Plastering

Plastering

J.B. TAYLOR

FIFTH
EDITION

PEARSON
Longman

Harlow, England • London • New York • Boston • San Francisco • Toronto
Sydney • Tokyo • Singapore • Hong Kong • Seoul • Taipei • New Delhi
Cape Town • Madrid • Mexico City • Amsterdam • Munich • Paris • Milan

Pearson Education Limited
Edinburgh Gate
Harlow
Essex CM20 2JE
England

and Associated Companies throughout the world

Visit us on the World Wide Web at:
www.pearsoned.co.uk

© J.B. Taylor, 1970, 1980, 1985, 1990

All rights reserved; no part of this publication may be reproduced, stored in a retrieval system, or transmitted in any form or by any means, electronic, mechanical, photocopying, recording, or otherwise, without either the prior written permission of the Publisher or a licence permitting restricted copying in the United Kingdom issued by the Copyright Licensing Agency Ltd, 90 Tottenham Court Road, London WIT 4LP.

First published in Great Britain by George Godwin Limited 1970
2nd edition 1977
3rd edition 1980
4th edition 1985
Fifth edition 1990

British Library Cataloguing in Publication Data
Taylor, J. B. (James Ball) *1915–*
  Plastering
  1. Plastering
  I. Title
  693.6

ISBN 0-582-05634-9

20 19 18 17 16
08 07 06 05

Printed in Malaysia, PP

Contents

PREFACE

CHAPTERS

1 TOOLS AND EQUIPMENT　　　　　　　　　　　　　　1

Steel and wooden hand tools; Simple scaffolding; Boards, stands, mixing bankers or containers

2 MATERIALS　　　　　　　　　　　　　　　　　　　13

Limes; Plasters; Cements; Aggregates; Strength and cohesion of plastering mortar mixes; Plastering mixes; Gypsum plaster/lightweight aggregate mixes, their uses and coverage; Plasterboards; Lathing; Metal trim; Newlath; Additives; Workshop materials

3 INTERNAL PLASTERING　　　　　　　　　　　　　　37

Methods of application; Trueness of plastering systems; Attached piers; Detached piers; External angles; Internal angles; Finishing coats; Keene's cement; Patching walls; Curved walls; Convex curved walls; Floating ceilings; Patching ceilings; Barrel ceilings; Lunettes *in situ*; Circular domes *in situ*

4 FLOORS, SKIRTINGS AND STAIRS　　　　　　　　　　68

Granolithic floor surface (concrete topping); Roof and floor screeds; Cement skirtings; Stairs

5 SPECIALIZED TECHNIQUES AND PROBLEM SOLVING　80

Compound walls; Insulation and acoustic plasterwork; Pattern staining; Damp-proofing; Waterproofing; Water level; Complaints and remedies

6 FIXING AND PLASTERING OF LATHING MATERIALS　90

Plasterboards; Lathing; Plasterboarding faults

7 DRY CONSTRUCTION METHODS AND PARTITIONS　101

Dry lining; Dry partitions; Laminated plasterboard partitions; Metal-stud partitions; Plasterboard coves

8 EXTERNAL PLASTERING 114

Pebble dashing; Patterns and decorative effects; Rough casting; Spatterdash; Tyrolean finishes; Plain floated cement work; Coloured cement work; Ashlar jointing; Recessed joints; Artificial masonry; Cement lettering; Scraped texture; English cottage texture; Old English cottage texture; Fan texture; Stipple; Finger whorl texture; Drapery texture; Depeter; Sgraffito; External cement features; Modern proprietary finishes

9 MOULDED WORK *IN SITU* 144

External cornice; Arches; Eccentric rule; Niches *in situ*; Classical orders of architecture; Entasis to columns — cutting arc method; Entasis to columns — semi-circular method; Columns *in situ*; Entasised fluted columns *in situ*; Repair and maintenance of moulded and enriched surfaces

10 BENCHWORK 190

Benches; Moulding techniques and materials; Waste moulds; Wax moulds; Gelatine moulds; PVC moulds; Cold cure rubber moulds; Flood, open or fence moulds; Skin moulds; Case moulds; Cornice case moulds; Insertion moulds; Run cores; Run cases; Run casts; Reverse mouldings; Beam casings; Hand lathes; Turning moulds; Fibrous plaster arches; Barrel ceilings — fibrous; Lunettes in barrel ceilings — fibrous; Circular domes — fibrous plaster; Fibrous plaster columns; Portland cement casts; Glass-reinforced plastics (GRP); Glass-reinforced gypsum

11 MECHANICAL PLASTERING 240

One-coat projection plastering

12 GEOMETRY OF ARCHES 244

Arches struck from one centre; Arches struck from two centres; Arches struck from three centres; Arches struck from four centres; Plasterers' oval

APPENDIX A 253

Abbreviated explanations and definitions of terms used in the plastering trade

APPENDIX B 260

Information sources and references

INDEX 262

Preface

In an age of mass production techniques, plastering remains essentially a craft, and the plasterer still requires a high degree of skill to cope with both the modern and traditional aspects of his craft.

Although most plastering can be carried out by a number of methods, the aim of this book is to describe good standard practices in a way which will be helpful both to the qualified craftsman and to the student taking courses up to and including City and Guild's Plastering Advanced level. It is also anticipated that the book, whose contents are laid out in sections, will be suitable for any structural pattern now being developed for N.V.Q. levels I to III in plastering.

The fifth edition has been revised and updated to include newer trends and techniques. Opportunity has been taken to include illustration and extensions to many items of work described in previous editions, plus new additional material.

New items included are the latest recommendations for floor laying techniques, changes in dry lining methods, a new revised section on bonding adhesives, plus information of materials recently introduced. These changes take note of the withdrawal of many manufactured materials such as Keenes cement, the elimination of sand/plaster mixes, plus the changes in cement based mixes.

In decorative plasterwork, explanations are included of running a niche *in situ* with a semi-elliptical hood. Also included are new drawing details with descriptions for forming compound case moulds for enriched cornices, alternative methods of forming flexible insertions, plus alternative methods of forming strike-offs. Plain and rebated joints to sides of casts are also explained with easy to follow sketches.

I am indebted to David Wickens, the Secretary of the Plasterers Craft Guild, for his advice and help in preparing this edition, also in the checking of the manuscript and drawings.

J.B. Taylor

CHAPTER ONE

Tools and Equipment

The list of tools required to be provided by plasterers in accordance with national working rule agreements is as follows:

Tool box	Level
Hawk	Pincers
Trowels	Plumb bob
floating	Punch
gauging	Rule
Floats	Saw
cross grain	Scrapers
straight grain	Screwdriver
Bradawl	Small tools
Chalk line	Snips
Chisels	Square
cold	Stock brush
wood	Tool brush
Dividers	Scratchers, templates, small
Files	special tools and other such
Joint rules	tools (customarily provided
Jointer	by the plasterer in the local-
Knives	ity) as and when required
Lath hammer	for particular work.

STEEL AND WOODEN HAND TOOLS

Wooden tools for plastering, including floats and darbies, are best made from yellow pine. This is because the timber does not warp easily when subjected to alternate periods of wetting and drying. Yellow pine also contains few knots, and this is an advantage because knots are tougher then the rest of the timber and therefore wear away more slowly leaving the knot projecting beyond the face of the float or darby. This in turn causes scratches on the finished floating surface. There is little difference between the summer and winter wood growth in yellow pine, and therefore the grain is even textured and wears away fairly smoothly without exposing raised grains of harder winter growth. Yellow pine is also light in weight and is a soft wood that does not splinter or scratch easily.

Tool boxes

Most plasterers prefer an army-type valise to carry and store their tools, but a tool box has an added advantage in that it is easier to lock. Sizes of tool boxes vary from 450 mm — 900 mm in length with a minimum sectional size of 375 mm x 200 mm.

Hawk

The hawk or handboard should have a carrying surface of approximately 300 mm square. Most purpose-made handboards are made nowadays from aluminium with a detachable handle as shown in Figure 1. Good quality, outdoor specification plywood with a wooden handle makes a good, cheap serviceable hawk. However, this type of handboard should be discarded immediately the plywood layers begin to separate or to break up at the edges, otherwise it will become difficult to clean, and small pieces of timber will become detached with the trowel and mix in with the plastering stuff. When used for skimming, these small pieces are a great inconvenience and so their accidental inclusion should be avoided.

Trowels

Skimming trowels are made from high quality steel, the blade being thin and flexible to ensure good contact on the finished plaster surface when laying down and trowelling up. Because the blade is so thin, the tang or rib of steel or alloy must extend almost the full length of the blade and is usually fastened to the blade with eight or ten rivets. The handles are usually single hang and include a variety of shapes including the banana shape as well as parallel and slightly tapered handles with circular sections (Figure 2).

Floating trowels have slightly thicker steel blades, shorter tangs and are fastened together with usually from five to seven rivets. Occasionally they are double hang by having an extra backstay at the rear of the handle, and this is an advantage when used for concrete or floor laying.

Flooring trowels are made with a longer blade and pointed nose (Figure 3).

Coving trowels for internal and external angles can be obtained in various sizes and styles. The curves vary from pencil round to bull nosed and most curved shapes are made in single or double hang style (Figures 4, 5).

Gauging trowels are made in different lengths, the most popular being 150 mm — 250 mm. They are general purpose tools, being used for gauging small quantities on the handboard for patching, tiling, etc (Figure 8).

Angle or twitcher trowels are used chiefly as an internal angle trowel but can also be used as margin trowels. They were used more frequently when lath and plaster ceilings were common, because these ceilings were often still soft after finishing. To prevent damage to the softer ceiling side of the angle when skimming the wall, the twitcher was used along the ceiling/wall angle.

STEEL AND WOODEN HAND TOOLS

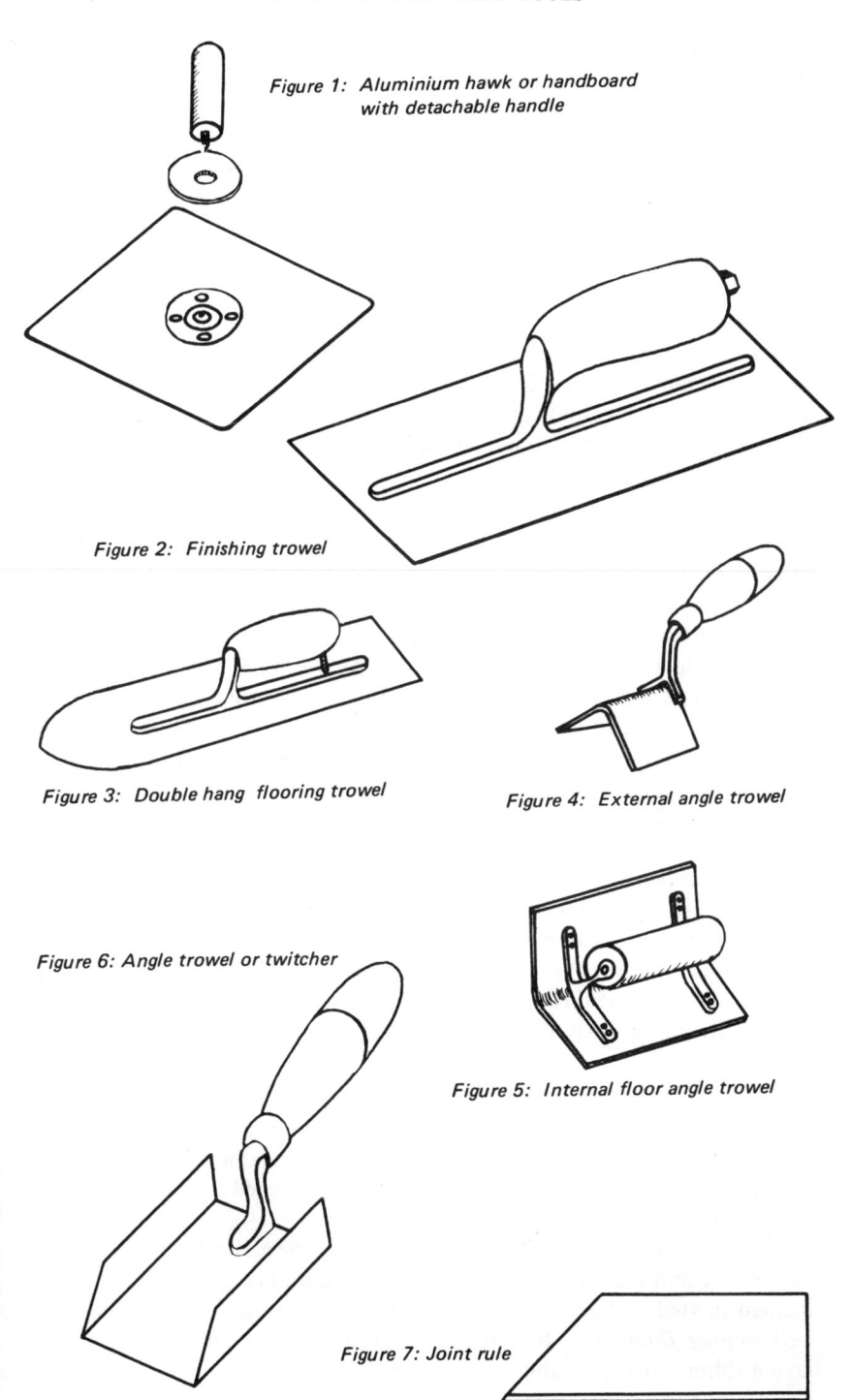

Figure 1: Aluminium hawk or handboard with detachable handle

Figure 2: Finishing trowel

Figure 3: Double hang flooring trowel

Figure 4: External angle trowel

Figure 6: Angle trowel or twitcher

Figure 5: Internal floor angle trowel

Figure 7: Joint rule

This avoided the use of a skimming trowel with the consequent risk of 'digging in' to the soft ceiling (Figure 6).

Margin trowels are made in many widths below 75 mm wide, and convenient sizes are 25 mm and 50 mm widths because these will suffice for the majority of margins. Narrower widths can often be filled in with the square end of a small tool, and wider margins with floats, trowelling off the latter with the nose of the trowel held at an angle to the run of the margin.

Small tools

These are used for touching up moulded or other fine or particular work. They are made from tempered steel to produce thin, flexible blades of various shapes. The most common are trowel or spade, leaf, spoon and square. Combinations such as leaf and square, or leaf and spoon can be obtained to give the plasterer his own preference for the type of work for which each shape is best suited (Figure 9).

Joint rules

These are used for extending lengths of plaster moulding of all types to form mitres, stopped ends and returns. The joint rules are made from thin sheets of inflexible steel with one end cut to a point with usually a 45° splay. The long side is brought to a fine edge by forming a rolled or swaged (bevelled) finish. Joint rules are unsuitable for moulded coloured cement work, partly because of the risk of staining which occurs as a reaction between the metal and the white or light colours of cement. Joint rules of boxwood or similar types of hardwood are more suitable for moulded cement work (Figure 7).

Busks and drags

These are sheets of flexible steel which are used to scrape over the surface of fibrous plaster casts to obtain a good surface, as well as for the cleaning off and finishing of joints (Figure 10). *Saw-toothed drags* have an edge similar to a saw. These are useful for scraping back unwanted or excessive thicknesses more easily and quickly than a smooth-faced busk.

Floats

Straight grain floats have a blade with the grain running its length. A convenient size is 275 mm x 112 mm and 15 mm — 18 mm thick. Wooden handles are usually nailed from beneath, the nails being punched deeper as the timber face of the blade is worn away.

Devil floats have three or four nails, usually at one end of the float, protruding about 1 mm from the face of the blade (Figure 11). This is to scratch the finished floated surface to provide a key for the setting or skimming coat.

Skimming floats are straight grained floats without nails. They usually have a thinner blade of about 12 mm thickness.

STEEL AND WOODEN HAND TOOLS

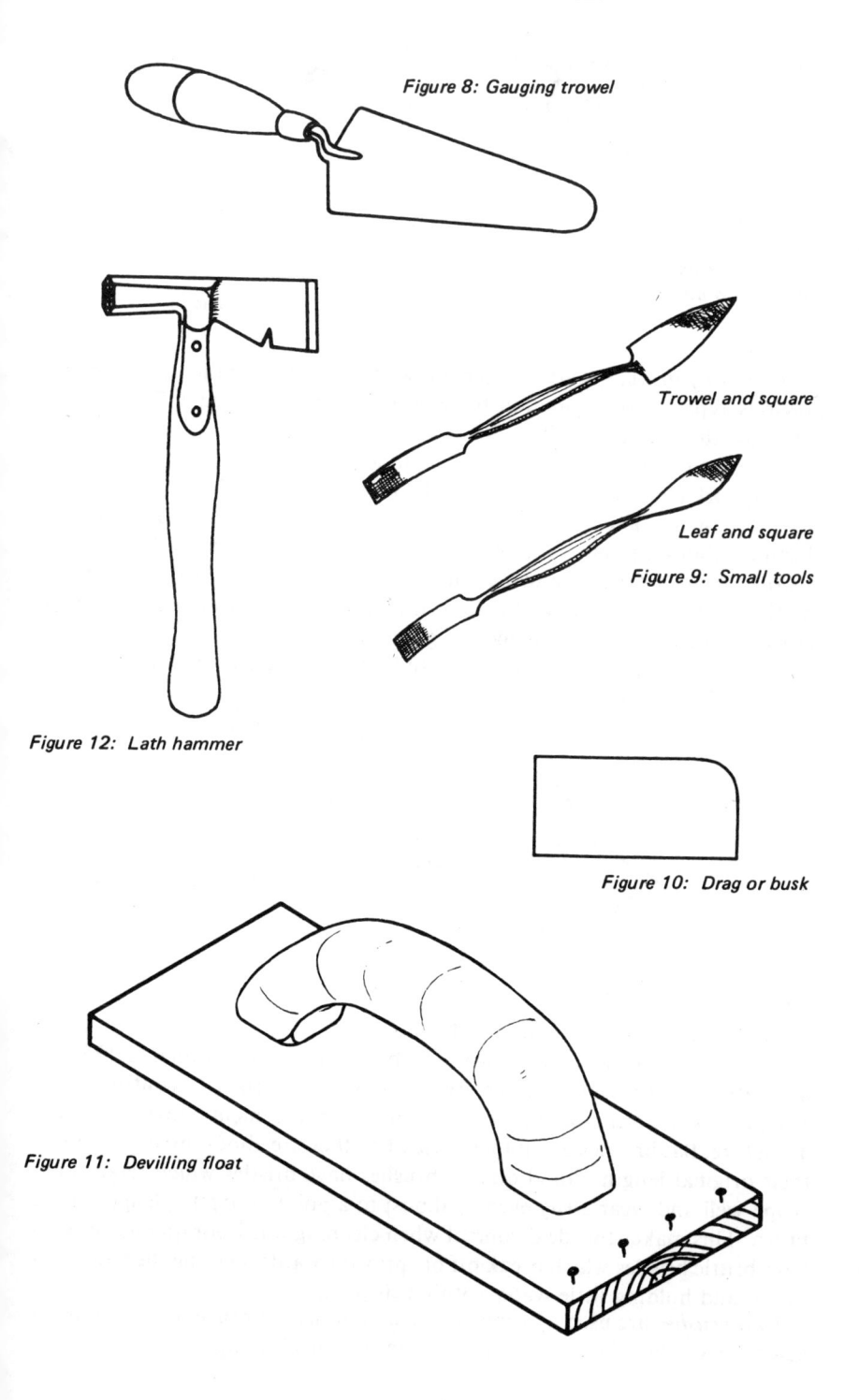

Figure 8: Gauging trowel

Trowel and square

Leaf and square

Figure 9: Small tools

Figure 12: Lath hammer

Figure 10: Drag or busk

Figure 11: Devilling float

Figure 13: Chamfered or feather edged float

Figure 14: Convex and concave floats

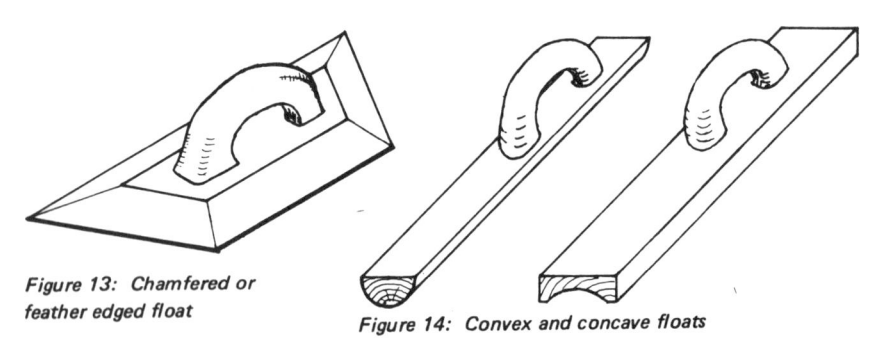

Fining or finishing floats for cement work are often smaller than the two previous types, and a common blade size is about 150 mm x 75 mm and approximately 12 mm thick.

Chamfered floats are invaluable for touching up V joints and moulded cement work (Figure 13). Curved special floats can be made to fit any given curvature on site and are useful for work such as arches, curved and coved skirtings, flutes and any type of moulded cement work (Figure 14).

Cross grained floats are made with the grain running across the width of the float. Because of the weakness of this type they are strengthened with a hardwood dovetailed stengthener (Figure 15).

Plastic floats have been recently introduced and have proved satisfactory for plain floated work.

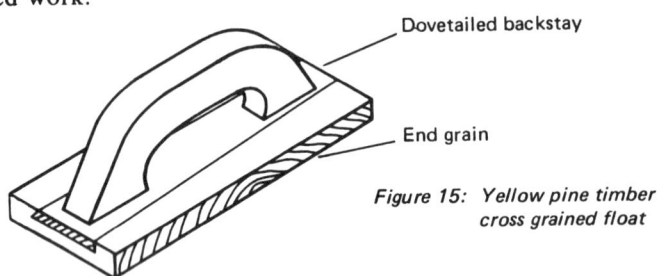

Figure 15: Yellow pine timber cross grained float — Dovetailed backstay, End grain

Brushes

Flat brushes are used for solid plastering when damping down walls, trowelling off and washing down. The best type is a good quality 175 mm decorator's brush. New brushes are too long in the bristle to control easily, especially when cleaning off woodwork or surplus cement above a skirting. Therefore, flat brushes are at their best when they are about three quarters of their original length. Good quality brushes have bristles which retain their shape well and wear away evenly, almost to a point, along the length of the brush. This makes for ideal control when cleaning off. Poor quality brushes have brittle bristles which break off or spray outwards, making them difficult to use and holding little water without dripping.

Tool brushes are used as a smaller version of the flat brush, particularly on benchwork jobs when damping down mitres or cleaning off after bedding

STEEL AND WOODEN HAND TOOLS

Figure 16: Splash brush

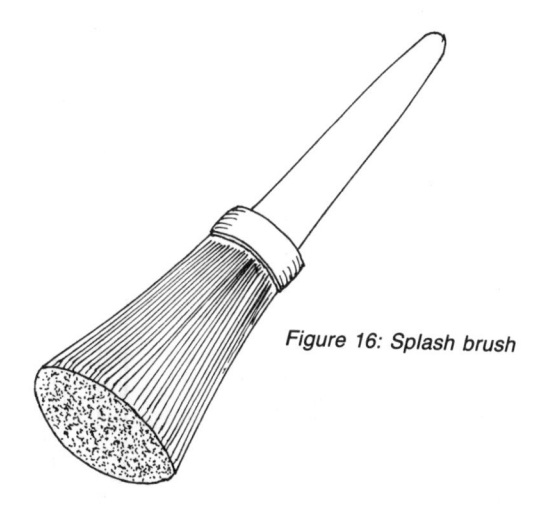

down pieces of moulding. Suitable sizes are 25 mm and 40 mm paint or sash brushes. Similar types of brushes are used for applying grease, shellac, etc, in fibrous plastering workshops.

Splash brushes are necessary when filling in casts and they are useful in many other moulding processes (Figure 16).

Levels

Spirit levels are necessary. Useful sizes, with or without plumb levels, are 225 mm and 750 mm. The longer length is more useful and accurate for the laying of floor and roof screeds, as well as tiling.

Water levels are often used and they are certainly quicker and more reliable once the technique of correct use has been mastered. A water level consists of a length of rubber tube with a transparent glass tube at each end. The rubber tube is filled with water until its level can be observed in each of the glass tubes. Unless there is an air lock in the tube, the water level will be the same in each tube and this will be the same even if the ends of the tube are held apart. Purpose-made water levels are available with perspex end tubes with valves to eliminate the risk of air locks, and screwed caps to prevent the escape of water when it is carried about between jobs.

Plumb bobs and gauges

These are used to plumb vertical dots in good class work.

Plumb bobs are often made of lead, at least 350 grammes in weight, with a hole bored through the centre. A string line is passed through the hole and knotted below.

Gauges are pieces of timber with shoulders. They are used in pairs which must be identical in size so that the plumb line, when suspended from the shoulder of the upper gauge, is an equal distance from the upper and lower dots when the plumb line is touching the shoulders of the upper and lower gauges.

TOOLS AND EQUIPMENT

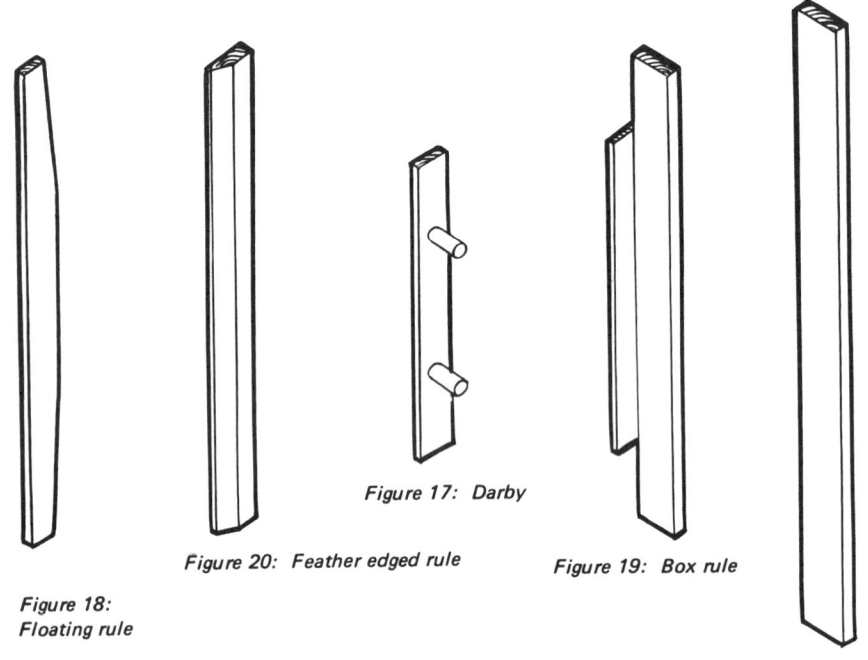

Figure 17: Darby

Figure 20: Feather edged rule

Figure 19: Box rule

Figure 18: Floating rule

Figure 21: Parallel rule

Raised panel

Sunken panel

Plinth

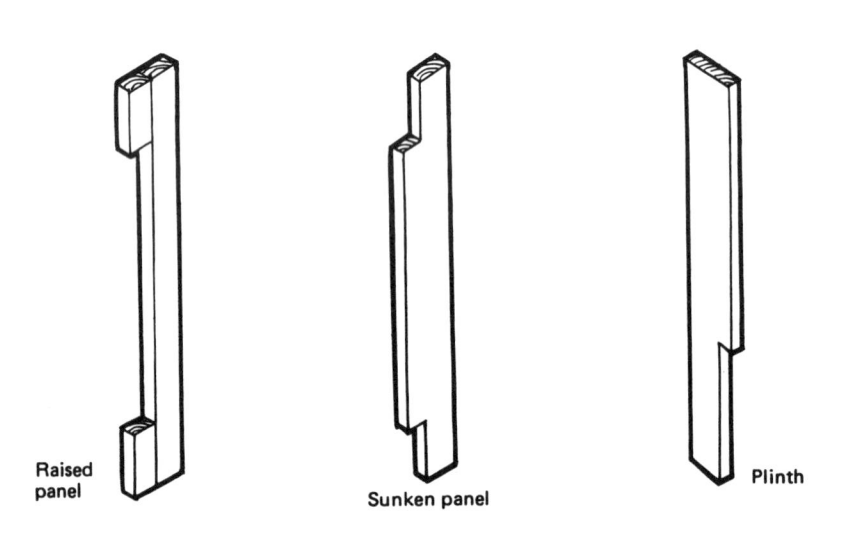

Figure 22: Rebated rules

Lath hammer

This is a double purpose tool having a combined axe and hammer head with a wood, steel or alloy shaft (Figure 12). The axe was more frequently used when wood lathes were common, but it is still excellent for knocking off old plaster and trimming edges. The notch in the axe is useful for withdrawing plasterboard nails, but it is unsuitable for large or oval headed nails.

Rules

The Darby is a short rule, approximately 1.05 m long x 100 mm wide and 15 mm thick, with two handles to the back (Figure 17). It is used to straighten floating coats without the use of screeds. The edge is also used to scrape off the high parts of the floating, and this surplus is lapped back into the hollows. The flat face of the darby is then pressed flat over the floating with pressure to squeeze it apparently straight and flat.

Floating rules are used for better class work (Figure 18). They are usually 2 m — 2.3 m long and with a sectional size of approximately 125 mm x 25 mm.

Box rules are floating rules with a catching board nailed at right angles to the back edge (Figure 19). They are used on ceiling work to save the droppings from falling onto the floor or the scaffold.

Feather edged rules have one long edge chamfered to make it easier to use in internal angles. Shorter versions are also used to staighten freshly applied skimming applications (Figure 20).

Parallel rules up to 4.5 m long and having sectional sizes up to 150 mm x 40 mm are used for floor and roof screeds. (Figure 21). Such rules should be checked frequently for parallel because of the excessive and uneven wear caused by ruling in sand or granolithic mixes.

Rebated rules with double rebates are suitable for forming raised or sunken panels. Single rebated rules can be used for ruling in parallel plinths (Figure 22).

Curved templates, or rules, may be cut to any given curvature and used to form screeds on circular or other curved work.

One coat projection plastering

Tools used for one coat projection plastering include:

Aluminium feather edged rules, either 'h' section or trapezoid feather edge section, are available in lengths of 1 m, 1.2 m, 1.5 m, 1.8 m and 2 m (Figures 23, 24).

Swiss float or two handed trowel has a length of 500 mm and a width of 130 mm (Figure 25).

Aluminium scratching tools for scraping internal angles have a maximum length of 450 mm and a width of 85 mm. The scraping blades project about 18 mm below the frame (Figure 26). An alternative corner plane is shown in Figure 27.

TOOLS AND EQUIPMENT

Figure 23: 'h' section

Figure 24: Trapezoid section feather edge rule

Figure 25: Swiss float (two handed trowel)

Figure 26: Aluminium scratching tool

Figure 27: Corner plane

Figure 28: Sponge rubber float

Figure 29: Internal angle trowel

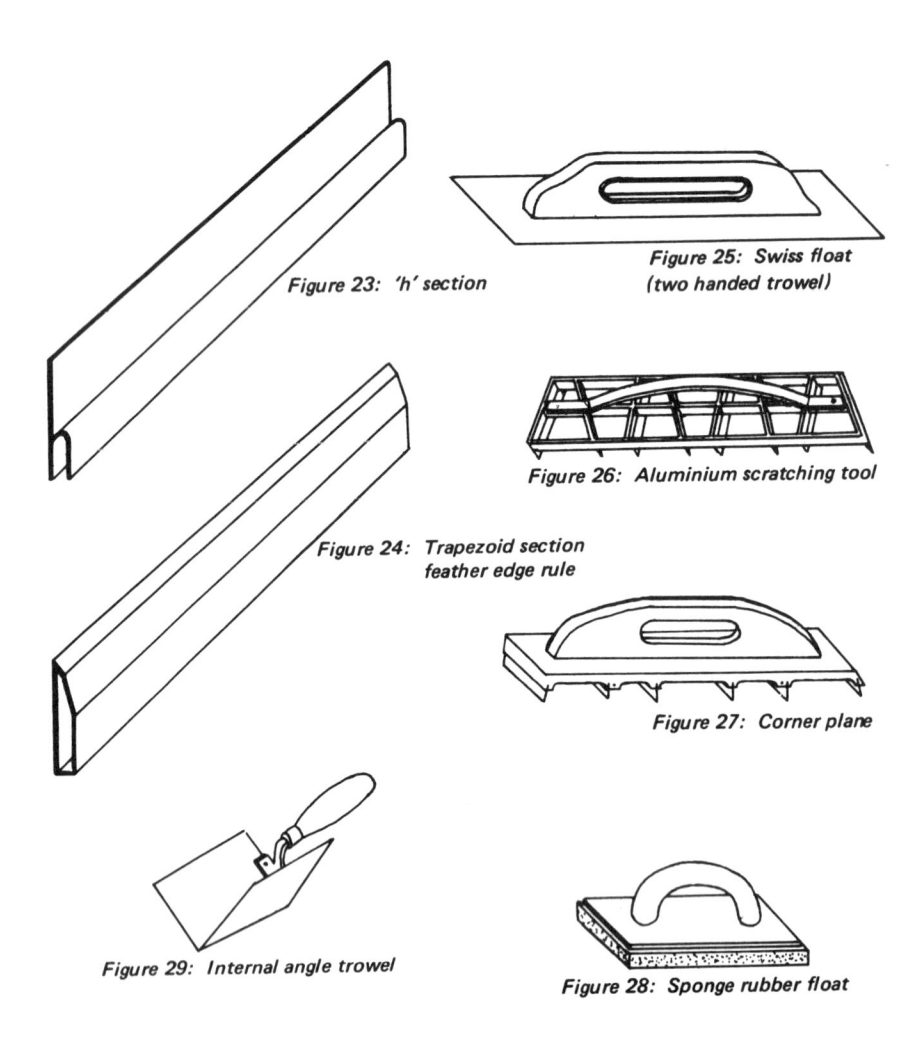

Sponge floats are approximately 240 mm long and 120 mm in width (Figure 28).

Power floats can also be used.

Square internal angle trowels can be used, particularly for wall/ceiling angles (Figure 29).

Angle or twitcher trowels are favoured for internal wall angles (Figure 6).

SCAFFOLDING AND EQUIPMENT

Figure 30: Timber split-heads

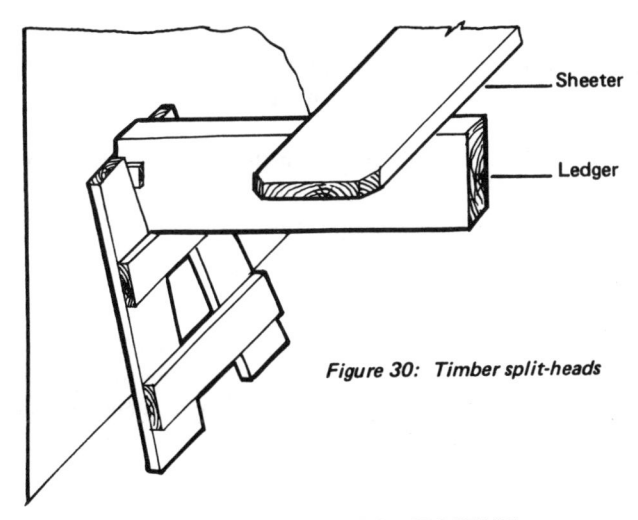

SCAFFOLDING AND EQUIPMENT

Interior scaffolding for normal room heights is often carried out by use of tripods or trestle stands. The practice of using timber split heads has largely disappeared (Figure 30).

Tripods

Tripods are three-legged steel stands with telescopic extensions, allowing them to be set to the correct platform height required. They can be obtained in three size ranges from 600 mm — 2.4 m platform heights (Figure 31).

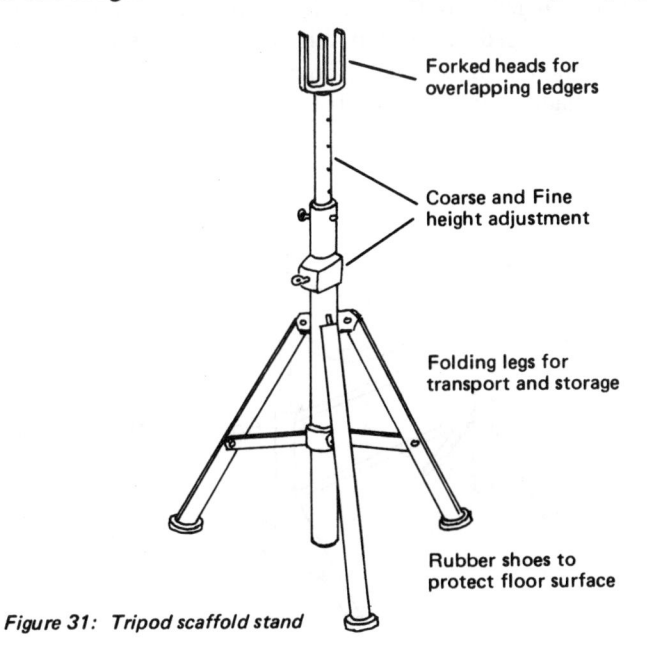

Figure 31: Tripod scaffold stand

Trestle stands

Trestle stands are adjustable for platform heights from 450 mm — 1.8 m (Figure 32). The following recommendations are advisable:

1. Guard rails should be used when the platform height exceeds 2 m.
2. Trestles over 3.15 m in height should be tied back to the building.
3. Maximum spacing between trestles for 50 mm planks should be 2.55 m.

Board and stand

The mortar board is often made of tongued and grooved boards nailed to strong battens or, alternatively, from 20 mm blockboard. A convenient size is from 900 mm — 1 m square. The stand is the frame to support the mortar board and can be folding or rigid. A suitable size is 600 mm square and is 675 mm — 750 mm in height.

Mixing bankers or containers

Bankers are mixing platforms and can be made from timber sheets, planks or, for sand mixes, merely created by spreading sand over a suitable clean space.

Waterproof troughs

Waterproof troughs are used for pre-mixed lightweight aggregate/plaster mixes. A shallow tray lined with sheet metal and about 2.4 m long and 1.2 m wide with 200 mm sides is excellent. The tray can be raised slightly at one end and clean water poured in. The plaster is emptied into the upper end of the box and then dragged into the water for mixing. It is important to clean out any remaining gauged material as this will accelerate the setting times of the mixes mentioned in Chapter 2, and reduce their ultimate strength. Bungalow baths made from galvanized sheet metal make excellent portable mixing containers.

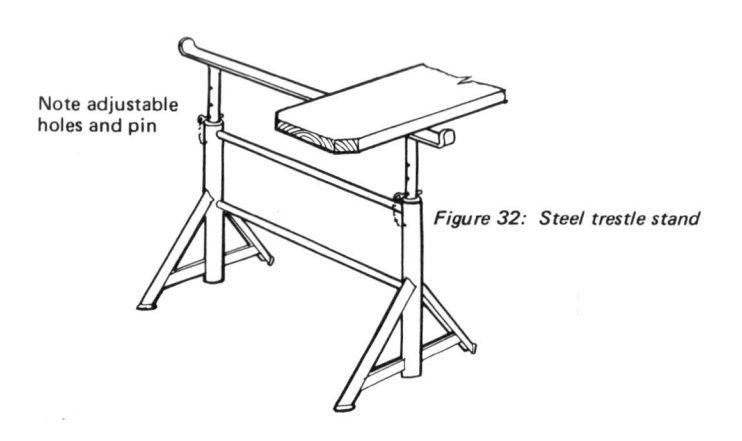

Figure 32: Steel trestle stand

Note adjustable holes and pin

CHAPTER TWO

Materials

Materials and mixes used in plastering applications are either a matrix or a matrix plus one or more aggregates. A matrix, or binder, is that part of a mix which has a setting action. The aggregate is the inert filler whose purpose may include providing special wearing qualities, better workability, regulation of shrinkage, cheapening of the mix costs, improved insulation, etc.

The main matrices, or binders, used by plasterers are gypsum plasters, limes and cements. Other types of binders, less used, include the transparent synthetic resins, epoxy, polyester and similar formulations.

Many of the traditional plastering materials are used only rarely under present site conditions. Their inclusion in the following descriptions is to provide a more comprehensive background knowledge of the types of plastering materials, traditional and contemporary, that can be used for a particular plastering job either in new or repair work.

LIMES

The raw materials from which building limes are obtained are limestone and chalk. Both of these materials are largely composed of calcium carbonate and were formed over a period of millions of years by calcium deposits from marine life in the oceans of the world. As a result of later earth movements, layers of limestone are to be found inland in many countries of the world.

The limestone or chalk is quarried, crushed and then heated in kilns to drive out the carbon dioxide content. The resulting material withdrawn from the kilns is calcium oxide, known also as burnt lime, lump lime, quick lime, or simply as lime. The material is then slaked with water to form slaked lime (calcium hydroxide), also known as lime putty.

Lime putty sets by a process of carbonation as a result of loss of water and also by combining chemically with carbon dioxide from the atmosphere. In doing so, it reverses the chemical actions which took place during the manufacturing process (Figures 33, 34).

Classification of limes

Most of the limestones or chalks quarried contain impurities such as silica, alumina, iron oxide and sulphur. These impurities are retained by the lime after manufacture from calcium carbonate, and their presence will greatly

MATERIALS

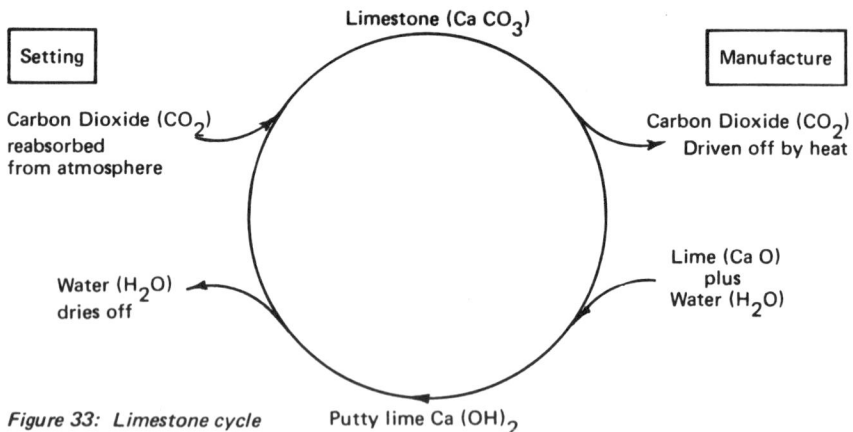

Limestone (CaCO$_3$)

Setting — Carbon Dioxide (CO$_2$) reabsorbed from atmosphere

Manufacture — Carbon Dioxide (CO$_2$) Driven off by heat

Lime (CaO) plus Water (H$_2$O)

Water (H$_2$O) dries off

Putty lime Ca(OH)$_2$

Figure 33: Limestone cycle
Figure 34: High calcium lime

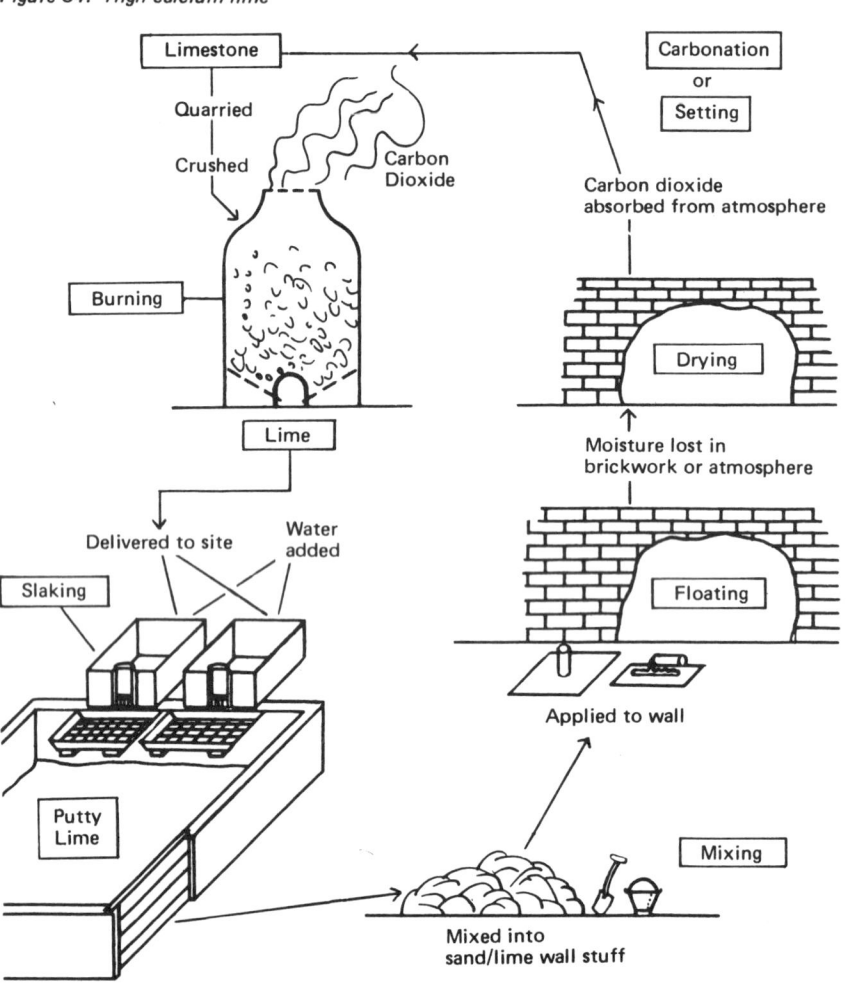

Limestone — Quarried — Crushed — Burning — Carbon Dioxide — Lime

Delivered to site — Water added — Slaking — Putty Lime

Mixed into sand/lime wall stuff — Mixing — Applied to wall — Floating — Moisture lost in brickwork or atmosphere — Drying — Carbon dioxide absorbed from atmosphere — Carbonation or Setting

affect the setting times, working properties and strengths of the resulting slaked limes. Limes of this type do not depend upon carbonation for setting and are capable of setting out of air or even under water. This property is termed hydraulicity.

The classification of limes is based on the degree of hydraulicity as follows:

1 *Non-hydraulic limes*
This is the purest type of lime containing up to 98 per cent calcium oxide. This produces a putty lime which has high plasticity and workability and sets slowly by carbonation. This is also known as high calcium lime.

2 *Semi-hydraulic limes*
These limes contain sufficient impurities to impart to them feebly or mildly hydraulic tendencies.

3 *Eminently hydraulic limes*
These contain a high proportion of impurities which give the resulting slaked lime properties similar to Portland cement. A well known type of hydraulic lime is called Blue Lias Lime.

4 *Hydrated lime*
This should not be confused with hydraulic lime. Hydrated lime is obtained by slaking high calcium lime under controlled conditions in a manufacturer's plant to ensure efficient slaking, after which the excess water is dried off. The resulting powdered hydrated lime is sold in 25 kg bags. Hydrated lime, soaked in water overnight, will produce a lime putty with improved yield and workability.

PLASTERS

Plasters used in the building trade are manufactured from gypsum and anhydrite. Production of plaster from anhydrite has ceased in Britain though limited use still occurs in other countries.

Gypsum is a fairly soft rock containing calcium sulphate and waters of crystallization. The rock is obtained mainly by drift mining, after which the crushed gypsum is fed into containers called kettles. The crushed material is then heated for about two hours at a temperature of 170°C until three quarters of the waters of crystallization have been driven off. The resulting material after grinding is known as hemi-hydrate (Class A) plaster, also known as plaster of paris.

When water is added to plaster, the reverse chemical action takes place, and the waters of crystallization reform to convert the plaster into gypsum again. Chemically, the plaster will accept back only the exact amount of water driven out during the calcining or heating process, any excess being left to dry out later. The growth and interlocking of the crystals is the setting action of plaster and any attempts to re-temper a setting mix means that the crystals are broken apart and will not re-form. This results in a weak final set.

Class A plasters set too quickly for normal solid plastering uses and a retarder is added by manufacturers to convert the plaster to *Class B* retarded hemi-hydrate plaster.

Class C plasters have had all the waters of crystallization driven off during manufacture from gypsum, and they are known as anhydrous gypsum plaster. These plasters are slow setting but have a harder finish than hemi-hydrate plasters.

Class D plasters are calcined at a higher temperature and have accelerators added to improve the setting time. The final set is harder than other plaster, and they are used in positions where damage to the plasterwork would otherwise occur, such as external angles, reveals, etc. Anhydrous hard burnt plasters Class D are known as gypsum cements, among which is Keenes cement. Anhydrous plasters often stiffen up after mixing and can be re-tempered without adverse effect on the final strength. They are unsuitable for use on plasterboards.

Class B retarded hemi-hydrate gypsum plasters are used as binders in a number of proprietary pre-mixed gypsum plaster/lightweight aggregate mixes used for a variety of plastering jobs. These are itemised on p. 23. Board Finish is another type of Class B plaster. Note neither Class C or Class D plasters are now being manufactured.

Properties of plaster

The plasters described must always be protected from contact with moisture before use. They must be stored under cover clear of the ground and other damp surfaces. If a small proportion of plaster is affected by contact with moisture, then this will appreciably shorten the setting time of the whole batch. This condition is variously known as 'starved' or 'perished' plaster, and is often characterised by the presence of lumps of partially set plaster in the bags.

Gypsum plasters should never be mixed with Portland cements because of the probable formation of calcium and sulphoaluminates which would disrupt the resulting work.

Plasters of different categories should not be mixed together.

Water used for mixing plaster should be clean, otherwise the impurities in the dirty water will affect the setting time and ultimate strength of the set material. Mixing machines, water barrels, mixing baths, mixing bays, buckets and any tools used in mixing must be kept clean to avoid contamination. Dirty water can reduce the setting time and the strength by 50 per cent quite easily.

Corrosion can occur when gypsum plasters are applied to unprotected ferrous metal surfaces. (Ferrous metal is a metal containing iron.) Plastering mixes may be neutral, acid or alkiline in reaction, or affected by the use of an accelerating salt or similar (giving an acid reaction).

The risk of corrosion being caused by plasters which are alkiline in reaction is slight, and therefore it is good practice to use lime in the mix when applying calcium sulphate plasters to metalwork. (An exception to this is that certain specifications forbid the use of lime in the plastering mix because of its adverse effect on subsequent oil-bound decorating application.) To combat the risk of corrosion when plastering on expanded metal lathing, a small amount of hydrated or putty lime should be added to Class B retarded hemi-hydrate plasters in contact with the metal. Proprietary metal lathing plasters have already been adjusted for anti-corrosion by the manufacturers.

Other metalwork which is to be covered by plaster mixes should be protected if possible by galvanizing, painting, or other similar methods. Initial protection of all metal surfaces is the safest way to reduce the risk of corrosion in damp conditions.

Plaster swells when setting, the amount of expansion varying with the proportion of water added. Stronger mixes with little water expand most.

Set plaster mixes have considerable resistance to fire. This is due to the chemical combination of water which must be driven off as steam before the heat from the fire can be passed through the plasterwork to the superstructure.

CEMENTS

Portland cement

Portland cement is an artificial cement invented by Joseph Aspden in 1824. He named his new cement Portland because of its resemblance to Portland stone when set.

The raw materials used in the maufacture of Portland cement are limestone or chalk and clay, or shale, roughly in the proportions of 78 per cent limestone: 22 per cent clay.

Portland cement is manufactured by two different methods, the wet and dry processes.

In the wet process, the separate materials are quarried, crushed, ground, mixed into a slurry and then combined together. At this stage, the slurry is tested for its chemical constituents and adjusted as necessary.

This corrected slurry is fed into a rotary kiln to be burned to a cement clinker. The rotary kilns are in the form of horizontal cylinders about 120 m long and 4.2 m in diameter. The kiln itself revolves once every minute. A slight incline is allowed in the length of the kiln to allow the material inside it to travel gradually through during manufacture.

Heat is applied at the lower end of the kiln, often by powdered coal and forced air. The temperature is approximately 1650°C at this portion of the kiln, becoming progressively cooler towards the far end.

In its passage through the kiln, the slurry is dried, heated and burned until the materials fuse into a new chemical combination known as cement clinker. This is now passed into a clinker store to cool. After cooling, it is mixed with 1-5 per cent of raw gypsum and then ground until at least 90 per cent of it will pass through a 170 sieve. This sieve contains 28,900 holes per square of 25 mm sides. The purpose of the gypsum is to prevent a flash set of the mixed cement.

In the dry process of manufacture, the raw materials contain only a small proportion of water, resulting in cost saving. Many of the newer plants use this method.

All cement manufactured in Britain is made to conform to British Standards Specifications, and this includes strict standards of fineness, chemical composition, strength and setting times.

Setting time is the period between the addition of water and the time when the mix stiffens. The initial set should not be less than 45 minutes, and the final set not more than ten hours.

Hardening commences after the final set and, providing sufficient moisture is available for continued hydration (chemical action between water and cement), will continue indefinitely. To ensure conditions as described above, all important cement work should be 'cured' by restricting the moisture loss or providing sufficient moisture for hardening to continue long enough to achieve the desired strength.

Other types of Portland cement

White Portland cement is manufactured from pure limestone and white china clay. Special precautions to prevent contamination by iron or other staining agents have to be taken during the manufacturing process.

Coloured cements are obtained by mixing pigments with white cement for light tones, and with ordinary Portland cement for darker colours. Coloured cements are also pre-mixed with special graded sands and only require the addition of water. This type is sold as Cullamix and has been designed for use with Tyrolean textured work, and for certain types of plain floated and stippled work.

Rapid hardening Portland cement hardens quicker than ordinary Portland cement due to finer grading and burning at a higher temperature during manufacture. The extra manufacturing costs are responsible for its higher price. Its setting time is no quicker than ordinary Portland cement, but after the final set, hardening takes place more rapidly.

The addition of 2 per cent of calcium chloride to ordinary Portland cement increases the rate of setting and hardening. Calcium chloride should not be added to quick setting cements. Due to faults arising from past misuse of calcium chloride, great care must be taken to prevent incorrect usage.

Weatherproofing and water-repellent cements are composed of ordinary Portland cement plus very finely ground materials which result in a more complete pore filling mix to give a dense impervious mass when set.

Masonry cement is ordinary Portland cement plus an addition of finely ground materials which help to plasticise the mixed cement. With this type of cement, a fatty mix with good workability can be obtained without the use of putty lime or other separate plasticiser.

High alumina cement is not a Portland cement and is manufactured from bauxite and clay. It is black in colour, it sets and hardens more quickly than ordinary Portland cement, and is also resistent to certain sulphates and weak solutions of acids. The trade names for high alumina cement are Ciment Fondu and Lightning.

Different grades of cement should not be mixed and used together, particularly Portland cement and high alumina cement because of the serious adverse reactions.

AGGREGATES

These make up the bulk of plastering mixes and are composed of inert materials. They are classified as coarse or fine aggregate. All aggregates passing through a 4.6 mm sieve are considered to be *fine aggregates*; the larger sizes are *coarse aggregates*.

Sand

Sand is a fine aggregate formed by the natural disintegration of rock, or it is artificially created by crushing stone or gravel to the required sizes. Only small amounts of crushed stone sand are used in the plastering industry, its use being confined to special cement work on the whole.

Natural sand has been formed over the centuries by the action of wind, rain, frost and running water to break down rocks into small particles. These small grains, varying in size, have been washed by the action of the seas and rivers and deposited in certain areas which have in turn been moved inland due to past changes of the earth's formation. This is the reason for large pockets or deposits of sand many miles inland.

The two main types of sand available at present are *pit sand* from inland quarries and *river sand* obtained by dredging. Sea sand is unsuitable because of the risk of efflorescence due to the presence of salt. Crushed stone often contains too much fines.

A good sand should contain a suitable proportion of large, medium and small sized grains. The reason for this can be seen in Figure 35.

Sketch of a sample of badly graded sand

Note—voids or spaces between the particles of sand

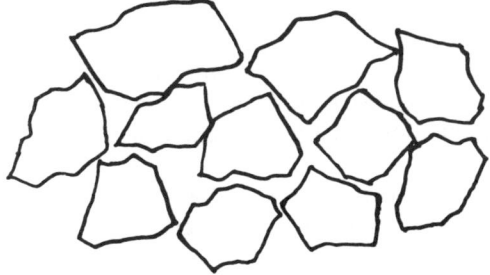

Figure 35

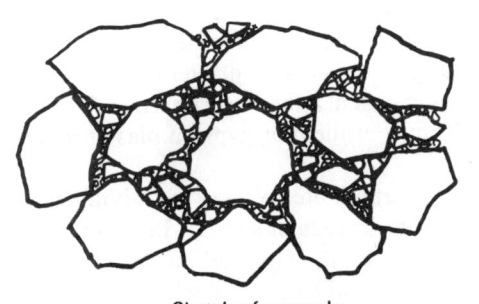

Sketch of a sample of well graded sand

Note—large voids filled with medium and small sized grains

If the sand is composed of large particles only, then there will be many spaces or voids between the grains. This type of sand would require a lot of lime or cement to make a strong, dense mix. As a result the mix would be expensive and, because of extra shrinkage, would not be strong enough.

A well graded sand has medium sized grains to fill in the larger voids and small sized grains to fill in the smaller voids.

The functions of a sand are:

1 To induce the mix to shrink uniformly during the process of setting and hardening, irregular shrinkage being a general cause of cracking.

2 To lower the cost of the mixed material by providing the biggest bulk of the mix.

3 To assist workability, particularly on thicker applications such as floating coats.

Expanded perlite

This is a lightweight material used in modern lightweight aggregate mixes with gypsum plaster or Portland cements. It is obtained from the mineral perlite and is a volcanic rock which is a type of Rhyolitic glass containing a small amount of combined water.

The perlite ore, after crushing to correct grade size, is softened by pre-heating and then expanded by greater heat to many times its original volume, due to the formation of steam from the combined water. The resulting material is a glass-like bubble of cellular structure which weighs as little as 70 kgs-130 kgs per cubic metre, depending upon the grading sizes.

Vermiculite

Vermiculite is the name given to a group of laminated minerals resembling mica in appearance. It is found in various parts of the world. The crude ore consists of thin flat flakes, each lamination or layer separated from its neighbour by microscopic particles of water.

When the flakes are heated quickly to temperatures of 700°C-1000°C, the contained water turns to steam which forces the laminae or layers apart. This is known as exfoliation, and the exfoliated or expanded vermiculite now consists of accordion-like granules containing millions of minute air layers. The high insulation value and lightness in weight is due to the latter property.

Exfoliated vermiculite is made in many grades with particle sizes from 1.5 mm-6 mm. It is used for lightweight aggregate/gypsum plaster mixes including concrete bonding plaster which has excellent adhesive properties. Particle sizes for plaster mixes are 1.5 mm-3 mm. Vermiculite/gypsum plaster mixes give excellent fire protection.

The larger particle sizes of exfoliated vermiculite are used for mixing with Portland cement for insulation work, roof screeds or similar. Because of its special properties, exfoliated vermiculite is easily damaged on floor or roof screeds and vermiculute/cement surfaces should be covered with a minimum of 13 mm of sand and cement as a protection on this type of screed.

Granulated pumice

Granulated pumice is another mineral of volcanic origin, containing millions of minute air cells. It is used in acoustic plaster mixes for its high sound-absorption qualities.

Granite chippings

Granite chippings are composed of crushed granite up to 9 mm in size and used as the aggregate in granolithic work such as floors, pavings, paths, etc. Granite is an igneous rock, formed by heat, and is an extremely hard rock. Its hardness and wearing properties are used to advantage on floors left uncovered as a wearing surface.

Carborundum chippings

Carborundum chippings, also known as carborundum dust, are a compound of silica and carbon. They are mixed into or sprinkled on granolithic floors, steps and landings for their extra hard-wearing or non-slip qualities.

Pebble-dashing materials

These include a wide variety of self-cleansing stones including white or cream limestone spar, Dorset spar (brown sea pebbles), and white or coloured marble chippings.

STRENGTH AND COHESION OF PLASTERING MORTAR MIXES

Hair was traditionally used to give strength and cohesion to plastering mortar mixes. The main types used were cow and goat hair. Nowadays vegetable or synthetic fibres such as flax, sisal, jute or nylon can be used.

PLASTERING MIXES

These are mixtures containing a material with setting properties, generally known as the *matrix*, or binder, and an inert material termed the *aggregate*.

Traditional sand/lime mixes

Three parts of sand mixed with one part of putty lime makes a plastering mortar and is termed *raw stuff* or *coarse stuff*. If beaten hair is mixed in, then the mix is termed *haired coarse stuff*. When raw stuff is required to set more quickly or to attain a higher initial strength, then it is mixed again with gypsum plaster or Portland cement to the strength required. It is then termed *gauged coarse stuff*.

The above mixes are used as undercoats, but sand and lime putty can also be used for skimming on sand/lime floating coats. The typical skimming mixture is composed of three parts lime putty to one part sand and the mixture punched through a fine mesh sieve. This skimming *raw stuff* also known as setting stuff, is gauged with a minimum of 25 per cent Class A or Class B gypsum plaster.

Gypsum plaster/lightweight aggregate mixes

The lightweight aggregates used are perlite and vermiculite, and the gypsum plaster used is a retarded hemi-hydrate Class B. These materials are pre-mixed and require only the addition of water.

The various grades of mixes available and the backgrounds for which their application is recommended are given in the table on page 23.

Additional specialist plasters

Lightweight aggregate/browning HSB for use on high suction backgrounds.

Acoustic plaster contains granulated pumice plus a Class B gypsum plaster. It is used as a textured finishing coat to provide sound absorption qualities for spray applications only.

Projection plaster is a blend of gypsum plaster with special additives to regulate the setting time, improve the workability, and to increase water retention.

Universal one coat hand applied, a retarded hemi-hydrate plaster with a small proportion of perlite aggregate and additional additives. The plaster is designed to be applied in one coat to normal plaster thicknesses, e.g. 13 mm to normal brickwork. The material should be mixed using a heavy duty drill and suitable whisk attachment. This plaster is an alternative to the normal one, two or three coat specification.

Thistle Hardwall, a gypsum pre-mixed undercoat plaster designed to be used in situations where higher impact resistance is required and to give better resistance to efflorescence.

Thin coat plasters are Class B finishing plasters in one coat work for fair faced aerated concrete blocks or similar types of background.

X-ray plaster is a special pre-mixed material composed of a retarded hemi-hydrate gypsum plaster with barium sulphate as the aggregate. It is supplied in undercoat for floating coats on walls, fibred for use on metal lathing and finish grade for skimming coats on the previous two grades. The barium sulphate has insulation properties similar to lead and X-ray plasters, suitable for use in X-ray theatres and certain laboratories, are used to provide insulation properties against electromagnetic radiation.

Portland cement mixes

Portland cement mixes vary in strength according to the finished strength requirements. For floor screeds, plinths, skirtings, and similar types of work the normal mix is sand: cement, 3:1. For granolithic work such as floors or pavings where excessive wear is expected, the mix is Portland cement: granite chippings, 2:5.

On other types of work the cement content is decreased when weaker strengths or greater suction will be advantageous. Mixes of six parts of sand

PLASTERING MIXES

GYPSUM PLASTER/LIGHTWEIGHT AGGREGATE MIXES, THEIR USES AND COVERAGES

Grade	Aggregate	Background	Approximate coverage $m^2/1000$ kg.
Lightweight aggregate/ browning plaster	Expanded perlite	Brickwork Coke breeze Concrete bricks No fines concrete Clay tile partitions Thermalite blocks	130 — 150 Average thickness 11 mm
Lightweight aggregate/ metal lathing plaster	Expanded perlite and exfoliated vermiculite plus rust inhibitor	Expanded metal lathing	60 — 70 Average thickness 11 mm (from face of lath)
Lightweight aggregate/ bonding plaster	Exfoliated vermiculite	Precast concrete *In situ* concrete Plasterboards Stonework Cork slabs Backgrounds treated with Synthraprufe or PVA bonding liquids	145 — 155 8 mm 150 — 165 100 — +10 11 mm 145 — 155 8 mm
Lightweight aggregate/ finish plaster	Exfoliated vermiculite	For use as a neat finish on all the above floating coats	410 — 500 2 mm

to one part cement, plus one part of lime putty are used when a certain amount of suction will be required from the rendering or floating coat.

Mixes of one part Portland cement, two parts lime putty and nine parts sand can be used on certain types of indoor or protected outdoor work. Note that the ratio of binder to aggregate is still 1:3.

Portland cement/lightweight aggregates

Mixtures of Portland cement with perlite and vermiculite are used for wall and floor applications in a variety of ways.

Pre-mixed Portland cement and perlite is made for backing or floating coats on walls, using traditional methods or machine application.

MATERIALS

Suitable backgrounds include brickwork, no-fines concrete, aerated blocks and clay pots. Coverages are 6.5 m^2 — 7.5 m^2 per 50 kg bag when applied 10 — 11.5 mm in thickness. The same mix is also suitable for use on expanded metal lathing and Twil-lath backgrounds. Coverage (two coats) is approximately 4 m^2 per 50 kg bag.

Cement/perlite lightweight screed mixes are also made to special requirements. The mixes of cement and perlite mentioned above are pre-mixed dry and sold in 50 kg paper sacks requiring only the addition of water on site.

Special gypsum based finishing plasters with additives suitable for application on the floating coats described are available.

Vermiculite is widely used in mixtures with Portland cement for insulated floor and roof screeds. Proportions of vermiculite: cement vary from 4:1 to 8:1. These mixes, when set, are not strong enough to resist breakdown and crushings by usage, so they are strengthened by covering the surface with a sand and cement topping during the laying process.

PLASTERBOARDS

These consist of an aerated gypsum core between layers of special fibrous paper. Insulated grades and thermal boards are available. Types of plasterboard available for plastering and dry lining are as follows — Gypsum wallboard and plank used for dry lining and partitions — Gypsum baseboard and gypsum lath used as a background for plastering only. The wallboard and plank types can be obtained with ivory finish and have tapered, bevelled or square edges. Gypsum laths have rounded edges on the long side of the lath. Baseboards have square edges. Plasterboards of these types are made in a larger variety of lengths and widths, and have thicknesses of 9.5 mm, 12.5 mm, 15 mm and 19 mm (the latter size for plank only). Gypsum laths are made in widths of 400 mm only. Plasterboard coves are made in two girth sizes, 100 mm and 127 mm. Both sizes have wall and ceiling members of 6 mm. The smaller size has a projection and depth of 67 mm. The larger size has a projection and depths of 83 mm.

LATHING

Expanded metal lathing

This is made from good quality steel plate, cut and expanded to form a network of diamond-shaped meshes. The size of the meshes is normally 6 mm, although 9 mm was previously available. The thickness of the metal varies from 0.5 mm to 1 mm. Sheets are manufactured in galvanised metal and are also available in stainless steel.

Rib lath
This is an expanded metal lathing stiffened by steel ribs 10 mm deep and formed in the same sheet of metal. This type is also available in sheets of 2.500 m by 700 mm.

Hy-rib
Hy-rib is steel lathing stiffened by rigid high ribs. It is made from a single sheet of steel, and contains six ribs spaced at 89 mm centres across the width of the sheet. The lengths vary from 2 m to 5 m in 1 m increments, and are 445 mm wide. Each rib is 21 mm high and is made to interlock with the first rib on the adjoining sheet. Hy-rib lathing can be used as the form work and as part reinforcements for concrete floors or roofs. The lathing gives an excellent key for plastering the soffit, or ceiling, when the concrete has set. Alternatively, hy-rib lathing can be used for suspended ceilings, partitions, fireproof constructions and other similar use.

METAL TRIM
Metal trims used for plasterwork include metal angle beads, screed beads, picture rails, casing beads and bell cast trim.

Metal angle beads
These have a straight true nosing attached to two expanded metal wings, all made from one sheet of galvanized steel, and supplied in lengths varying from 2.4 m to 3 m (Figure 36).

Thin coat beads
These are made in two sizes, 3 mm and 6 mm, to provide a straight protective external angle for thin coat work at the thicknesses quoted. They are especially valuable for use on plasterboard one coat work external angles. Lengths are as for metal angle beads. The wings are pierced steel with scalloped edges (Figure 37).

Plaster stop beads
Plaster stop beads are used to provide a straight protective finish against opening or abutments such as door openings. They are made in four depth sizes for plaster thicknesses of 10 mm, 13 mm, 16 mm and 19 mm (Figure 38).

Perforated thin coat stop beads
These are also available for 3 mm and 6 mm thicknesses (Figure 39).

Plasterboard edging bead
This is a reversible dual effect bead which protects the ends of plasterboard and assists in fixing, and is provided for 10 mm and 13 mm plasterboards, in lengths of 3 m (Figure 40).

MATERIALS

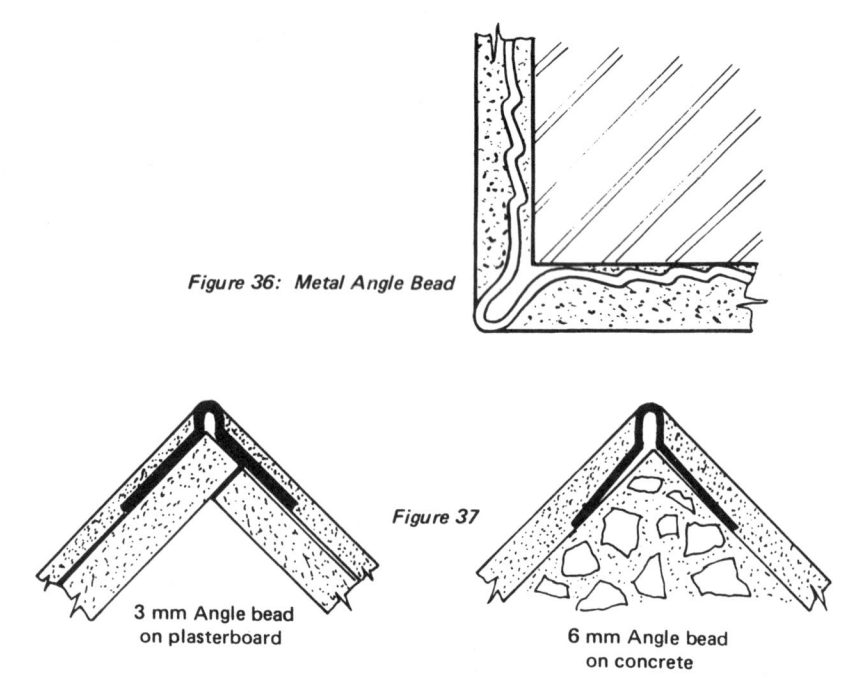

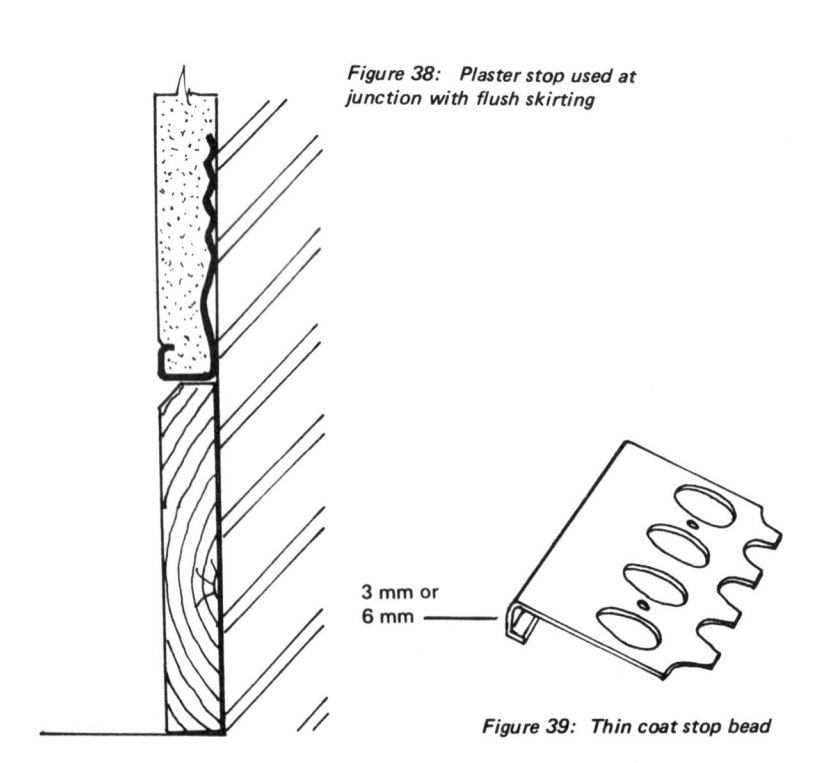

Figure 36: Metal Angle Bead

Figure 37

3 mm Angle bead
on plasterboard

6 mm Angle bead
on concrete

Figure 38: Plaster stop used at junction with flush skirting

3 mm or
6 mm

Figure 39: Thin coat stop bead

METAL TRIM

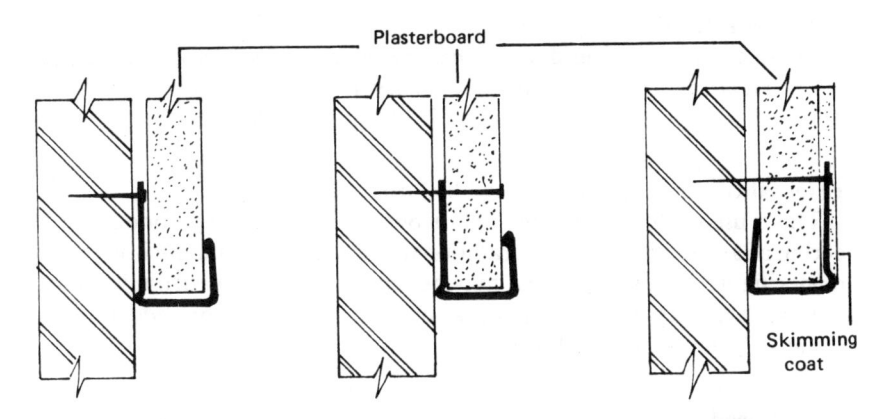

Figure 40: Plasterboard edging bead. Note alternative fixing methods

Sectional views showing method of fixing

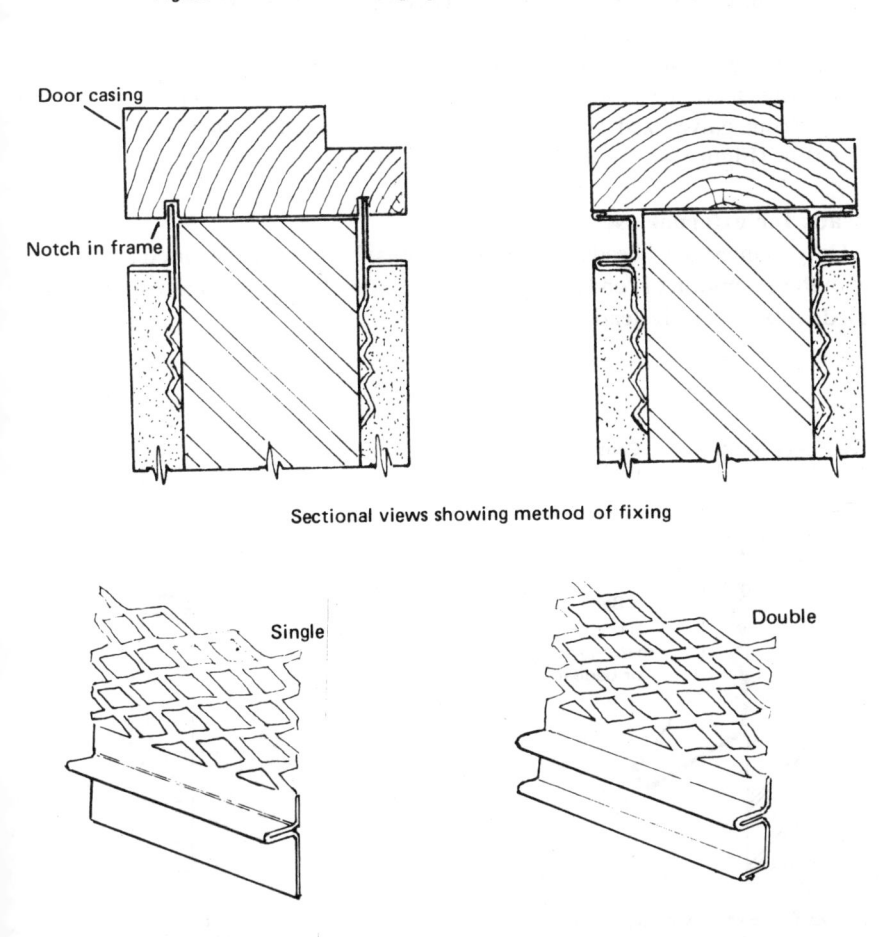

Figure 41: Expamet architrave beads

MATERIALS

Architrave beads

These are used around door frames, acting as a plaster stop, and also providing an attractive shadowline effect. They are available in two types (Figure 41), and are supplied in lengths of 2.300 m and 3.000 m.

External render stop (also termed bell cast stop)

This is used to form a neat projection over window openings for outside rendering as an alternative to the traditional bell casts. Its main use is in external rendering where a stopped end may be required. It is made from galvanized steel with an expanded metal wing for fixing with nails or bedded. This is made in lengths of 3.000 m (Figure 42).

Screed bead

This is used to provide a division between two types of plasterwork on one surface, such as a plastered wall with a flush cement skirting. The flush bedded screed beads provide an excellent bearing surface when floating, but skimming clearance should be made before the setting coat is applied. This is made from galvanized steel with two expanded metal wings in lengths of 3.000 m (Figure 43).

Beads for external use

Only stainless steel beads and special external beads should be used externally with the exception of render stop. It is a common fault to see galvanised bead used, this generally leads to problems later.

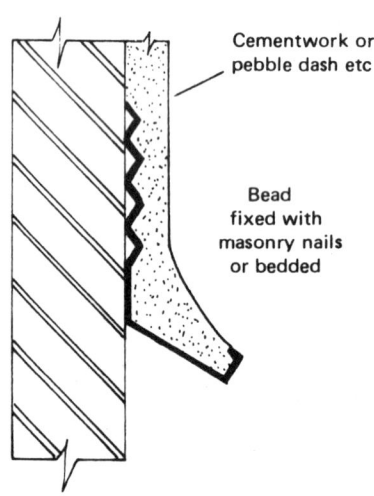

Figure 42: External render stop (Bell cast bead)

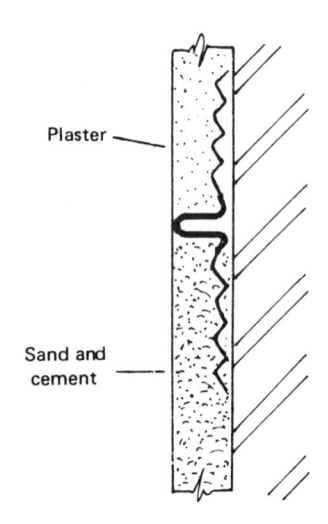

Figure 43: Screed bead used for a flush skirting

METAL TRIM

Corner mesh

This is expanded metal lath 100 mm wide, bent at right angles to provide 50 mm wings and for use in internal angles to reduce the risk of cracking. It is available in lengths from 2.500 m.

Strip mesh

Strip mesh is similar to corner mesh, but it is supplied flat, and has a normal width of 100 mm. It can be used either as a reinforcement strip in plasterwork or as a protective covering to service pipes or sunken conduits.

Twil-lath

This is composed of interwoven slotted paper with a welded wire mesh covering. The slots and wire covering provide key and reinforcement strength to plaster and cement mix applications. It is made in three types: G (general), SS (stainless steel), and AX (for damp conditions), each in two grades, 400 and 600.

All types are light to handle, can be stapled or nailed to timber joists or firrings, wired to metal runners in fireproof construction, and can be bent and fixed to curved surfaces if required. All Twil-lath 400 types are made for spans of 400 mm, Twil-lath 600 is made for spans of 600 mm. G lath 600 is made in sheets 2.42 by 0.71 m, and has 1.6 mm welded wire mesh 38 mm by 50 mm and stiffener wires (3.0 mm) at 150 mm spacings. The paper is absorbent chipboard paper, slotted to provide key for plastering applications.

AX400 has a similar specification to G lath 600 but it includes a special waterproof backing paper with overlaps and lacks a stiffener wire. It is suitable for use with cement applications to provide backings for tiles, terrazo and other cement finishes on internal work. It can also be used for sprayed plaster, including acoustic finishes.

AX600 heavy duty is similar, but there is a stiffener wire of 3.0 mm incorporated in it.

SS400BP and SS600BP have sheet sizes and wire mesh similar to other types, but in stainless steel. They have a chipboard slotted key paper with a waterproof breathing paper backing, and are used for all types of external applications. Stainless steel staples should be used in external fittings to timber. Rustless wires or clips must be used when fixing to metal constructions. The sheets are crimped 6 mm at 160 mm centres. The water resistant paper acts as a damp-proof membrane, and reduces the risk of dampness.

W or BP, AX400 and AX600 can both be obtained in grades for wet or exterior conditions, the grade description is followed by W, e.g. AX400W. Also available in a grade which has a breather paper attached for timber framed houses, the grade is followed by BP, e.g. AX400BP.

MATERIALS

NEWLATH

Newlath is manufactured in high-density polythene 0.5 mm in thickness formed into a pattern of raised studs linked by reinforcing ribs. The studs project 8 mm from the face of the wall when fixed and this allows air to circulate freely around behind the lath. On the face and attached to the studs is a polythene mesh which provides a key for plastering. Newlath is obtained in rolls 1.5 m in width and 10 m in length. The lath should be fixed at maximum centres of 300 mm using Newlath plugs, sides should overlap 100 mm. This lath is designed to be used over damp backgrounds to provide a waterproof base for plastering. A small gap should be left at the base and top of the wall, this will allow air to circulate behind the lath.

ADDITIVES

Plasticizers

These are materials which are added to plastering mixes and mortars to assist their workability. They are mostly based on resins and they have the property of entraining millions of microscopic air bubbles in the air. The air bubbles in turn act as ball-bearing cushions to overcome the friction of the sand or other aggregate.

These air bubbles remain in the mortar after it has set, and this is an advantage in certain cement mixes where they help to break up the capillary channels and therefore reduce or prevent the penetration of moisture. The air bubbles also, in certain instances, provide expansion chambers for any expansion of water in freezing conditions.

Other types of plasticizers are straightforward lubricants based on artificial soap-like liquids which act as greasy lubricators to overcome the friction of the aggregates.

Bonding adhesives

These materials are similar to glues in that they adhere well to smooth surfaces, and also to applied plaster coats. Many of these adhesives are based on polyvinyl acetate (PVAC) and manufactured in accordance with BS 5270. Other adhesives are manufactured from polymers based on acrylic resins and styrene-butadiene rubber (SBR), the latter now being the most commonly specified bonding adhesive for external use. Styrene-butadiene rubber is not affected by moisture. All the adhesives are recommended for use on most smooth dense surfaces including concrete, facing bricks, and even tiled surfaces. Sound, painted brickwork surfaces can also be bonded with adhesives without hacking, but because the paint to background soundness is difficult to determine it should be used with caution.

The application of bonding adhesives varies according to the type and brand used. All background surfaces must be dust free, clean and sound before other treatment is attempted. Bonding adhesives are applied using one of the following methods.

- (a) *Brush:* using a diluted priming coat, followed by a coat undiluted and the plaster applied before the adhesive dries.
- (b) *Stipple:* the adhesive is mixed in with a cement slurry and a small proportion of sharp sand added. The slurry is scrubbed and stippled into the surface to form a key.
- (c) *Key coat:* a slurry coat applied and keyed with a notched trowel forming a key.

A completely different type of bonding adhesive is a bitumen-style solution extracted as a by-product of coat. It has excellent adhesive properties to most smooth surfaces and also has the additional advantage of being waterproof. Thus, it can be used in damp situations to prevent the penetration of moisture from the background and yet provide the base for an application of a gypsum based plaster which will in turn be resistant to the formation of condensation. The application of this bitumen type solution is by brush. While it is still wet, the surface should be blinded with dry sand. The waiting period before application of the floating coat should be twenty-four hours.

Note: all bonding adhesives should be used in accordance with the manufacturers' instructions.

Glass fibres

Glass fibres are incorporated in certain plaster and sand/cement mixes to improve the strength and abrasion resistance of these materials. Chopped strands of glass fibres are included in proprietary pre-mixed sand and cement rendering materials, available with either white or ordinary Portland cement as the binder.

Special chopped strands of glass fibre are also available for inclusion with gypsum plaster mixes, the finished work being known as GRG (glass-reinforced gypsum). Such mixtures are the modern counterpart of haired plaster mixes, but the increase in strength is incomparable.

Strict cleanliness should be observed in handling glass fibres, and in certain cases, the use of gloves may be advisable.

WORKSHOP MATERIALS

Shellac

Shellac is used to seal the plaster surfaces of models or moulds. Set plaster contains thousands of voids and these act as capillary tubes when coated with liquids, oils or pastes, etc. During the preparation of a plaster mould it is important to prevent this suction before the grease or oil is applied to ensure that the subsequent cast will free readily, after setting, from the mould.

Shellac, as used in the workshop, is made by dissolving shellac flakes in methylated spirits. The thickness of the solution can be regulated by varying the proportion of shellac flakes to that of the spirits. A very thin solution is ideal for the first coat on damp plaster, and mixtures of increasing thickness used in building up a dense surface coat. At least three coats are normally used, but more may be necessary for certain types of work, i.e., plaster piece moulds for cement castings.

Thick or burnt shellac is a very thick treacly solution used for bedding dry pieces of plaster moulding. This saves keying and the soaking of plaster surfaces, and is a good, quick method in certain circumstances. It is also used in some workshops as a glue to stick on the paper profiles to sheet metal as a guide for cutting and filing.

Thick shellac is made by dissolving the shellac flakes with the minimum possible amount of methylated spirits. Burnt shellac is made by burning off the excess methylated spirits from a normal solution of workshop shellac.

Shellac flakes are made from the secretion of the lac, an insect found mainly in India. It is transformed into an amber-like material with a scaly covering, by the action of the insects on the tree surface. The dried secretion is collected and further processed by washing, heating and squeezing, until it emerges in the familiar flaky form known as orange shellac flakes.

Oils and greases

These are used to provide a separation film between model and mould, and between mould and cast. A typical plasterer's grease can be made by heating tallow in a container on a hot plate until melted, and then adding an equal amount of machine oil or a similar type of oil. The type and amount of oil used can be varied according to the temperature at which the grease is to be used. On hot days less oil will be required, or alternatively, a paraffin oil which is less greasy, can be used.

Oils which are used neat for special types of moulds, such as gelatine, etc, include Colza, Rapeseed and linseed oils.

Size

Most natural glues will, when dissolved in hot water, slow down the setting action of plaster if used in the gauging water. Gelatine skins from the melting pot are often used for making up size in the workshop, but fresh glue cake or glue size can be used.

The glue should be melted in a container with hot water until completely dissolved. To ensure that the solution does not set back into a jelly when cool, a small amount of lime putty should be dissolved in the solution before cooling. If new gelatine is used, it is likely to give off an offensive smell due to its organic nature, but this can be avoided by dosing the solution of size with a small amount of carbolic acid.

Gelatine

This is a superior type of glue which is made from collagen and is obtained from cartilage, skin, sinews, or the bones of cattle. It is supplied in cake, powdered or slab form. The best quality cake gelatine is brittle.

When preparing the latter type for use in flexible moulds, the cake gelatine should be soaked in water until flexible. The excess water can then be drained off and the gelatine heated in a water-jacketed container. Gelatine melts at the boiling point of water, 100°C. When completely melted, a small amount of carbolic acid should be added to the new gelatine to kill off any living organisms. The latter would, if allowed to develop, cause a fungus or mould to grow on the surface of the cooled gelatine over a period of time.

Alum

Alum is used for accelerating the setting action of plaster, and also for the pickling, seasoning or hardening of gelatine mould surfaces.

The alum crystals (aluminium sulphate) should be dissolved in hot water and, when the solution is cool, a small amount can be added to the gauging water for quickening the set of the plaster. This quicker setting action is useful when gelatine moulds are used in hot weather. The longer a case is in contact with the gelatine, the greater the chance of mould damage due to the heat derived from the chemical action of setting plaster. Crushed alum will also dissolve directly, but more slowly, in the gauging water.

PVC (polyvinyl chloride)

This is a flexible moulding compound which is thermoplastic and is termed a hot melt compound. It is a modern alternative to gelatine and is a non-organic substance with properties similar to rubber. When compared with gelatine it has the following advantages:

1 Strength: it can be used in thinner mould thicknesses hundreds of times without noticeable deterioration.
2 Density: it does not require oiling or greasing, and no surface hardening or pickling is required.
3 Non-organic: it can be used over long periods without risk of undue shrinkage or warping.
4 Water and normal heat resistance: it is not affected by humidity, dampness or normal heat ranges.

Unfortunately PVC has the following disadvantages:

1 It has a high gelling point (120°C—170°C, depending upon the type), and special heating appliances need to be used for melting.
2 The high melting temperature would destroy the normal shellac model preparation, and special treatment for plaster surfaces is required.

3 Mould surface detail may be inferior in some cases, due to the special difficulties described above.

4 Greater care needs to be used in handling the hot melt. Excessive amounts of the fumes given off can be harmful.

Moulding wax

This consists of a mixture of approximately equal parts of beeswax and resin. It is prepared by melting in a container on a hot plate, but the molten wax must not be allowed to boil. The resin gives rigidity to the wax when cool, and its proportion in the mix can be varied to suit the requirements of the finished wax mould. This type of mould is used when clean, sharp, fine detail is required from a mould with no undercut parts.

Scrim

Scrim is made from hessian or jute, and has a mesh of 4.5 mm — 6 mm. It is made in widths of 850 mm in rolls of up to 450 m long, and this type is suitable for fibrous plastering reinforcement. For solid plastering, the normal size of the rolls of hessian scrim is 87 mm wide and 90 m long. Scrim is also termed 'canvas' by plasterers.

When used in solid plastering work, the purpose of scrim is to give strength to the joints in plasterboard ceilings and partitions, and to prevent the formation of cracks along these joints.

In fibrous plasterwork, scrim is used to provide a fibrous core to thin plaster coats, giving strength to an otherwise relatively weak material. Scrim is also used to seal and strengthen the joints between casts when fixing.

Other types of scrim used in solid plastering work are metal, plastic, cotton and paper. These are mainly 75 mm wide strips and, in the case of cotton, used for thin coats, although its strength is much less than that of hessian.

Sheet iron and zinc

These are used for making the profiles for running moulds. Zinc is a fairly soft non-ferrous metal which is easy to cut and file. Because of these reasons, and the fact that it does not rust, it was much favoured by plasterers in the past who had to keep stock running moulds for long periods. The price of zinc nowadays is high compared with that of iron, and as a result, sheet iron is mainly used in industry today.

French chalk

This is a smooth, finely ground powder which is used as a dry lubricator in certain instances during moulding and casting processes. It has properties similar to those of talc.

Clay

Clay is used as cottles or fences, and for other processes in model making and casting in benchwork. It should be kept pliable by storing in an air-tight container. Small batches can be stored by wrapping them in a damp cloth inside a closed polythene bag.

Fibreglass reinforced plastics

These are being used increasingly in industry for a wide variety of constructions. Many production techniques are used, some quite complicated, and involving the use of machines and engineering facilities. However, certain hand-production contact moulding methods are similar in technique to fibrous plasterwork moulding and casting systems.

Materials used for fibreglass hand lay up and similar types of moulding include the following:

Resins: Thermosetting polyester resin, manufactured in a comprehensive range for specialised work, is the type most suitable for hand lay up methods. Two grades of resin are used in casting. A *gel coat* is used first to give a better surface finish to the cast and to hide the raised fibre pattern effect which may otherwise be noticeable. The *lay up resin* is used for the follow on application. (The use of gel coat and lay up resin are similar in technique to firstings and seconds in fibrous plaster castings.)

Polyester resins are normally pale viscous liquids which should be stored before use in cool conditions in closed containers. They have a limited life unless stored under the conditions stated, but when stored correctly will last for at least six months.

Gel coat resins are thicker in consistency and contain thixotropic properties which are useful when working on vertical or irregular surfaces during moulding or casting processes.

Setting action of these resins is achieved by mixing in small quantities of specially developed catalysts. For gel coats, 2—3 per cent (22 cc—34 cc catalyst to 1 kg of resin), and for lay up resin, 0.50—1 per cent (7 cc—11 cc catalyst to 1 kg of resin).

The setting action is normally too slow for practical purposes, and resins are usually supplied pre-accelerated. This is a safety measure because catalysts and accelerators when mixed together are explosively reactive.

Glass reinforcement: most of the laminates used as reinforcement are based on glass fibre chopped strand mat or similar types. Chopped strand mat consists of fine glass strands about 50 mm long, laid down in a random pattern and held together in the form of a mat by resinous binders.

They are produced in rolls with standard widths of approximately 1.05 m and 1.425 m. Surfacing mat is a thin fibrous veil to provide a resin rich surface. It is used on the surface of the building panels, and translucent sheets for improved weathering properties. If surfacing tissue is used, the gel coat may be omitted and the glass tissue placed on the mould surface to be followed by the chopped strand mat at the lay up stage.

Glass filaments are wound on mandrels or bobbins in the form of roving for use in spray contact moulding. The roving is chopped into strands by a chopper unit and projected on to the recently applied resin from a spray gun. Chopped glass strands of various lengths up to 50 mm are also available for use in other types of glass reinforcement techniques, including GRG (glass reinforced gypsum), and for use with Portland cement/sand rendering mixes. mixes.

Plaster moulds used for fibreglass casting should be sealed with shellac and coated with wax polish. Special release agents, mostly based on solutions of polyvinyl alcohol in methylated spirits or water, should be applied to the mould surface before coating.

Tools and brushes used in casting should be cleaned immediately after use by washing out in a suitable solvent such as cellulose thinners, trichlorethylene or acetone. Parting agent brushes can be cleaned in soapy water.

CHAPTER THREE

Internal Plastering

The purpose of plastering the walls and ceilings inside buildings is to provide a surface that is sound and of good appearance. In most cases it also contributes to the sound and thermal insulation of these buildings and in many instances to its fire protection.

Backgrounds

These are the surfaces to which the first coat of plaster is applied and they include brickwork, stonework, coke breeze blocks, concrete bricks, hollow clay tiles, *in-situ* and precast concrete, no-fines concrete, wood wool slabs, Thermalite blocks, cork slabs, expanded metal lathing, Twil-lath, Newlath, plasterboards and plaster lath.

It will be seen that the number of different types of backgrounds, each with its own properties and characteristics regarding suction, mechanical key, strength, etc, require differing techniques when plastering on them.

Preparation of backgrounds

Most background surfaces in new buildings require no preparation before plastering and these include plasterboards, expanded metal lathing, wood wool slabs and similar blocks. Brickwork, if new, clean and with reasonable key, is suitable for immediate application. If however, the brickwork is dirty, dry, smooth or composed of dense facing bricks then it may require brushing, raking out of joints, damping, hacking or coating with a bonding adhesive as necessary.

In-situ and precast concrete are certain to require one or more of the following: hacking, cleaning, spatterdashing or coating with a bonding adhesive. In extreme cases when adhesion is difficult to achieve, the surface may have to be covered with expanded metal lathing which is nailed to plugs in the concrete or on to timber battens.

Composite backgrounds containing different materials are often a source of cracking or loss of adhesion. Concrete or timber lintels over doorways and windows can cause trouble unless treated. Timber wallplates and frames to partitions should be covered by plasterboard, expanded metal lathing or scrim.

Smooth, painted surfaces may be treated with a bonding adhesive if the paintwork is sound. Water soluble paint and distempered surfaces are not suitable for bonding adhesives and hacking thoroughly is the only suitable method apart from covering with expanded metal lathing.

Plain interior plastering

This can generally be described as either one, two or three coat work. The number of coats required is decided by the conditions on site. These will include considerations of background, thickness of application, surface finish, sound or thermal insulation requirements, fire precautions, waterproofing, strength, hygiene and cost.

One coat work is mainly confined to skimming, or setting, coats on plasterboard and lath. Other examples of one coat work include skimming coats on straight backgrounds such as precast or vibrated concrete, and on various types of fibre building board and cork slabs. The reasons why one coat work is advised for the backgrounds given include cost, bonding risks and weight.

Cost is reduced by the amount of material used and the speed of completion. The risk of bonding failure is less with thin coats than the thicker two or three coat work. A reduction in the overall weight of the finished plasterwork is an advantage in certain types of structures, such as suspended floors, ceilings, etc.

A modern technique of one coat hand applied plastering is the use of a gypsum based plaster with perlite and other additives to control the setting time and give other desirable properties. The plaster is supplied in 33 kg bags under the trade name of Snowplast and Thistle Universal.

It can be applied to most backgrounds including concrete, brickwork, plasterboard and expanded metal lath. The latter will need a pricking up coat applied and allowed to stiffen before a further one coat finish is applied. Dense concrete should be treated with a bonding adhesive. High suction surfaces should be primed also with a dilute solution of bonding adhesive to reduce excessive suction.

Mixing of the plaster can best be achieved by adding the plaster to water in a large tub and using a heavy duty drill with a whisk attachment.

The mixed plaster should be applied to the surface to the following recommended thicknesses. Plasterboard 3 − 5 mm; concrete 5 − 8 mm; bricks and blockwork 10 − 13 mm. Greater thicknesses will require dubbing out applications. The applied plaster should be ruled off to an even surface and allowed to stiffen after which a large spatula or trowel can be used to improve the flatness. When the surface is firm it is scoured with a damp sponge faced float. Overwetting of the surface and the working up of too much fat should be avoided.

After a further period of setting the surface should be given an initial and then a final trowelling when hard. Overpolishing should be avoided.

Two coat work is carried out on a wide variety of backgrounds including brickwork, clay tile blocks, coke breeze, masonry, concrete, plasterboard and lath, etc.

If the background to be plastered is fairly even then good results can be expected from two coat plastering. However, if there are several thick places in the brickwork or stonework these are best filled in before the floating coat is applied. This filling out of thicker places or hollows is known as *dubbing out*.

Normal two ooat work is carried out by applying a floating coat of approximately 13 mm thick, straightening, rubbing up and keying. This is followed later after the floating coat has set by a thin skimming or finishing coat of 3 mm or less.

Three coat work is essential for successful results on lathwork and is preferable for high class work on brickwork and similar types of background.

The purpose of the first or rendering coat work is to even out the suction for the floating coat. This enables any irregularities in the background, and any differences in thickness and suction to be reduced or eliminated. The differences in suction are often caused by different materials in the background, such as concrete lintels, stone cills, timber wallplates, etc.

Besides removing the risks of shrinkage cracks developing due to insufficient or irregular suction, the extra thickness of three coat work increases the sound and thermal efficiency and fireproof protection of the structure.

Three coat work on brickwork should not normally exceed 19 mm. Thicknesses of finished plasterwork on metal lathing should be about 13 mm for lightweight gypsum plaster and 13 mm for other types of plaster (thicknesses measured from face of laths).

METHODS OF APPLICATION

The *rendering coat* should be applied about 9 mm thick roughly straightened with the trowel and then scratched deeply for key with a wire scratcher. This should be allowed sufficient time to shrink or set, according to the type of material used, before the application of the floating coat.

Floating on most of the building sites is straightened with a darby, filled in with the trowel and rubbed up. This often results in inferior work with hollow or bulging wall surfaces and bent angles. A far safer method when using a darby to straighten the floating is to form the angles first with the aid of a long floating rule. The angles can then be used as a guide or screed for the darby and much straighter floating will result.

A better, though slower, method is to use floating screeds. This can be accomplished by vertical or horizontal screeds. In the simplest method a band of the floating mix about 75 to 300 mm wide is placed across the width of the wall at a reasonable height above the plasterer's head, and of average thickness. This is straightened with a straight edged floating rule about 2.1 m long. A similar screed is placed about 150 mm up from the floor. After the screeds have stiffened the wall space between the screeds is filled in stages working from the right. When a patch about 600 mm wide has been applied the rule is worked over this portion keeping the edges tight to the upper and lower screeds. The surplus floating is ruled off exposing any hollows which are filled in and the process repeated along the wall until it is completed.

INTERNAL PLASTERING

TRUENESS OF PLASTERING SYSTEMS

It must be realised that two coat work in plastering can only correct minor irregularities or deviations in the background. If however the background is reasonably straight and plumb and any grounds or linings to be worked to are fixed correctly, then the finished plastering of not less than 13 mm thickness should be reasonably smooth and straight to within a deviation of not more than 3 mm in 1.8 m (1 in 600).

This in in accordance with British Standard Code of Practice for Internal Plastering (BS 5492).

In the *Plumb, Dot and Screed* method a much more accurate floated surface will be obtained with little extra work.

Firstly a small pat of floating material is placed at the top right hand corner of the wall about 150 mm from the ceiling and return wall. Into this pat of material is bedded a 75 mm lath so that its exposed surface is about 12 mm from the face of the wall. A similar lath is bedded at the top left hand corner of the wall. At the base of each wall under each lath is bedded another lath and this is fixed plumb beneath the upper one. It is done by using a plumb bob and line held suspended from the upper dot by a gauge. A similar shaped gauge is held in front of the bottom dot which is now adjusted backwards or forwards until the shoulder of the gauge just touches the plumb line as shown in the sketch (Figure 44).

When both pairs of corner dots have been plumbed, intermediate dots are lined in about 2.1 m apart. Screeds can now be formed between the dots and when complete can be used for ruling in as described previously (Figure 45).

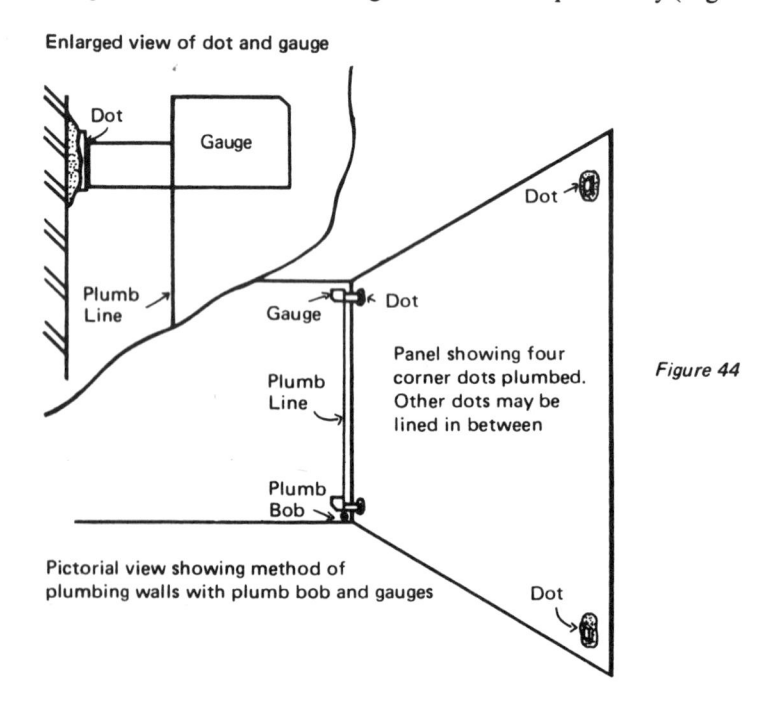

Figure 44

TRUENESS OF PLASTERING SYSTEMS

Figure 45

Plumb, Dot and Screed Work

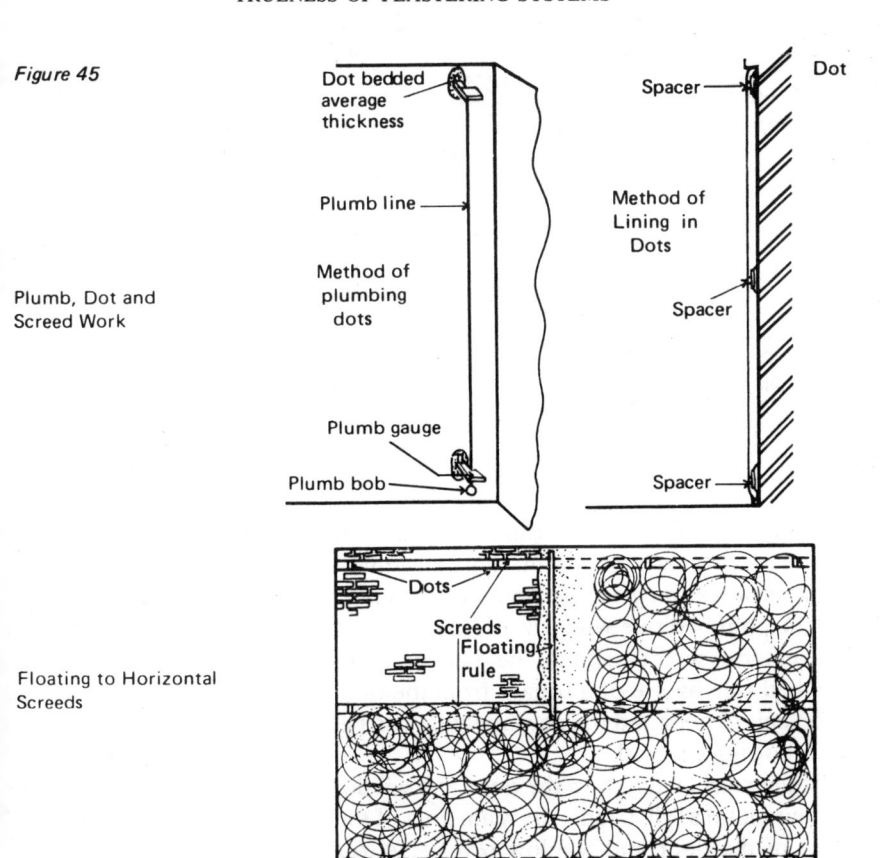

Floating to Horizontal Screeds

A convenient method of lining in dots is to use three equal thicknesses of timber, or similar material, as spacers. The method is to place one spacer on each of the corner dots and hold the line tightly over these. A dot is bedded in the position required between the corner dots and tested for being lineable by sliding the third spacer on top of the middle dot up to the line. Some plasterers prefer to use three nails as spacers but three pieces of 6 mm or 12 mm timber are easier to use (Figure 46).

When floating a wall with attached piers corner dots are plumbed at each end of the wall as previously described for a straight wall. The projection of the piers is measured and three gauges are cut having this length of shank to the shoulder of the gauge. Additional dots can now be lined in each recess by holding a line from gauges on each corner dot and using the third gauge to position each inside recess dot in turn. Dots on the face of the piers are bedded to the line.

From these dots the screeds can be formed and later the wall recesses and pier faces can be floated in and ruled off. All these wall surfaces will be plumb and lineable.

INTERNAL PLASTERING

Lining in dots

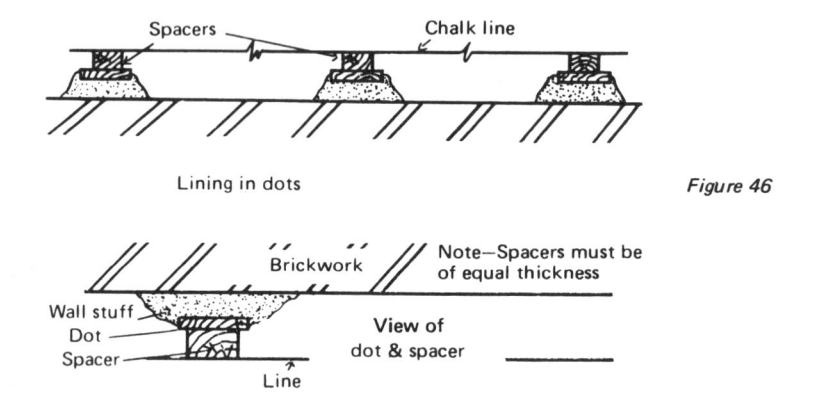

Figure 46

View of dot & spacer

Note—Spacers must be of equal thickness

To form the returns rules can be nailed, or held, plumb to the face of the piers, checking each pier for margin and equal widths. If the returns are only 110 mm or less they can be floated and ruled in off the floated recess wall and the plumbed rule with a scaffold square. If the returns are too wide for this method screeds should be bedded horizontally to the face of the plumbed rule by use of a square from the floated recess wall (Figure 47a).

Reveals to door and window openings can be floated plumb and square by the following method.

If the external angles to the door and window openings are to be formed without metal angle beads, the main area of the wall is floated first up to the edges of the door and window reveals. A rule is held or fastened to the floated wall surface, plumb and linable with the door or window frame, allowing a floating thickness of about 12 mm. The rule should first be positioned by squaring at the base from the frame, having first decided on a suitable margin. This margin, wherever possible, should be kept the same on each window frame throughout the building.

The rule is plumbed by sighting through to the window frame or squaring from the equal margin at the top of the frame.

A floating gauge is cut to the required margin as shown in Figure 47b. The gauge is then used to rule in the applied floating, bearing off the rule and window frame as shown in the sketch.

With this method, all the reveals in the building can be floated plumb, square, and the margin on each frame will be parallel and the same width throughout.

The method of forming external angles to reveals with metal angle beads is similar to that described previously, except that the metal beads will have to be fixed before the main wall area is floated. Extra care will have to be taken to ensure straightness of the beads between upper and lower squaring points if this method is used, and also the width of the reveal itself will have to be checked for parallel.

TRUENESS OF PLASTERING SYSTEMS

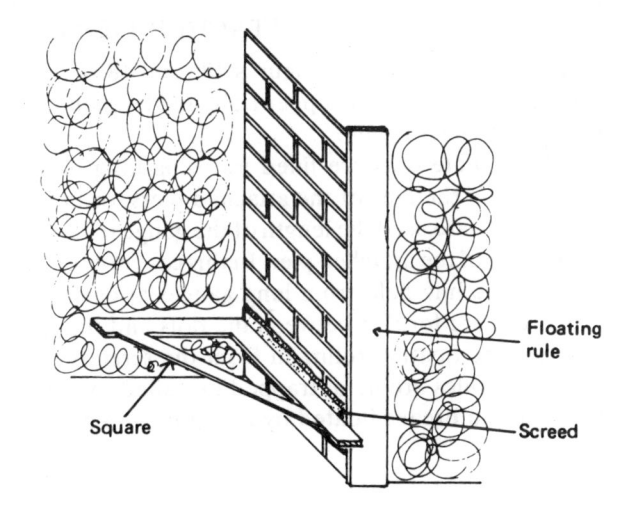

Figure 47a: Squaring breaks or reveals

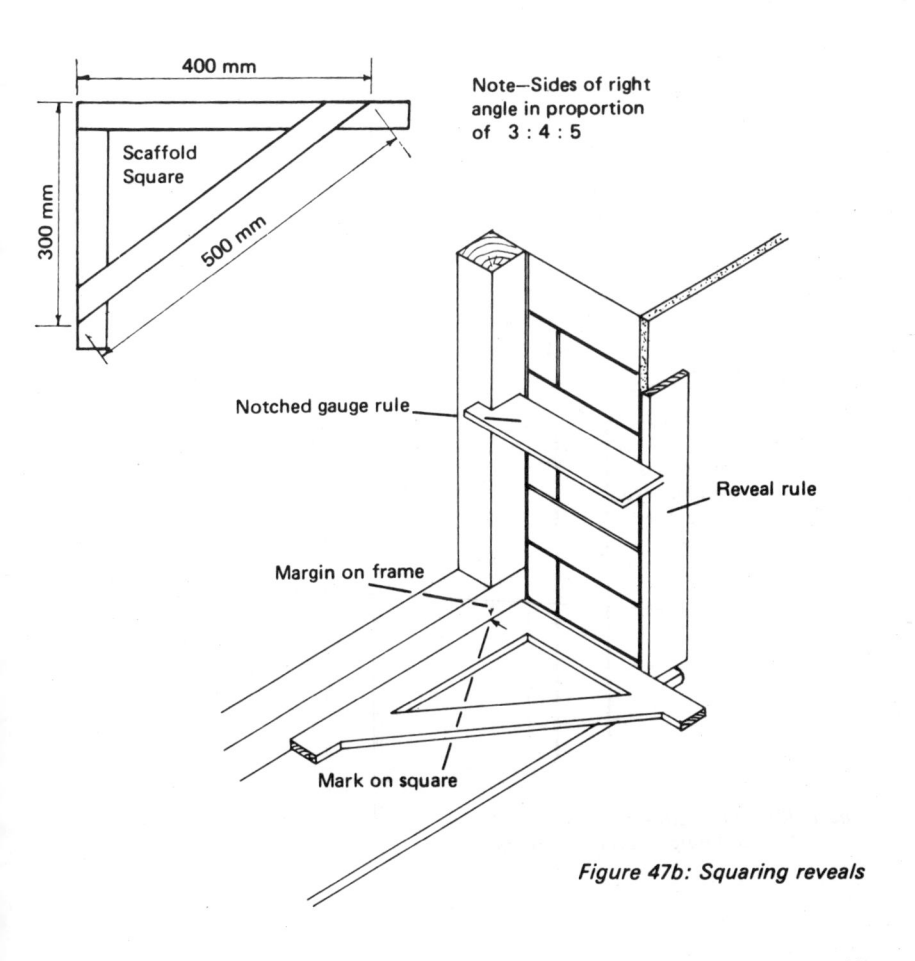

Square — Floating rule — Screed

Scaffold Square
400 mm
300 mm
500 mm

Note—Sides of right angle in proportion of 3 : 4 : 5

Notched gauge rule
Reveal rule
Margin on frame
Mark on square

Figure 47b: Squaring reveals

43

INTERNAL PLASTERING

ATTACHED PIERS

Two methods are shown (Figures 48a and 48b). In each case a dot is bedded average thickness at the upper right and left hand ends of the wall. Dots are plumbed beneath each.

Intermediate dots are lined in, using long or short gauges to establish positions on the face of the attached piers, or in the recessed walls. If long gauges only are used the length of the gauge to the shoulder should equal the projection of the attached pier. In this case dots on the face of the pier will be fixed to the line held from long gauges at each end of the wall recesses.

Screeds are formed between the dots and the faces of the piers and wall recesses are floated, ruling in from the screeds.

Timber rules are held, trapped or nailed to each side of the floated pier face and the breaks floated by ruling in with a square off the main floated

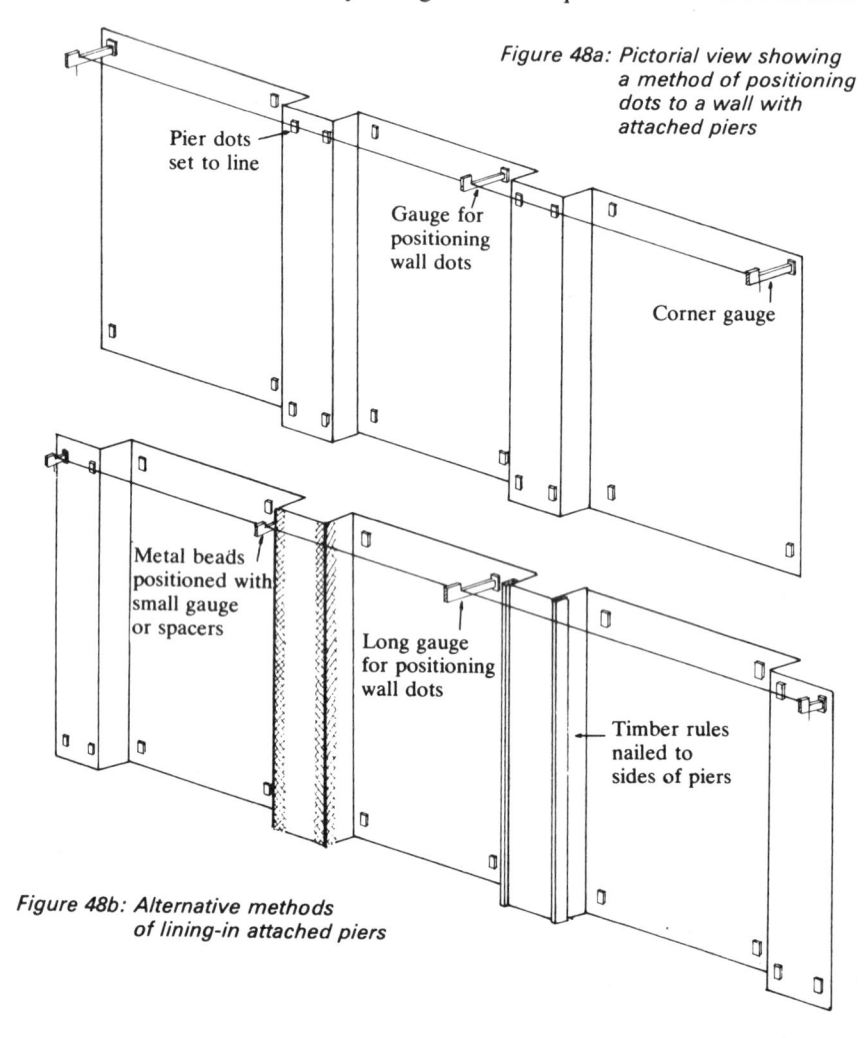

Figure 48a: Pictorial view showing a method of positioning dots to a wall with attached piers

Pier dots set to line

Gauge for positioning wall dots

Corner gauge

Metal beads positioned with small gauge or spacers

Long gauge for positioning wall dots

Timber rules nailed to sides of piers

Figure 48b: Alternative methods of lining-in attached piers

EXTERNAL ANGLES

wall and bearing against the rule (Figure 48a).

Alternatively metal beads can be bedded to the line at top and base, straightening and plumbing between these points.

DETACHED PIERS

In a room with a row of detached piers or square columns the following method can be used. Decide on a floating thickness on an end column, mark the floor, note the width between marks and repeat the floor marks adjacent to the other end column. Check that the widths are similar and then snap parallel chalk lines (Figure 49a). Draw lines at right angles from the chalk line on each side of each column, allowing floating thickness and checking margin widths in each case.

When hard angles are being formed, timber rules can be nailed to the two external angles of each pier or column, plumbed from the long parallel lines. The area between the rules can then be floated, the rules taken off, re-fixed on the opposite side and this side can then be floated to the rules. Margins should be checked top bottom and middle for accuracy.

The side walls are formed in a similar manner by nailing rules plumb from the right angled cross lines, again checking for margins and squareness.

Figure 49b shows a method using metal angle beads. In this case the outer beads must be plumbed up from the intersection of the long parallel line and the outer right angle cross line. Other beads are fixed from the external angle mark on the floor to a chalk line at the top, checking the margins for width.

EXTERNAL ANGLES

These can be formed, planted or run. External angles which can be formed include square arris, pencil round, 15 mm round, bull nose and splay (Figure 50). Planted or fixed angle beads are usually metal nosing beads with perforated metal or expanded metal lath wings. Timber angle beads are rarely used these days. Run angle beads include a variety of moulded shapes, the most common being sunken ovolo and double quirked bead.

Forming a simple square arris to an external angle can be carried out as follows. The angle should first have been floated plumb and square with the aid of a floating rule. 50 mm x 12 mm planed rules should be tacked down one side of the external angle allowing skimming thickness on the side to be skimmed first. The rule should be plumbed and straightened, after which the finishing coat can be applied in a band 75 mm wide and tight to the rule. When the material has been laid down and trowelled off it should be cut off neatly from the rule which may now be removed. The return wing is now skimmed tight to the arris and finished in the same way as the first half of the angle. The sharpness of the angle itself should be softened down by rubbing over with the back of a wet trowel when finishing.

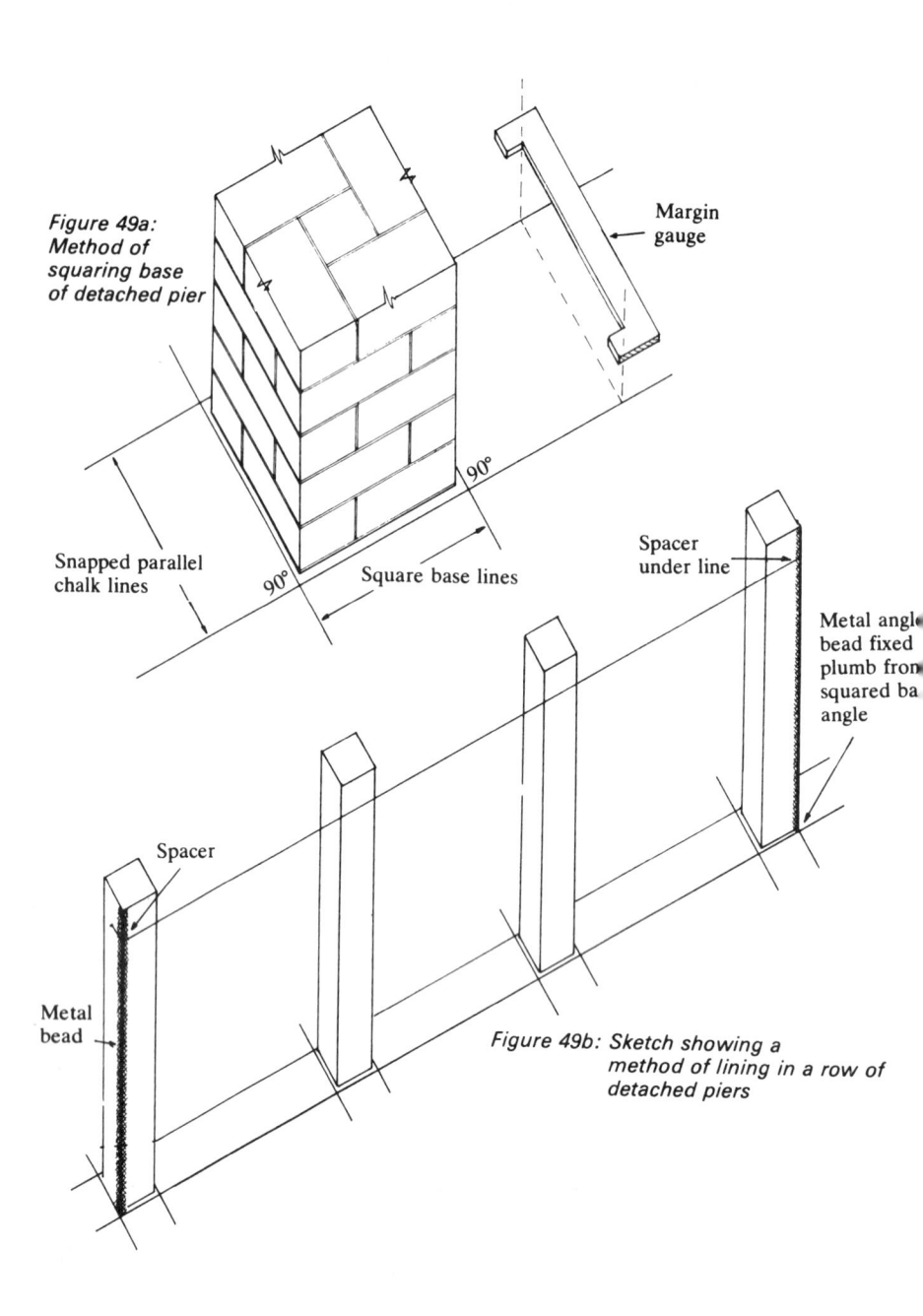

INTERNAL PLASTERING

Figure 49a: Method of squaring base of detached pier

Margin gauge

Snapped parallel chalk lines

Square base lines

90°

90°

Spacer under line

Metal angle bead fixed plumb from squared base angle

Spacer

Metal bead

Figure 49b: Sketch showing a method of lining in a row of detached piers

EXTERNAL ANGLES

Figure 50: External angle sections

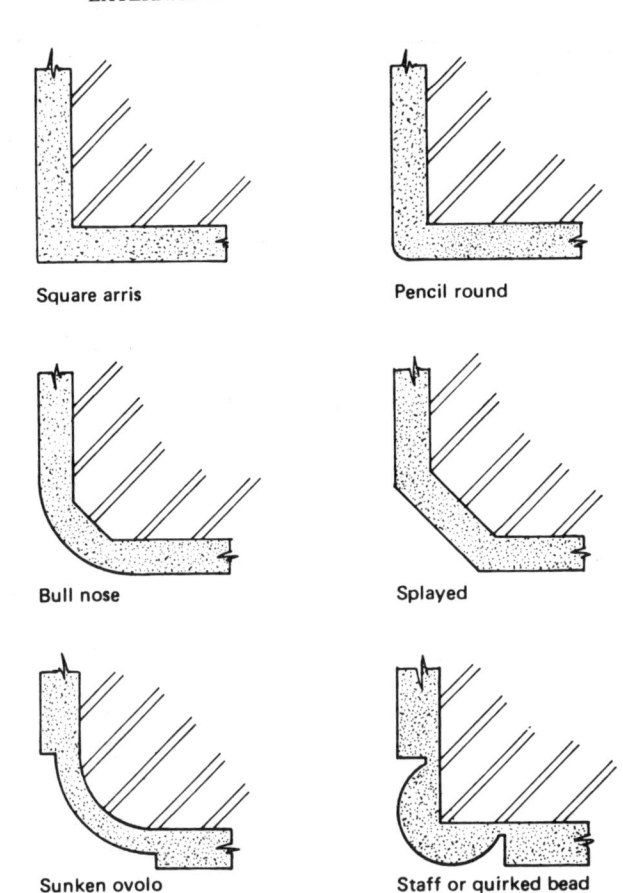

Square arris

Pencil round

Bull nose

Splayed

Sunken ovolo

Staff or quirked bead

Because of the weakness of this shape of angle, strong plastering materials should be used and in good class work the floating coat would be sand and cement and the finishing coat, a hardened plaster such as Crystacel or Herculite.

Pencil round angles are usually formed square in both floating and skimming applications and rubbed round to the desired shape with darby, straight or cross-grained floats.

Larger radius angles are usually turned with the aid of straight edge and darby in the case of floating coats. The skimming is either applied freehand with the trowel and straightened with darby and float or else a special angle trowel is used.

If a number of bull nosed angles are to be formed in one room it is advisable to cut a template in the form of a quadrant and a screed formed top and bottom of each external angle (Figure 51a).

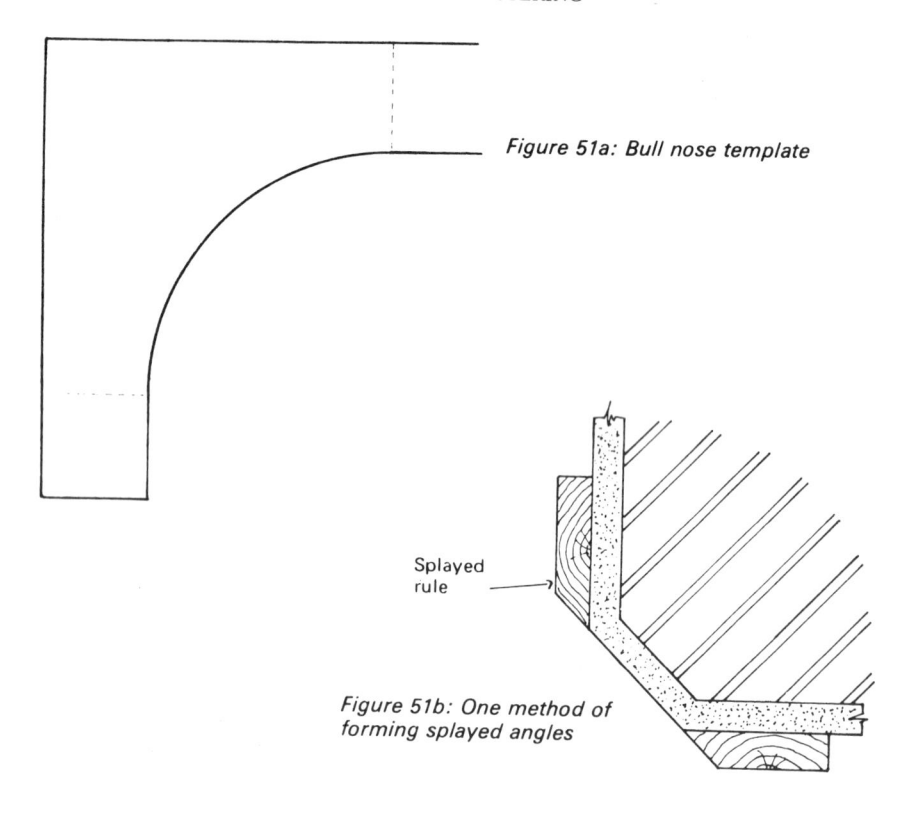

Figure 51a: Bull nose template

Splayed rule

Figure 51b: One method of forming splayed angles

Splayed angles can either be formed or run. One method of forming splayed angles is with the aid of two splayed rules. The floated splay can be formed by measuring back from the extended angle on each wing, marking the extent of the splay and cutting back to the mark. This can now be rubbed up and keyed. The rules are now fixed as shown in sketch, care being taken to check for margin, and then the face of the splay is filled in to the rules (Figure 51b).

Metal angle beads and trims

Metal angle beads are used to form straight arrisses which are resistant to normal damage. They are supplied in a variety of lengths ranging from 2.400 m − 3 m. If required, lengths can be joined together by inserting a short length of 4.064 mm diameter galvanized wire into hollows of the two bead ends to be joined. This will ensure true alignment and continuity of the nosing. (Figure 52).

Fixing the beads can be carried out by one of the following methods:

1 Plaster dabs are applied on each side of the external angle at intervals of about 600 mm. The wings of the metal bead are then pressed into the plaster dabs, and the bead straightened and plumbed by use of a straight edge and plumb level while the plaster dabs are still soft.

EXTERNAL ANGLES

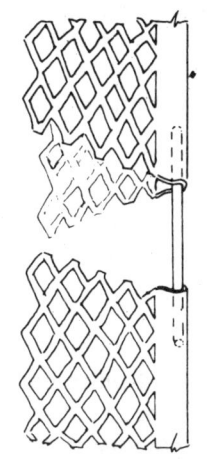

4.064 mm diameter
galvanised wire or nail

Figure 52: *Joining lengths of angle bead*

2 A thin coat of the plastering material is applied to the full length of each angle wing, instead of plaster dabs. While this material is still soft, the bead is bedded, straightened and plumbed as before. Extra applications over the wings may be required to consolidate the firmness of the bedding.
3 30 mm galvanized nails are nailed through the wings at each side of the angle. The nails should first be driven to grip the metal lath wings only until straightening and plumbing has taken place. Any adjustments can be made by tightening the nails or moving the lath and bead to the correct alignment. Plaster dabs or full length applications are then pressed firmly through the meshes of the lath wings on to the backgrounds to ensure a strongly embedded angle.

Cutting of the beads to the desired length is best carried out by sawing through the bead nosing with a hacksaw and cutting each wing with tinsnips.

When plastering to metal angle beads, the floating coat should come just below the level of the bead nosing, and any surplus on the nose itself wiped clean whilst the plaster is still soft. The skimming coat should be slightly proud of the angle bead to protect the nosing when trowelling. Scraping off the zinc coating on the bead must be avoided.

Perforated thin coat beads for use on plasterboarded or other external angles requiring thin coat applications only, are made for finishing thicknesses of 3 mm and 6 mm.

Plaster stop beads are used to provide a strong neat finish to openings and abutments. They are available in four sizes for plastering thicknesses of 10 mm, 13 mm, 16 mm, and 19 mm.

Fixing is carried out by nailing through the holes in the stop beads, or bedding on to plaster dabs. When fastening to metal lathing, the bead can be wired direct to the lath with 1.219 mm diameter galvanized wire. Examples of the use of plaster stop beads are shown in Figure 38.

INTERNAL PLASTERING

Thin coat stop beads are also available for thicknesses of 3 mm and 6 mm.

Expansion joints in a building frame can be covered by two stop beads as shown in Figure 53. To cover the expansion joint, a thin steel plate or similar material can be used under the beads to cover the actual joint, but it should only be fastened to one side of the expansion joint.

Movement bead is specially manufactured for expansion joints but is limited to about 10 mm maximum movement. This bead is basically as shown in Figure 53 with two stop beads back to back but the bead has a PVC extrusion which bridges the beads and is exposed when plastering or rendering is complete. The bead is fixed using the same methods as other beads.

Plasterboard edging bead is a reversible, dual effect bead which provides fixing and protection for the edges of plasterboards. It can be fixed by nailing in different ways as shown in Figure 40. The edging beads are made in lengths of 3.048 m, and for use with 10 mm or 13 mm plasterboard.

External render stop bead is used over window and doorway openings, or at the lower edge of external renderings to form a firm, straight projection. It is sometimes called *bell cast stop bead*.

The fixing of the render stop bead can be carried out by use of masonry nails through the mesh near to the solid portion of the bead and into the

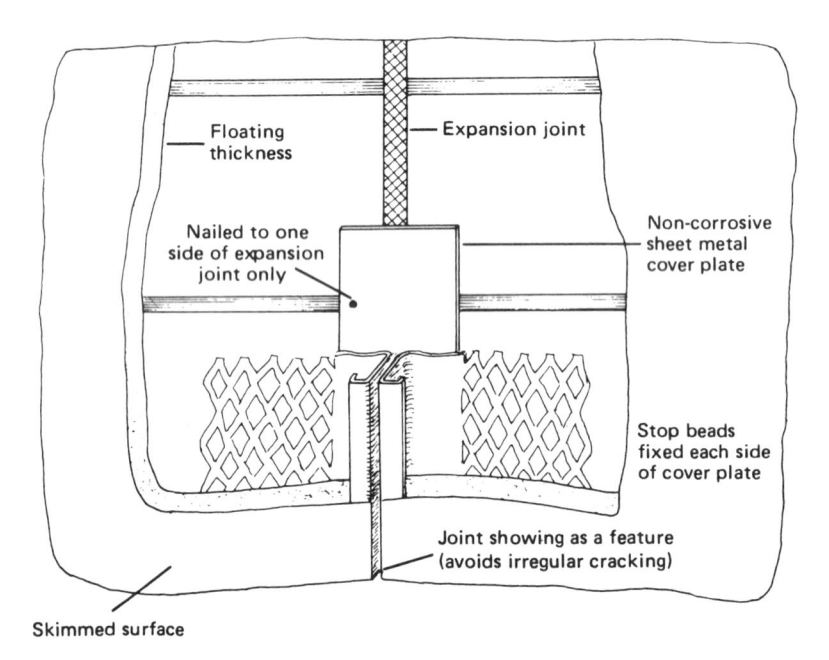

Figure 53: Treatment of Expansion Joint

brickwork. It can also be fixed by bedding into cement mortar dabs (Figure 42).

Screed bead is used in positions to provide a neat division between different types of plastering applications. It can also be used as a skirting bead, either flush or projecting (Figure 43). Fixing is again carried out by nailing or bedding into cement mortar dabs.

The running of moulded angles will be described in the chapter on moulded work.

INTERNAL ANGLES

These are normally finished square but occasionally they are coved for reasons of hygiene or their decorative value. Moulded internal angles will also be described in the chapter on moulded work.

Square internal angles are usually formed in the skimming coat with the trowel, cleaning out any skimming 'fat' with a wet flat brush. If one side of the angle is set and finished first the risk of 'digging in' or damaging is lessened. Occasionally both sides of the angle may have to be finished together and when using soft, weak finishes a special square internal angle trowel can be used to trowel both surfaces together at the same time. A twitcher may also be used for the same purpose (Figure 6).

Small coved internal angles can be formed by using a coved trowel. Strongly gauged skimming material is applied in the internal angle and then wiped out with the coving trowel. As the material sets it is tightened in and trowelled off with the main wall surface.

Instead of a coving trowel a circular bottle or cylindrical tin can be used.

Larger coved internal angles are formed by cutting a template to the desired size and using this to wipe out the floated core of the angle. This type of coved angle is best skimmed with the trowel.

FINISHING COATS

These are termed the setting or skimming coat and vary according to the type of floating coat on which it is to be applied and the desired strength of the finished surface.

In general the finishing coat should not be stronger than the floating, or danger from shelling off the skimming coat is a possibility. This can be likened to a strong brittle icing sugar layer on a soft sponge cake.

Most present day skimming mixes are retarded hemi-hydrate gypsum plasters, either neat, or mixed with small additions of perlite and/or vermiculite.

The skimming mix is applied in an even coat with the laying trowel starting from the left hand side of the wall and working to the right. Each trowelful should overlap the previous one keeping the toe of the trowel tight to avoid gathering on a double thickness at the overlap. In normal skimming techniques this is followed by a very tight coat of a similar or weaker mix and as this begins to set is trowelled up. The trowelling up is accomplished by going over each part of the wall methodically at least twice with a clean trowel lubricated with water. The trowel is held at an angle of about 45 degrees or greater to the wall surface and the wall is traversed using a fair amount of pressure. Water is applied to the wall surface 150 mm in front of each trowel stroke by the use of a flat brush, until a good, clean slightly polished surface is obtained.

On better-class work the angles are wiped out in the scratching or first application coat, with the aid of a feather-edge rule. In the laying down or tightening coat a straight grained skimming float is used and this ensures a straighter final surface. This is because although a thin steel trowel will bend to any uneven surface a wood float cannot bend and hollows are exposed and can be filled out. Angles in better-class work are sometimes scoured out with a cross-grained float. Trowelling up the main wall surface is as described previously though extra trowelling up may be required if a thicker application has been necessary.

Floating coats that have been allowed to become thoroughly dry, lead to difficulties such as loss of workability, peeling and blistering.

As each stroke of neat Finish plaster is applied to very dry porous floating coats, the water will be immediately sucked out by capillarity into the voids of the floating coat. This means that the overlap of the next trowel stroke will mean a double thickness application unless extra pressure is used or extreme care in avoiding the overlap is taken. These difficulties will be repeated in the laying down coat, particularly if any uneven surface needs to be straightened due to double thicknesses on the overlaps. In the trowelling-off stage the surface sometimes peels away in tiny flakes when the trowel is edged to finish the surface. This is due to the plaster drying without setting and further time wasting can occur in making good and re-trowelling when the skimming has hardened.

Blistering occurs occasionally when pockets of moisture are entrapped and later replaced by air as the water is sucked away. Pressure from the trowel causes localised pockets or blisters to develop and again time is required to assist or allow the air to escape and tighten the surface back to the floating coat. A likely cause of blistering on dry floating is when water used to kill the suction has not been sufficiently absorbed into the background.

All the disadvantages mentioned can be avoided by skimming with neat Finish plasters whilst the floating is still 'green'. This means as soon as the floating coat has set and in this condition all the voids will still be full of excess water from the gauging mix. Problems of suction therefore will not arise, there will be no loss of workability and a reliable rate of drying and setting can be assured. The exception is cement based undercoats that should be allowed to dry out to ensure that any shrinkage has taken place.

Another method of course is to damp down the floating with water, care being taken to avoid the risk of blistering occurring in the skimming coat later.

New materials have been developed recently to increase the viscosity of the mixing water and this helps to resist the capillary attraction on high suction backgrounds. Plasticity is therefore retained much longer and the original workability of the mix is maintained until normal setting takes place. This type of material is known as a *bonderiser*.

KEENE'S CEMENT

Keene's cement, an anhydrous plaster, was placed in a class of its own in BS 1191 and called a cement. The reason being the nature of the plaster which was extremely hard and its ability to be brought to a near perfect surface. The plaster was used for areas which may have suffered damage, i.e. lower parts of a wall (the dado) and external angles. Many squash courts were plastered in Keene's cement on a cement and sand backing coat. There is no comparable site plaster although by using one of the number of types of hard class 'A' plasters (Herculite, Crystacal, etc.) by retarding, or mixing with lime, a reasonable attempt to match the original is made.

PATCHING WALLS

A wide variety of techniques may have to be used for the many different types of patching or repair work. In each case it is good practice to ensure the area surrounding the patch is sound, the piecing scraped clean and trimmed to a reasonably straight or regular line, the background is keyed or will provide satsifactory adhesion, the area of background is free of dust and the surface has been damped if necessary to prevent excessive suction.

Small holes which are of, say, 75 mm or less can usually be treated in the following manner. Any wallpaper covering the piecing should be scraped clear of the patch to a distance of 25 mm or greater. Usually it will be sufficient to damp the patch's background and then to fill in the patch with a neat plaster or a lime putty and plaster mix. As the plaster application begins to set a further 'tightening' coat can be applied leaving the application slightly proud of the original plastered surface. When this coat has set, the patch can be trowelled off by use of a clean plastering trowel and wet flat brush. The blade of the trowel should be at an angle of approximately 60° to the wall surface and sufficient pressure should be used until the plaster patch is level with the original plastered surface. The piecing should show clearly without any 'gatherings' on the original surface. At this stage the outside edge of the piecing should be washed with clean water so that when the patch and surrounding area have dried no surplus material or dirty marks are left on the plaster surface.

With larger wall patching areas, the surrounding plastered edges should be tested for soundness by tapping with a lath hammer. Loose or unsound plastering should be trimmed back to a solid piecing. The piecing itself is best cut to a reasonably straight or regular curved line, using a lath hammer or bolster chisel and a heavier hammer. (Irregular piecings are more difficult to patch and the finished work often looks unsightly.) The background is keyed, if necessary, by raking out mortar joints, hacking or applying a bonding agent. Any wallpaper or covering on the line of the piecing is scraped clear, after which the wall is usually brushed clear of dust with a stiff broom. At this stage the background and piecing edge are damped sufficiently to kill any excessive suction anticipated. Often the old defective plaster which has been knocked off the wall is left on the floor until damping has been completed. This ensures a clean dry floor surface when the plaster rubbish has been removed and before patching starts.

Materials used for the new patch will depend upon the requirements of the particular situation. Often plastering material of similar type as the original plastering is used. If, however, the original has become defective because it was unsuitable for the particular situation, then a material more suitable will be used as a replacement.

Assuming the material to be used is a gypsum plaster/lightweight aggregate plastering mix, then one floating coat should be applied to the background and straightened by ruling off the existing plastered edges. A skimming clearance of approximately 2 mm − 3 mm should be scraped from the floated surface particularly at the piecing edge. The floated surface is then rubbed up and the piecing washed clean. When the floating coat has set, the skimming mix is applied to bring it in line with the plastered edge, after which a laying down coat is applied over the patch surface. The laying down coat should be slightly 'proud' of the piecing. In the trowelling off stage the surplus projection is trowelled back to the piecing, with the blade half on the old and half on the new plaster surface, until the piecing shows through clean and level. Most plasterers prefer to use the nose of the trowel to complete the final treatment to the piecing and to ensure a perfect joint without 'gatherings' or 'slacks'. Finally, the outside edge of the piecing is washed clean of any accumulated 'fat' or plaster.

CURVED WALLS

In rooms where the centre for the curve can be located and positioned on the floor, the work can be carried out as shown in Figure 54.

Firstly the centre can be located on the floor by bisecting any two chords of the curve, the intersection of the bisectors being the given centre. One practical method of doing this is to place a parallel straight edge against two points of the curve. At a position exactly half way along the straight edge place a scaffold square. The right angled edge of the square will bisect the chord and if this line

CURVED WALLS

is extended as shown in Figure 54 it will pass through the position of the curve centre. This is repeated on another part of the curved wall and the intersection of the bisectors positions the centre.

If the floor is made of timber then a nail can be driven into the centre point. If the floor is concrete a centre block with a nail through its mid point can be bedded temporarily with strong gauged stuff.

A timber batten is cut to the length of the radius of the curve and a pivot point fitted at one end to fit over the nail at the centre. Dots can now be bedded at intervals of about 1.8 m on the bottom of the curved wall using the batten radius rod and a square from the level floor.

Dots are plumbed from each of the lower dots, and upon completion any intermediate dots required are lined in from the upper and lower ones. See Figure 55 elevation.

A template is cut to the correct curvature about 2 m long, and this can be marked by use of the radius rod on to a length of timber and cut to shape. The template can be used to rule in the horizontal screeds working from each pair of dots in turn. When the screeds are hard the wall can be floated and ruled in off the horizontal screeds using a normal floating rule, care being taken to ensure that the rule is always used vertically otherwise 'digging' in will occur.

In situations where the centre point for the curved wall is inaccessible, and the depth of the curve is not too great, the method shown in Figure 56 can be used. A cardboard template is cut in the form of a semi-circle with radius equal to the rise of the curve from the springing line. The diameter is divided into a number

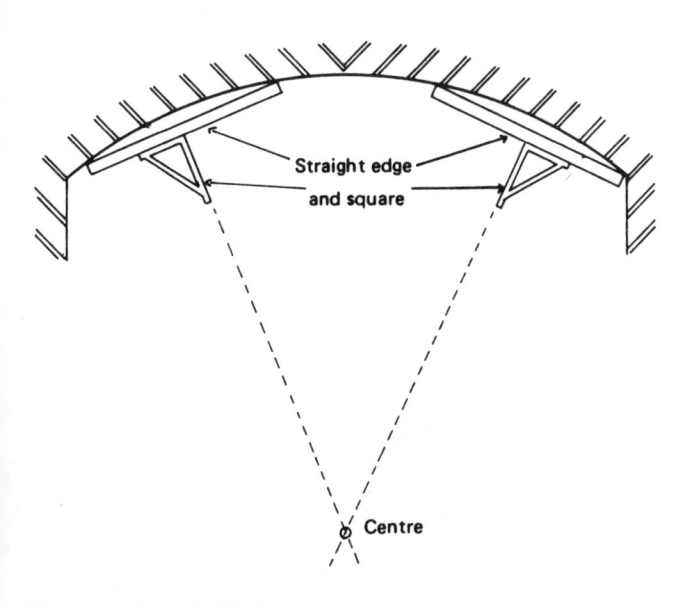

Figure 54: Method of finding centre for curved walls

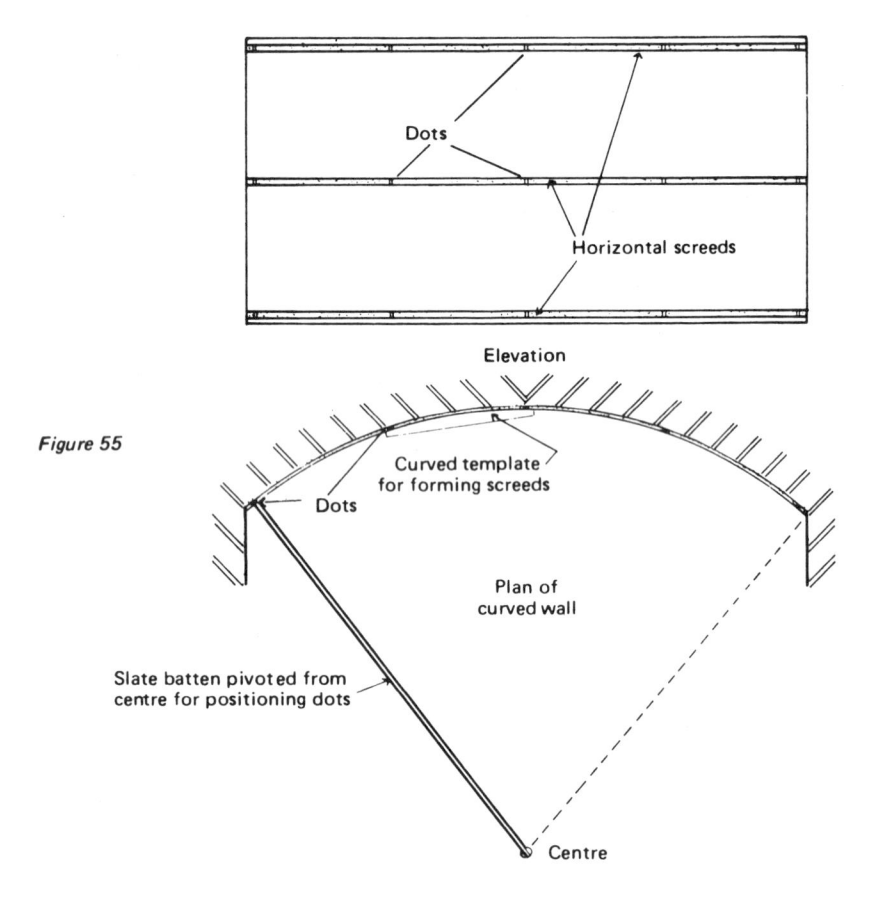

INTERNAL PLASTERING

Figure 55

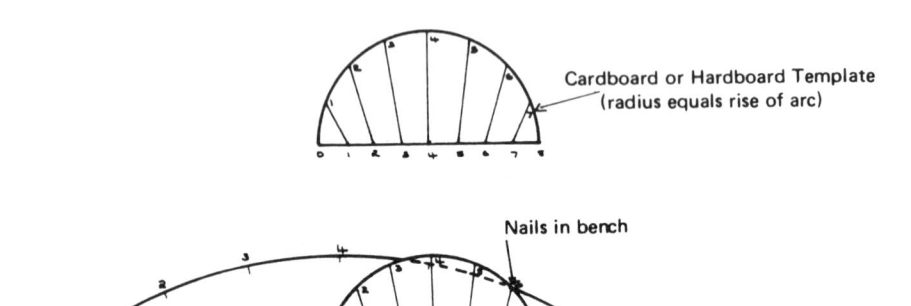

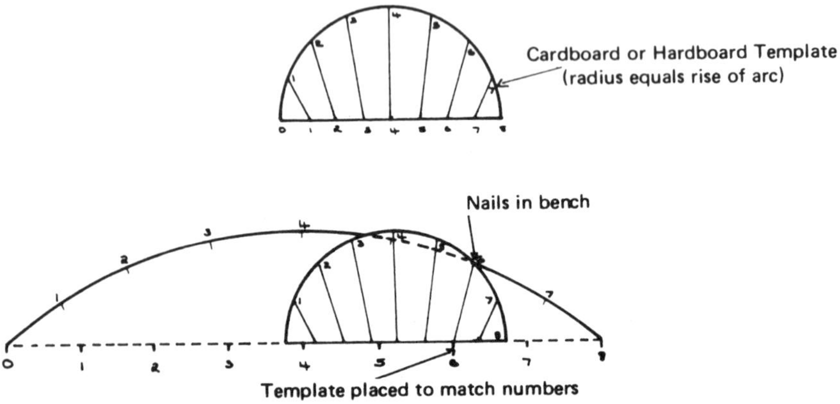

Dots

Horizontal screeds

Elevation

Curved template for forming screeds

Dots

Plan of curved wall

Slate batten pivoted from centre for positioning dots

Centre

Cardboard or Hardboard Template (radius equals rise of arc)

Nails in bench

Template placed to match numbers

Figure 56: Method of setting out a curve with inaccessible centre on a bench

of equal parts and the curved position of the template into a similar number of equal parts.

The span of the curve is marked on the floor and divided into the same number of equal parts as that of the template.

Assuming the number of parts to be eight as shown in Figure 56 then the template can be placed with its diameter on the line of the span chord. The mid base point 4 on the template should first coincide with point 4 on the span. The position of the upper point 4 on the template will establish point 4 on the floor curve and can be fixed by driving in a headless nail or marking with a pencil. Points 3 and 5 can be established in a similar manner by matching the numbers on the template with the numbers on the floor. It will be necessary to fold under, or cut the top off, the template to enable each succeeding pair of numbers nearer the base to fit inside the space available.

When setting out segmental curves for benchwork the template can be allowed to overlap as shown in the sketch.

CONVEX CURVED WALLS

In situations on site when the curve is convex instead of concave it is usually necessary to find out the radius. This may be found by reference to site drawings. Assuming this is not possible the radius can be obtained by placing the mid-point of a long straight edged rule against the wall and measuring an equal distance at each end of the rule to the wall. These distances must be at right angles to the rule. The length of the rule is the span of an arc of the convex curve and the distance measured from the end of the rule to the wall is the rise of the arc from a chord on that curve. The radius can be calculated by reference to the sketch in Figure 57.

By calculation

$$a \times b = c \times d$$

$$\therefore b = \frac{c \times d}{a}$$

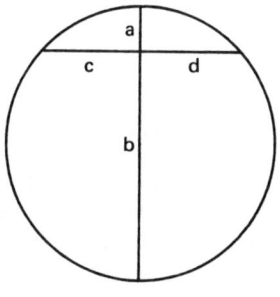

Figure 57

INTERNAL PLASTERING

For example if the straight edged rule was four metres long and the equal distance from the end of the rule to the wall was one metre, then the calculation would be as follows:

$$a = 1\text{ m} \quad c = \frac{4}{2} = 2\text{ m} \quad d = 2\text{ m}$$

$$\therefore b = \frac{c \times d}{a} = \frac{2 \times 2}{1} = 4\text{ m}$$

Diameter of circle = a + b = 1 + 4 = 5 m

Radius of circle = $\frac{5}{2}$ = 2.5 m

The curve can now be drawn to the correct radius on a floor space using a radius rod or strong tautly held line. A line is drawn from the centre to the arc at its mid point. The centre point of a straight edge is placed at right angles to the newly drawn radius to form a tangent to the curve. Ordinate rods are nailed at right angles to the rule, the ends of the rods just to touch the curve.

On a convex curved wall dots are placed by bedding the end dots to average thickness and then positioning the others by using the ordinate gauge previously made as shown in Figure 58. The procedure is then as described for the concave curved wall. Curved screeds are formed between the dots by the use of a curved rule cut from timber to the radius.

If the radius can be found from site drawings, or by the method previously described, an alternative method can be used. A scale drawing is made of the convex curve with a tangent as shown in Figure 59. Ordinates are included at suitable spacings.

The procedure on site is then similar to that described previously when using the ordinate gauge. In this case the mid-point of a straight edge rule is placed against the convex curve and the ordinates are measured at right angles from the rule using scaled measurements from the drawing.

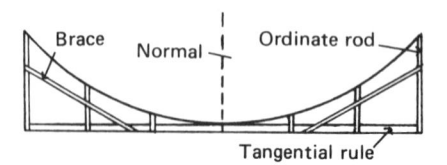

Figure 58: Ordinate gauge for convex curve

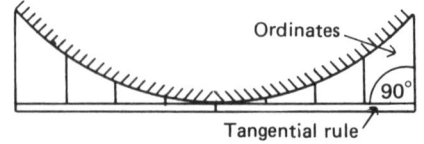

Figure 59: Ordinates found by scaled measurements

FLOATING CEILINGS

Assuming that the ceiling to be floated has already been rendered and that the walls have been floated plumb, the following method can be used.

Snap a level datum line around the room about 225 mm from the ceiling. If the room contains beams of a greater depth than this the datum line can be adjusted to any convenient level below the soffit of these beams.

FLOATING CEILINGS

Select a point on the ceiling about 150 mm in from the wall which appears to be the lowest point. At this position bed a piece of lath as a dot, and with a scaffold square bearing off the floated wall, tap the lath securely in place. If the side of the square bearing against the floated wall is plumb then the upper part must be level, therefore the dot will be level (Figure 60).

Having positioned this first dot a mark should be made on the square coincident with the datum line on the wall. Now further dots can be bedded on the ceiling and each one adjusted until the line on the square is level with the datum line. In this way each dot will be bedded level and at an equal distance above a level datum line around the room, therefore the dots will be level with each other.

Screeds can be formed between the dots and intermediate screeds if necessary are also formed by lining in new dots.

The ceiling can then be floated and ruled in to a true and level surface.

If the ceiling is intersected by a beam it can be floated plumb, square and linable by the following method.

A chalk line is snapped on the floated ceiling surface on one side of the beam. The line must be parallel with the beam and about 225 mm away.

The depth of the beam at its lowest point is measured and pieces of lath are cut 9 mm longer than this measurement. Starting at one end of the beam the first lath is bedded tight to the ceiling angle and square off the level ceiling by using a scaffold square as in the sketch. This means that when one side of the scaffold square is in contact with the level ceiling then the other part of the square in contact with the lath must be plumb. Therefore the lath must be plumb.

A mark is now made on the scaffold square where the datum line crosses. Laths can now be bedded along the beam at a convenient distance apart by using the scaffold square off the level ceiling until the mark on the square coincides with the datum line (Figure 61).

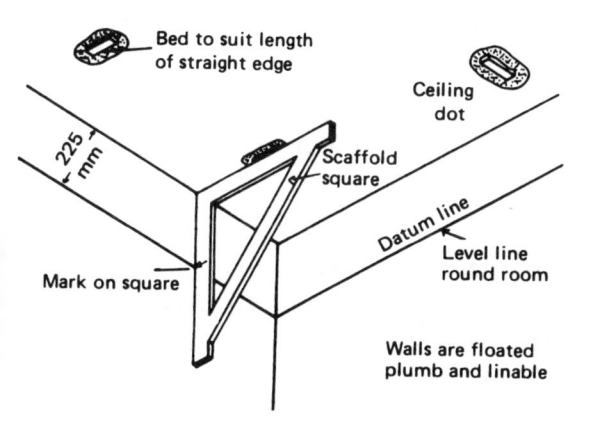

Figure 60: Levelling ceilings

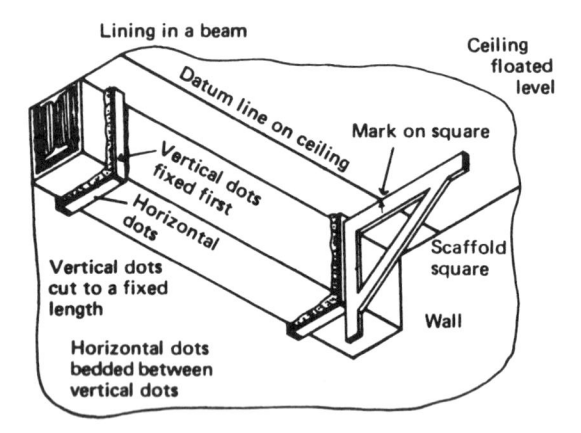

Figure 61: Lining in a beam

To position the laths on the other side of the beam a lath is bedded square off the ceiling at one end. The position of the datum line at one end can now be marked from the original on the square.

A gauge can now be cut to the width of the beam soffit at this end. Using this gauge another lath can be bedded square from the ceiling at the other end of the beam, and a mark placed on the ceiling against the mark on the square. A datum line can now be struck on the ceiling on this side of the beam, and if the width of the soffit at either end of the beam is the same then the two datum lines will be parallel.

Laths can now be bedded on the second side of the beam as described for the first.

All laths on both sides of the beam should now be plumb and linable. Also they are of the same length, so if the tops of the laths are all touching the ceiling then they are all level. Therefore the bottom of the laths will be level and these can be used to position new laths to bed across the soffit of the beam. The new laths for the soffit will have to be cut to fit between the laths bedded to the beam cheeks.

The beam can now be floated and ruled in plumb, level and linable using the laths as screeds, removing the laths and filling in on completion.

PATCHING CEILINGS

Plaster patching on wood lath ceilings can be carried out by different methods. Assuming the patch is approximately one metre square, the repair can be made good by nailing a plasterboard, cut to the size and shape of the patch, to the ceiling joists. The surface of the plasterboard is skimmed to bring it in line with the original ceiling plastered surface, making good the piecing as before described. Where sufficient thickness is available, the ends of the board can be scrimmed to avoid the possibility of a cracked 'joint'. If the thickness available will not permit the use of a plasterboard nailed to the existing wood laths, then these are removed carefully on the centre line of the outside joints exposed in the patch area. This will mean that inevitably some making good to the full thickness of the original plastering will have to be carried out on the perimeter of the patch before skimming can take place.

In situations where the use of plasterboard is not possible or desirable, the patching can be done in the following manner. The surrounding edge of the patch can be tested for soundness and any loose unkeyed or defective plaster removed by hand, the plasterer standing underneath the sound part of the ceiling. Any broken or missing laths should be replaced at this stage. Dust should be removed from the laths by sweeping with a broom away from the operative. The piecing only should be damped. The replastering can be done with a lime/sand/gypsum plaster or vermiculite/gypsum plaster mixes. The rendering or pricking up coat should be applied to the perimeter edge first and then diagonally across the laths away from the operative. It is important that sufficient material goes through the lath joints to form keys above the laths and, at the same time, retaining sufficient under the laths to form a continuous bond in one application. The area should be floated out to the full thickness with the same material as the 'pricking up' coat. It may be possible to fill out to the full thickness in one operation but often the pricking up coat has to be keyed and allowed to set before the floating coat is applied. The floating coat is straightened, scraped back for skimming clearance and rubbed up as described previously for wall patching. Skimming coats on lime/sand/gypsum plaster undercoats should be lime putty and plaster. Skimming coats on vermiculite/gypsum plaster should be lightweight aggregate/gypsum Finish plaster.

Patching on plasterboard ceilings can be carried out by removing the broken plasterboards to the centre line of the outermost exposed joist in the patch area. A new plasterboard or boards are cut to fit and nailed in the space exposed by the removed boards. The replacement plasterboards are then skimmed with Board Finish plaster and the piecing made good as described previously.

An easier method is to cut a piece of replacement plasterboard about 100 mm wider and longer than the hole in the plasterboard ceiling. (The width of the replacement board must be narrow enough to pass through the hole in the ceiling.) To fix the plasterboard patch it is first necessary to clean the upper part of the patch perimeter clear of dust. Two small holes are drilled into the replacement plasterboard and a length of string threaded through as shown in Figure 62. The upper side of the patch perimeter edge is coated with a neat plaster mix and the replacement board is bedded into the soft mix to cover the gap. This is achieved by threading the replacement board up through the hole in the ceiling, turning it to cover the whole of the patch and then pulling the replacement board down tight to the bedding coat with the string. (An alternative method to the use of string is to use two roundhead nails and to pull the

INTERNAL PLASTERING

Figure 62: Patching plasterboards

Stage 1

Hole in plasterboard ceiling

Plasterboard cover piece

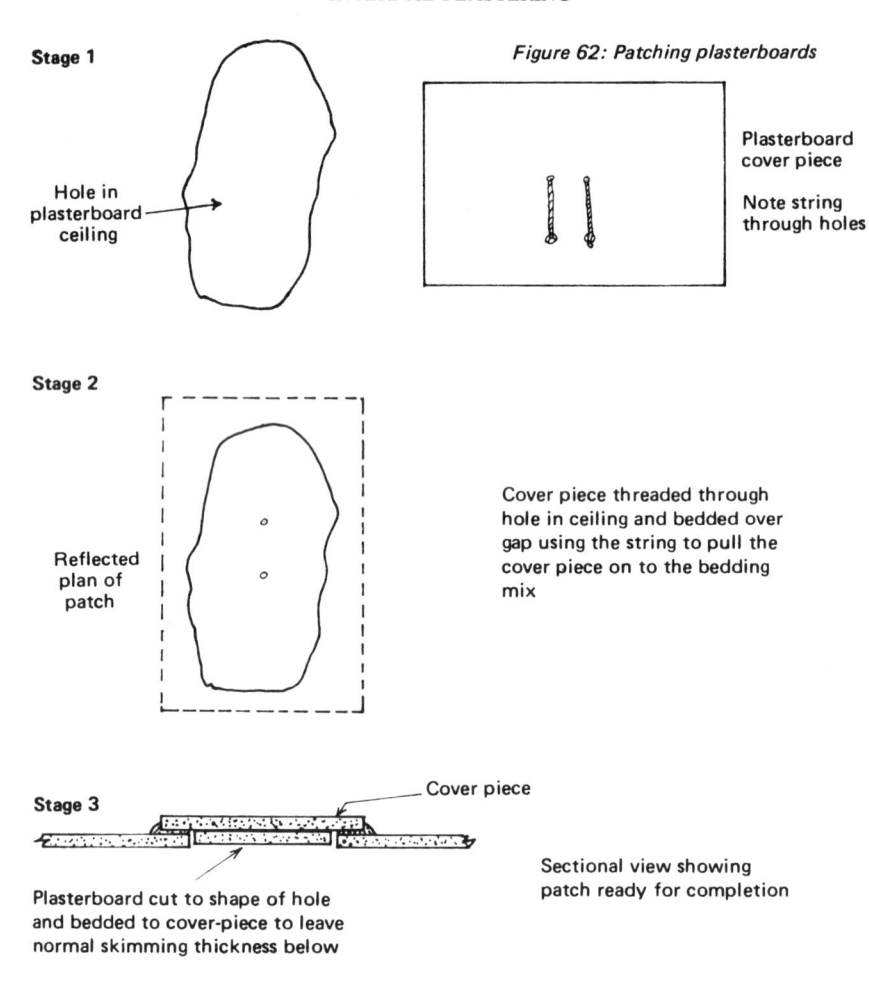

Note string through holes

Stage 2

Reflected plan of patch

Cover piece threaded through hole in ceiling and bedded over gap using the string to pull the cover piece on to the bedding mix

Stage 3

Cover piece

Plasterboard cut to shape of hole and bedded to cover-piece to leave normal skimming thickness below

Sectional view showing patch ready for completion

board down by gripping the nail shanks.) When the bedding coat has set, a further piece of plasterboard can be cut to fill the recess, and this is bedded tightly against the replacement board leaving a normal skimming thickness to make good afterwards. Another variation on this method is to bed the recess filling plasterboard to the larger replacement before bedding the combined boards in position to cover the gap in the ceiling. The string used for pulling the replacement board down on to the bedding mix is removed before skimming and making good the piecing as described previously.

BARREL CEILINGS

For barrel ceilings having a small span and rise, screeds can be formed by turning from a pivot point using a gig stick with a square at the traversing end (Figure 63a). A headless nail for the pivot point can be driven half its depth

BARREL CEILINGS

Figure 63a: Barrel ceiling solid plastering

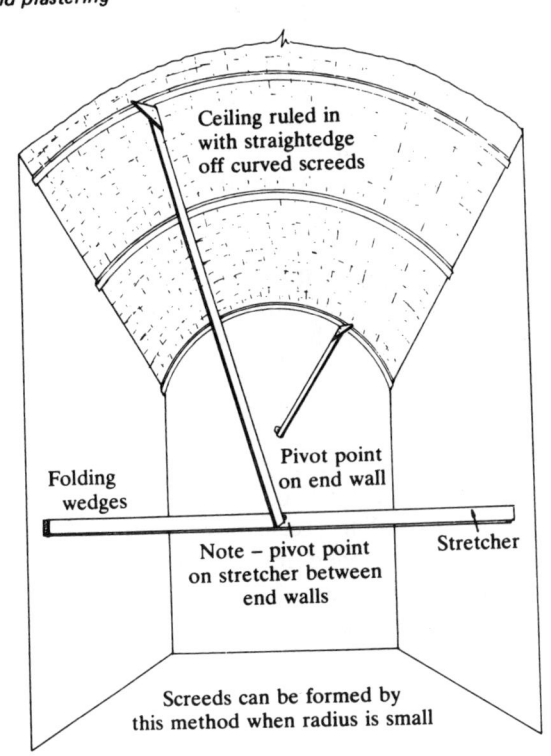

- Ceiling ruled in with straightedge off curved screeds
- Pivot point on end wall
- Folding wedges
- Stretcher
- Note – pivot point on stretcher between end walls
- Screeds can be formed by this method when radius is small

Figure 63b: Alternative method of floating medium to large barrel ceilings

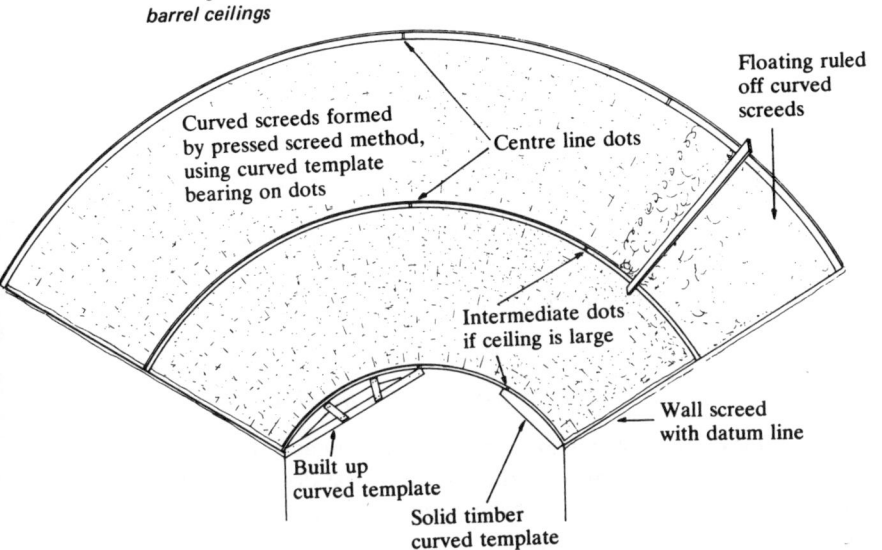

- Curved screeds formed by pressed screed method, using curved template bearing on dots
- Centre line dots
- Floating ruled off curved screeds
- Intermediate dots if ceiling is large
- Built up curved template
- Solid timber curved template
- Wall screed with datum line

63

into a timber plug on the side walls. Intermediate screeds can be turned pivotting off a timber stretcher wedged between the walls. This is repeated at suitable intervals along the barrel convenient for straight edge length. When the screeds have set the bays are floated by filling in and then ruling off the curved screeds with a straight edge floating rule. Skimming is of course carried out at right angles to the side walls.

Larger barrel ceilings can be carried out by positioning dots for the screeds and using curved templates to rule in from the dots, or by the pressed screed method. In the latter method a band of the floating material is applied between the dots on the curved surface. The curved template is then pressed tightly to the two dots to obtain the desired full curvature and surplus material at the sides of the screed removed.

When the barrel ceiling screeds have hardened the bays are filled in and ruled off as previously described.

Where the centre can be found on each end wall the dots can be positioned by pivotting a long batten, line or steel tape to each dot in turn. Intermediate dots along the barrel ceiling are bedded to a line from the dots previously positioned on each end wall (Figure 63b).

Where the centre is inaccessible a scale drawing, of say 1:10, can be made of the barrel section and divided into a number of ordinates. These measurements are then converted to full size, measuring up the height of each ordinate from a springing line on each outside wall to establish positions for each dot in the curve. Procedure is then as described previously.

LUNETTES *IN SITU*

Lunettes are openings into domed or barrel ceilings. Often the lunettes are curved window head openings to allow entry of natural light into rooms. The barrel ceiling would otherwise restrict the entry of natural light from the windows.

To form true plastering surfaces to curved lunettes various methods can be used. Paired pressed screeds can be formed by curved templates and then ruling off these with a cantilevered straight edge. Semi-circular headed lunettes can be formed as shown in Figure 64. Extra side bracing may be required to provide rigidity when turning larger-radius lunettes, especially on those intersecting into a segmental barrel ceiling.

CIRCULAR DOMES *IN SITU*

Small domes, circular on plan, can be run *in situ* as indicated in Figure 65a. A headless nail is driven into the crown of the dome and a screed placed around the ceiling perimeter of the dome. The mould is pivoted around the nail, the other bearing point being the screed.

CIRCULAR DOMES *IN SITU*

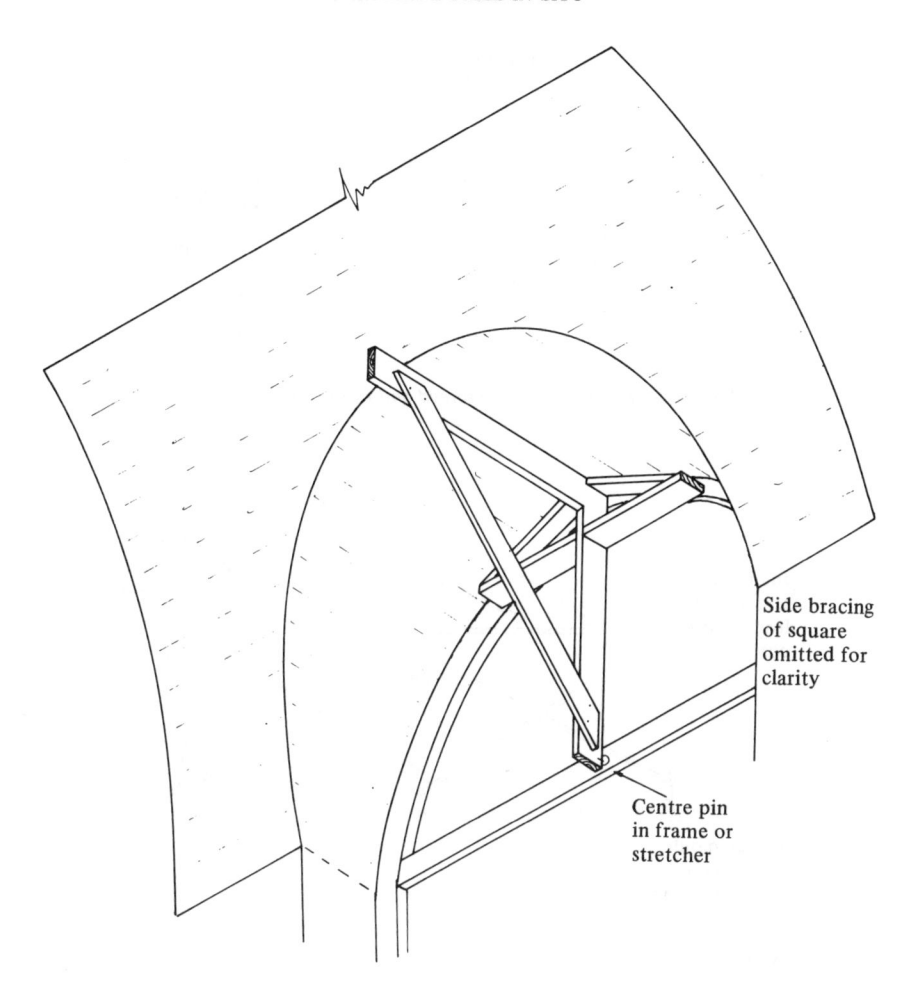

Side bracing of square omitted for clarity

Centre pin in frame or stretcher

Figure 64: Forming a lunette in a barrel ceiling

Pivot point

Dome core

Skimming member

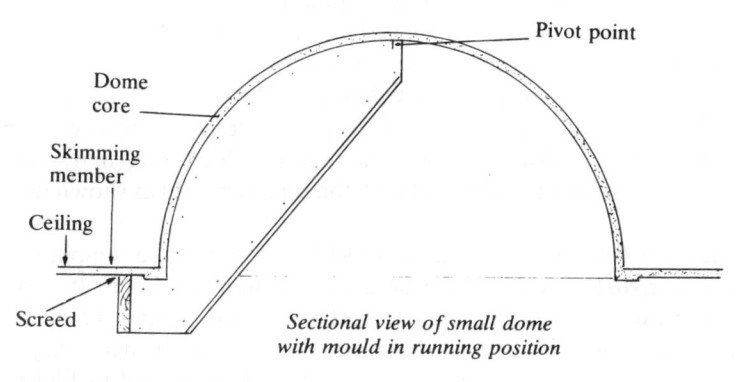

Ceiling

Screed

Sectional view of small dome with mould in running position

Figure 65a: Circular domes in situ

65

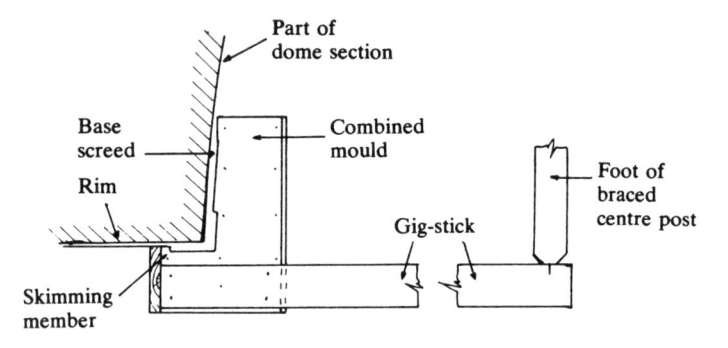

Figure 65b: Method of forming base screed and rim for large domes

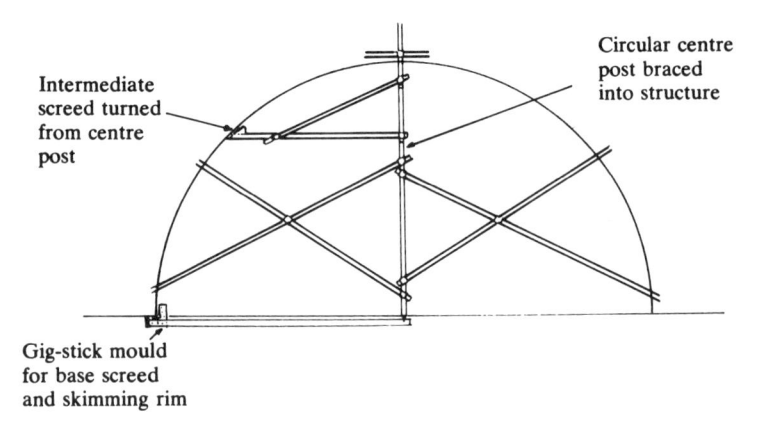

Figure 65c: Section through dome

In certain cases it is possible to turn a mould fixed to a 75 mm x 75 mm timber centre post, pivoted top and bottom. This will of course depend upon the height of the dome crown above a stable base.

Medium and large sized domes will need to be floated from screeds and the method used will depend upon size. A base screed can be turned from a centre post. The finished plaster plain or moulded rim, and skimming member, can also be incorporated with the screed mould as shown in Figure 65b.

In medium sized domes a curved blockboard or framed template, cut to the correct radial curvature, can be drawn around the dome. It is pivoted from the crown and bearing against the base screed around the perimeter.

On larger domes, one or more horizontal screeds will be necessary. These can sometimes be turned from a circular post as indicated in Figure 65c.

Alternatively, dots can be positioned from the centre post and horizontal screeds formed between the dots with a curved template cut to the correct radius for that particular part of the dome. Ruling in between horizontal screeds can be carried out with a curved template used radially in line with the centre point.

The base rim plastered finish can be run when the main floating has been carried out, before removing the struts of the centre post. The top plate screed profile is removed from the mould before the rim is run.

When the centre post has been removed, and the holes where the struts have been are made good, the dome can be skimmed to the finishing member on the run rim. Skimming is done either by using an old flexible trowel, or springing the centre rivets. Domes of smaller curvature may have to be finally finished with a flexible busk.

CHAPTER FOUR

Floors, skirtings and stairs

GRANOLITHIC FLOOR SURFACE (CONCRETE TOPPING)

Concrete floors are surfaced with a wide variety of materials including tiles, terrazzo, magnesium oxychloride, cement/sand, polymer modified cement, cold setting resin and granolithic. The techniques of laying granolithic are constant but the systems vary according to the design of the floor. There are three systems used for granolithic toppings. They are Monolithic, Bonded and Unbonded. The last two are both considered to be separate constructions. The main additions in BS 8204 Part II is the greater consideration given to floor preparation and the stating of minimum thickness as well as recommended thickness.

Monolithic system — a floor that is laid no more than three hours after the concrete is placed. The thickness recommended is 15 mm ±5 mm. The floor should be laid in bays, joints following the concrete base. Monolithic floors are considered to be trouble free due to this system of laying.

Bonded system is the most common system, the floor being laid on a hardened concrete base which will require preparation before laying. The soundness of the bond is reflected in the floor preparation carried out. The floor should be laid to a minimum thickness at any one point of 25 mm and to achieve this a design thickness of 40 mm is recommended. The floor should be laid in bay sizes not to exceed 20 m^2 for a thickness of 20 to 30 mm; where a substantial area of the floor is greater than 30 mm the bay size should be reduced to 15 m^2. The length to width ratio should be 3 to 2.

Preparation

Monolithic system, new concrete — should have the surface laitance removed by spraying with water, to expose the aggregate, unless the concrete and granolithic are being laid together. A perfect bond can be obtained with this system of floor. The bonded floor requires careful preparation and the extent of the preparation will have a direct influence on the floor bond. If a fully bonded floor (one which has a complete bond) is demanded the surface of the concrete has to be removed by mechanical means, i.e. scrabbling, planing, shot blasting (beyond the scope of normal plastering work and an expensive

GRANOLITHIC FLOOR SURFACE

process). If a perfect bond is not considered critical the surface will still need thorough cleaning with the laitance being removed. After the required concrete surface treatment the floor should be brushed clean and soaked with water. Cement grout (neat cement and water slurry) should be brushed into the surface no more than 20 mins prior to laying. A bonding agent can be added to the grout to aid adhesion but this should not be considered as alternative to concrete preparation. Unbonded floors only require a separating layer of waterproof paper, polythene or similar material to be laid on the concrete base with joints lapped 100 mm.

Laying

The floor should be divided into bays of appropriate size for the type of construction used and bays filled in draught board fashion allowing alternate bays time to harden.

Granite chippings 6 mm — 9 mm in size should be mixed with Portland cement at 3: 1 by weight or 5: 2 by dry volume. If the granite grading is too coarse, clean sharp sand can be added up to 20 per cent of the aggregate content of the mix.

Dots should be placed at each corner of the bay to the correct levels and screeds formed between the dots (Figure 66). The width between each screed being less than the length of the straight edge to be used by at least 150 mm. The actual traversing of the rule should be done on timber screeds, except when using the set surface of finished bays. Figure 66 also shows a method of levelling a floor from a datum line on the wall.

Screeds formed with the surfacing mix are unsatisfactory because when used the same day as laid the traversing rule will dig in. When this type of screed is used on the following day the shrinkage rate will be uneven and the screed having completed its initial shrinkage will finish up 'proud' or projecting above the rest of the floor.

The granite and cement mix should be laid and ruled off in strips of about 600 mm, on top of freshly supplied grout as stated previously. A good fair finish should be obtained with the rule and this can be trowelled lightly to smooth out the marks of the rule. This is repeated to the end of the bay, the timber screeds removed and filled in as necessary.

The final trowelling should take place within 10 hours of the mixing of the materials and the actual time will vary according to the temperature, suction and other site conditions. It is often possible to suspend a plank above the recently laid granolithic surface when trowelling off, but if this is impossible two large boards, one for each foot, to spread the load, will save excessive damage to the surface. The boards can be moved in sequence to the farthest point of the floor. Each reach is then trowelled off, returning the boards step by step, back to the starting point.

Curing

It is essential that sufficient moisture is available for the continued hydration, or setting action, between the Portland cement and water for up to seven

FLOORS, SKIRTINGS AND STAIRS

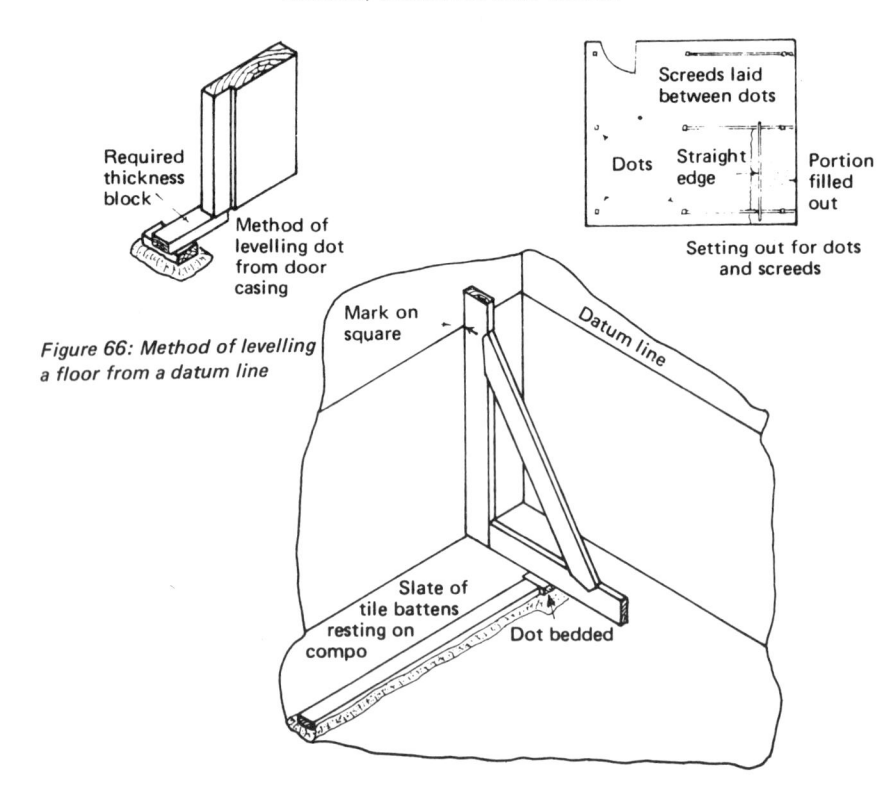

Figure 66: Method of levelling a floor from a datum line

days after laying. This can be carried out by restricting the moisture loss of evaporation from the mix by covering the finished topping with building paper, polythene sheets, damp hessian or a layer of damp sand.

Grinding

Occasionally the surface finish is ground by power-operated grinding machines, the floor being kept wet during this operation. The material ground off is removed by squeegee and the surface flushed clean with water. Any pin holes or blemishes should be filled in with neat cement paste and allowed to harden for three days, keeping the surface moist meanwhile. A final grinding to remove the film and give the floor its final polish is made and the floor surface thoroughly washed to remove all residue.

Surface treatments

Non-slip and wear-resisting materials can be added to the granolithic mix or sprinkled on and trowelled into the surface whilst still soft. Carborundum chippings or commercial floor grit A graded between No 20 and 26 are often used. When mixed into the topping material between 1.5 kg — 3 kg per square metre should be used for a 20 mm thick topping. Care should be taken to ensure that the mix is not too soft otherwise the grit will sink to the bottom of the mix, and in this position is useless.

When sprinkled on the surface up to 2 kg per square metre of abrasive can be used and again the granolithic surface should be stiff enough to resist the deep penetration of the grit otherwise its value will be minimised.

The idea of using abrasives as a non-slip surface is that the softer granite chippings will wear quicker than the abrasive, and in time the latter will project slightly above the surface. These irregular projections prevent slipping and also contribute greatly to the wearing qualities of the floor.

Ferrous aggregates have been used for their hard wearing qualities, particularly in factory floors to resist breakdowns of the surface by wheeled trolleys. They are sold in various proprietary brands in coarse, medium and fine gradings. Among the trade they are termed iron filings. Their use on constantly wet floors is not recommended.

Hardeners are materials applied to the surface of granite floors to prevent dusting. This latter defect is due to the breaking up of the surface, and the very fine dust particles resulting from this cause considerable inconvenience especially in factories containing machinery or paint shops. The cause of dusting is due to bad practice in the laying of the granolithic topping including badly graded or poor mixes, too much water in the mix and too rapid drying conditions.

The hardeners used include silicate of soda (P84 grade), also zinc and magnesium silica fluoride.

Before application of the hardener the floor surface should be thoroughly cleaned and this may require scrubbing with soap and water, detergents or the use of wire brush.

Silicate of soda should be diluted with four times its volume in water and spread over the floor surface evenly with a watering can. This can then be scrubbed into the surface with a soft brush or mop. This should be followed twenty-four hours later with a second application and the strength of the solution increased to three parts of water to one of sodium silicate. A third or even fourth application may be applied increasing the ratio to two parts water to one of the silicate. The number of applications required is determined by the rate at which the hardening solution is absorbed into the floor. Any unabsorbed hardening solution left on the surface after the final application should be washed off, otherwise a white deposit is left which is difficult to remove later.

Zinc or magnesium silica fluorides may be used separately as hardeners, or a mixture of one part zinc silico fluoride to four parts magnesium silico fluoride gives good results. The hardening solution should be made up of 250 grams of the mixed silico fluoride to 4.5 litres of water for the first application and then increasing this to 1 kg for further coats.

It should be applied as described for silicate of soda, again until there is little or no absorption and washing down on completion.

Due to the danger of the possible formation of hydrogen fluoride gas during this latter treatment, adequate ventilation should always be arranged. Care should be taken to prevent contact of the solution or gas with the eyes or open cuts of the operator.

FLOORS, SKIRTINGS AND STAIRS

ROOF AND FLOOR SCREEDS

These are usually carried out in sand and cement although roof screeds requiring thermal insulation properties use mixes containing aggregates of greater insulating value than sand.

Roof and floor screeds are placed to form a good surface to receive a final surface when set. Floor coverings include wood blocks, vinyl and rubber overlays, a wide variety of tiles including thermoplastic and cork, etc. Roof coverings include different types of waterproof toppings including asphalts, cold or hot bitumen solutions used in conjunction with roofing felt, etc.

The screeded surface may have to be finished level in the case of most floor screeds, or laid to falls to clear the surface of water and direct it to gulleys or rain-water outlets with roof screeds.

A typical mix for sand and cement floor or roof screed is three parts of sand to one part of Portland cement. When laying cork, or similar types of tile or floor covering that need to be tacked down with hardened steel pins, then the mix can be as lean as six parts sand to one cement.

When positioning the original dots before laying the screeds it is essential to allow for the thickness of the subsequent covering. A datum point must be fixed and dots levelled or lined from this position allowing for the thickness of the wood blocks or tiles, etc to follow. Top left sketch, Figure 66.

The procedure for laying screeded floors or roofs is as described for granolithic work except that the consistency of the laying mix should be as near semi dry as possible and also that trowelling off as described for granolithic floors, is unnecessary.

The latest recommendations also advise that screeds need not be laid in bays of restricted size provided that the proportion of length to width is not greater than 3:2, and unless they contain underfloor warming cables or are to receive an *in situ* finish. It is considered that the cracking which can occur in large bays is less undesirable, and easier to make good, than the curling which can, under certain circumstances, form around the perimeter of each bay. Thicknesses of 12 mm — 25 mm for monolithic construction, 40 mm minimum for separate construction, 50 mm minimum for unbonded construction and 65 mm for unbonded construction with heating cables, also floating screeds laid on compressible sheets, are also recommended.

Certain surface requirements insist on a trowel finish and trowelling of the semi dry surface as the work proceeds is normally adequate, although a much better finish can be obtained by 'tightening in' as the floor hardens.

When laying roof screeds to falls the dots are positioned by starting first with the lowest point usually the rain-water outlet. Here a dot is bedded to the minimum thickness of the screed and above the level of the outlet. Subsequent dots are 'levelled' or lined in allowing an agreed fall to ensure the discharge of all rain-water. In many cases the specified rate of fall is 1—100. One method of obtaining this is to nail a 30 mm fillet of timber under one end of a 3 m straight edge and placing this end on the lowest dot. The correct height of the new dot can be found by 'levelling' the straight edge with a spirit level

(Figure 67). This can be repeated until the highest part of the roof screed has been reached.

```
30 mm fillet                    Spirit level
nailed to rule                                          3m rule
                                                        New dot
Original dot
```

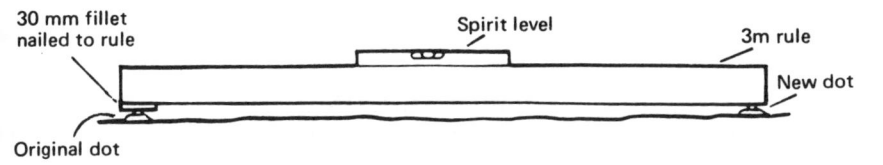

Figure 67: Sketch showing method of positioning dots allowing for a fall of 1 in 100

The procedure of placing the screeds and filling in, is as described for floor screeds.

Roof and floor screeds can also be laid with a mixture of exfoliated vermiculite and Portland cement. The advantages of this type of screed compared with the more usual sand and cement topping include good thermal insulation properties, good fireproof qualities and lightness in weight. Vermiculite/cement screeds are roughly one-fifth of the weight of normal concrete.

The vermiculite used is 'concrete aggregate grade' and should be mixed for roof screeds at 8 volumes of vermiculite to 1 volume of Portland cement.

Floor screeds should be slightly stronger at 6 volumes of vermiculite to 1 volume of Portland cement.

Mixing exfoliated vermiculite and Portland cement

When mixing by hand the vermiculite and cement should be mixed dry to an even colour and then mixed wet by *spraying* on water and mixing until uniform distribution has been achieved.

If machine mixing is used the water can be poured into the mixer first. Into this is added the cement and this is mixed to a slurry. The vermiculite is now added and mixed for approximately 1½ minutes.

Another method for machine mixing is to mix the vermiculite and cement dry in the mixer, and then add the correct amount of water until uniform distribution has resulted. This should again be mixed for about 1½ minutes.

Excessive mixing causes compaction of the vermiculite and should be avoided.

When laying vermiculite floor and roof screeds normal practice should be observed.

Vermiculite is easily compressed and because of this vermiculite roof and floor screeds should be protected from damage after laying by covering the surface with a protective sand and cement topping coat.

These protective topping coats should be applied while the vermiculite/cement screed is still 'green'. The mix used should not be stronger than 4 to 1 sand and cement, and it is an advantage to use a plasticiser in the gauging water.

The thickness of the protective topping coats for floor screeds should be at least 18 mm and for roof screeds a minimum of 9 mm.

Adequate time should be allowed for shrinkage and drying to take place before the waterproof covering is applied to roof screeds, otherwise troubles such as blistering and ceiling stains can occur due to the entrapped moisture.

CEMENT SKIRTINGS

These are made from 3 parts sand to 1 part of Portland cement. The normal practice when plastering a room which is to receive cement skirting, is to leave the floating coat off at the base of each wall to the desired height of the skirting.

The sand and cement mix is applied to the bare brickwork at the base of the wall to the thickness of the floating coat and scratched with a wire scratcher. As soon as this material stiffens a further coat is applied evenly about 12 mm thick. To do this the trowel should be loaded with sufficient material to coat one trowel width to the correct thickness and the desired height. When applying this, extra pressure is required at the base of the skirting easing off the pressure towards the top. This is to ensure that an even thickness is distributed from bottom to top of the skirting. It is also essential to tuck in the finish of the trowel stroke tight to the floating coat, otherwise the skirting is easily dragged down by its own weight.

When this coat has been applied it should be straightened with a rule or darby and then rubbed up with a float as soon as possible. A chalk line should now be snapped across the face of the skirting to the desired height and the surplus cut off with a trowel. If a square top edge is required a convenient gauge can be made with two legs nailed to a straight edge but this is only advisable when the floor is straight and level. When a splayed top edge is required the top can be cut off at an angle of 45 degrees, this is much better than cutting off square and filling in with a float.

The top edge should be rubbed up with a float, taking care to work away from the arris when finishing. If the suction is keen the float should be dipped in water from time to time to prevent loose sand being rubbed to the surface.

To finish off the skirting the trowel should be used to tighten in any loose sand, but smoothing should not be attempted.

One of the chief difficulties in forming a cement skirting is when the suction from the background is inadequate. In this case the stuff 'hangs' soft and although in many cases the surface stiffens slightly the portion in contact with the wall remains soft, and when extra weight or pressure is applied the whole skirting sags off.

If it is convenient, a simple method of overcoming this difficulty is to give the background one coat only, scratch it deeply with a wire scratcher and leave until set. This can then be second coated in the normal way, possibly on the following day.

Another method is to use a cement accelerator, but these are expensive and not readily available.

To finish a cement skirting on a background without suction and using ordinary sand and cement the following method can be used. Mix the sand and cement dry and then divide the mixture into two halves, re-mixing one half only with water. This wet mix should be applied fairly soft in a thin coat and immediately coated by throwing or spreading the dry mix over it. The dry mix will absorb some of the moisture from the wet coat for a time. When this ceases a further coat of the wet mix is applied to be followed if necessary by more of the dry mix. Usually one application only of the dry mix will be required but in severe cases of dampness two or even three may be required.

Semi-dry mixes can also be padded on to reasonable thicknesses but care must be taken that sufficient moisture will be available to cure the finished surface. This type of mix has a weak final strength when set unless cured correctly and it is also difficult to achieve a good surface finish with sharp arrisses as quickly as the other method.

Coved cement skirtings are formed at the base of walls to make cleaning of floor angles easier. This is particularly useful in buildings where the floors are cleaned by swilling with water in such places as abattoirs, etc.

A simple coved skirting can be formed by squeezing sand and cement into the wall/floor angle with a quadrant shaped coving trowel or bottle. If both the existing wall and floor surfaces are old and smooth it is unlikely that the cove will adhere successfully.

A better method is to combine the cove with a new cement skirting in which case the cove will adhere perfectly to the skirting. At the junction with the floor however the tapered feather edge of the cove will be very weak and likely to break off in normal usage.

Cleaning, hacking, or coating the floor with a cement bonding adhesive will lessen the risk of failure and a channel cut into the floor at the edge of the cove is a safer though more expensive job.

If the cove is to be formed in a room where the floor has to be surfaced with a granolithic topping then a timber batten can be bedded at the position of the coves limit and removed after the floor topping has set. The cove can now be formed in the normal way with no risk of failure at the feather edges (Figure 68).

STAIRS

Cement toppings for stair flights and landings can be carried out in sand and cement if they are to be covered with vinyl or other protective covering later. If the topping coat is the final surface finish then a mixture of 4 parts granite chippings, 1 sharp sand and 2 parts of Portland cement can be used. Carborundum chippings trowelled into the treads assists the wearing qualities and also forms a non-slip surface.

The setting out of each individual flight of stairs varies tremendously with the type and situation. If the steps abut on to a wall on one or both sides, no stringer or raised edging to the flight will be required. This is also the case of certain open-type flights without stringers on either side.

FLOORS, SKIRTINGS AND STAIRS

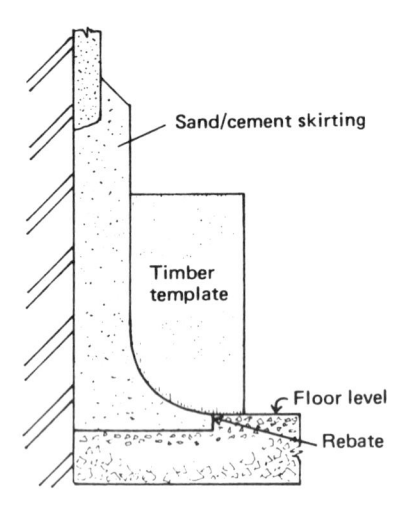

Figure 68

Sand/cement skirting
Timber template
Floor level
Rebate

To form cement toppings to these steps the total rise of the staircase should be divided accurately into a number of equal sized risers. (The riser is the vertical side of a step and the tread is the horizontal upper part.)

The total go, or total horizontal distance of the flight is also divided into a number of treads of equal sizes. Note that there will be one tread less than the number of risers in each staircase flight.

Temporary stringers or side boards may have to be positioned along either side of the flight. A nosing line can be snapped along each stringer or wall surface.

One method of marking out the individual steps is to use a storey rod. This is made from a length of timber batten and on this can be marked accurately the distance of each riser, and into each mark a nail can be driven.

The storey rod should be held or fixed vertically at the lower end of the flight with its base resting on the floor surface. The position of each tread can now be found by levelling through with the aid of a parallel straight edge and spirit level from each nail to the intersection of the rule with the nosing line (Figure 69a).

A number of riser boards, cut to the exact height and width of each riser can now be nailed to the correct position through the timber stringers, or wedged if the sides or stringers are brick or concrete (Figure 69b). The riser boards can be fixed at the top of the flight, working down from top to bottom in sequence. If the reverse way is attempted only three or four steps can be topped, each day. An advantage of this way however is that the riser boards can be fixed on top of a finished tread.

Before applying the granolithic mix the concrete steps should first be cleaned, damped and grouted with a cement slurry. The topping mix is first chopped down between the riser board and the concrete and well tamped down, after which the tread is filled in and trowelled up.

On the following day the riser boards are removed and the face of the riser

STAIRS

Figure 69a: Section showing method of setting out treads and risers

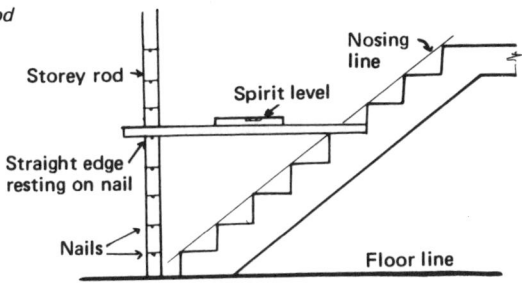

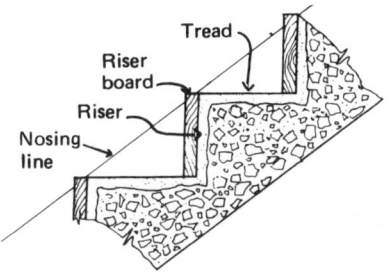

Figure 69b: Section of concrete staircase

rubbed down with a carborundum stone and water. Any holes or slack places can be filled in at this stage with a strong mix of sand and cement 1:1 and the surface rubbed up tightly with a wood float. The sharp arris of the step nosing is rounded with a carborundum to prevent damage and smarten the appearance.

Many plasterers prefer to have the front lower edge of the riser boards cut to a splay as shown in Figure 70. The reason for this is that when working downwards from the top of a stairs flight it is considered an advantage to be able to trowel the tread as near as possible to the riser previously placed.

When working upwards on a flight, a broader riser board is less likely to make an indentation, and the same is true when riser boards are positioned by wedging. The narrow wedge shaped lower edge of the riser board can however, be prevented from penetrating or marking the lower tread if extra care is taken, and this is easier when the riser boards are fixed by nailing through the stringer.

A narrow indentation is, of course, easier to make good when touching up, but it is best to avoid penetration whenever possible.

Special hard wearing and non-slip strips are sometimes inserted in the tread near to the step nosing. They are mainly strips of carborundum or materials with similar properties. The strips are usually bedded *in situ* as the treads are filled in, but in certain cases false strips are placed and the non-slip strips inserted later by bedding with a strong sand/cement mix. Other tread nosing strips of special types of rubber or toughened fibrous materials are sometimes fixed by plugging and screwing after the treads have hardened.

FLOORS, SKIRTINGS AND STAIRS

Figure 70: Riser board (narrow base)

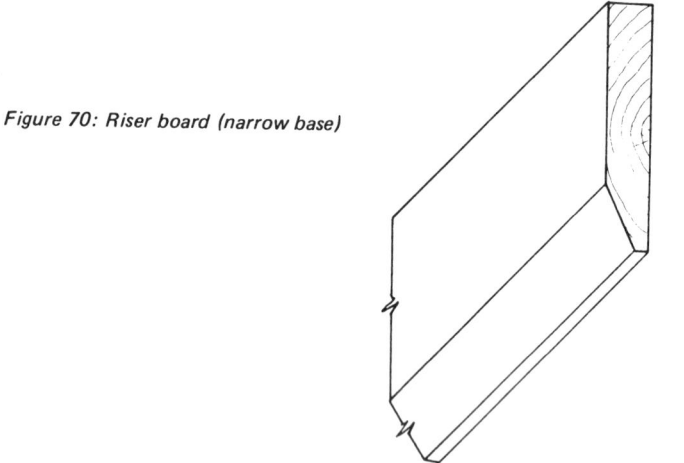

If a granolithic stringer or edging is to be formed to the sides of the staircase then a formwork box has to be constructed and fixed to the edge of the steps for the full length of the flight. This is filled in and finished as described previously.

Precast stair finishes are formed by bedding special slabs on to a concrete staircase to form a series of individual steps. The treads and risers may be cast and fixed separately, or combined treads and risers in separate units. Setting out for the treads and risers or step units is as described previously. The slabs are bedded to the concrete base with a strong sand/cement mix after grouting both the concrete and back edge of the slab to be bedded.

Materials used for manufacturing the slab units vary considerably, but many are of the terrazzo type containing marble of adequate hardness, other natural stone of similar characteristics, Portland cement and colouring pigments or coloured aggregates.

The units are cast in moulds which may be of steel, fibreglass or wood, the filled in moulds being compacted and vibrated to ensure strong dense casts.

The casts usually consist of a terrazzo facing only, the backing being composed of mixtures of crushed stone, gravel, sand and Portland cement. A minimum facing thickness of 12 mm of the terrazzo mix should be used for type A surfaces which are specified for staircase tread units. Minimum overall thicknesses for fully bedded precast treads is 40 mm for lengths of up to 1.5 m, 45 mm for 1.5 — 2 m, and 50 mm for 2 m — 2.5 m.

They should be made to within plus or minus 3 mm tolerance in thickness and plus or minus 1.5 mm tolerance for length and width up to 1.5 m. This is increased to plus or minus 3 mm tolerance for lengths up to 3 m.

The surface treatment of the casts is by grinding. This is carried out when the units are sufficiently hard to withstand grinding without dislodging the surface aggregate. Any imperfections are made good at this stage by grouting with a neat cement paste coloured to match the original.

After a further period of hardening, the excess grouting coat is removed by further grinding with a stone no coarser than No 80 grit. This should not take place sooner than three days after grouting.

Non-slip inserts can be set into the treads if specified, or provision made for fixing later.

Casts should be cured carefully to give maximum possible hardness of the finished product. The units should not be delivered or fixed for at least twenty-eight days after manufacture.

CHAPTER FIVE

Specialised techniques and problem solving

COMPOUND WALLS

Backgrounds composed of different materials, ie brickwork and concrete, often create problems of cracking in the finished work due to differences in the thermal movements of the two background materials. Expanded metal lathing fixed across the joint will reduce the risk of cracking. In cases where a brick wall contains a narrow linable concrete column, a better method is to cover the concrete with building paper and cover this in turn with expanded metal lathing fixed to the brickwork only (Figure 71).

If excessive movement is expected, it is usually best to form an expansion joint by bedding two metal casing bends at the intersection of the different background materials. This will certainly prevent the formation of unsightly irregular cracking (Figure 53).

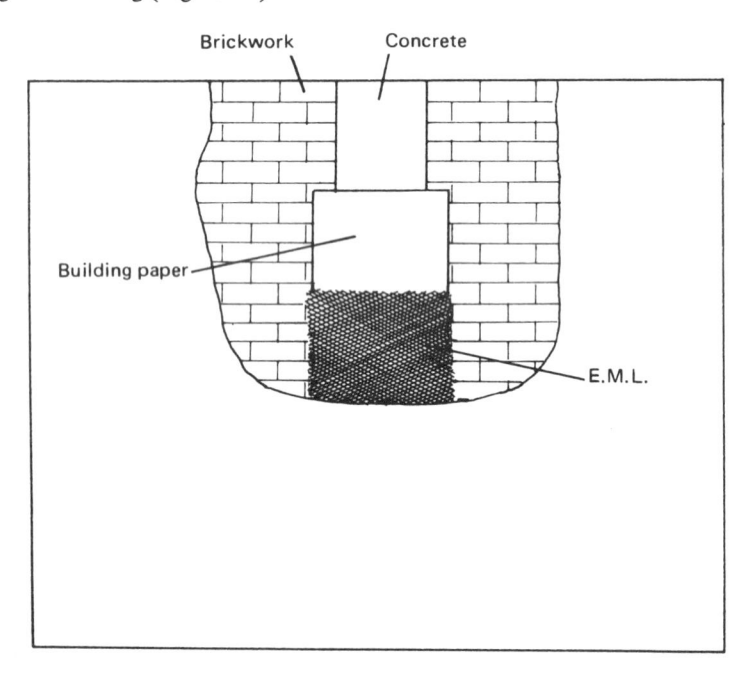

Figure 71: Compound wall

Thermal insulation
Heat transfer in buildings is by conduction, convection or radiation.

Conduction
This is the direct transmission of heat through a material, and the rate of conduction will depend upon its density. Metals and similar dense materials have a high conductivity. Gases and cellular materials have a low rate of conduction. The conductivity of a material is measured in 'k' values which is the amount of heat passing through 1 m² of the material of 1 m thickness for 1°C difference between the inner and outer surfaces. Insulation against heat losses will therefore depend in part on the resistance of the materials used in the structure. The resistivity of a material is the reciprocal of its conductivity, $1/k$, and this resistivity multiplied by the thickness of the material gives the resistance (R).

Plaster mixes using cellular or exfoliated aggregates such as perlite and vermiculite have a high resistance to heat conduction. Concrete or dense sand/cement mixes have a low resistance to the transfer of heat.

Convection
This is the transfer of heat in liquids or gases by circulation. Warm air is less dense than colder air, and the warmer will therefore tend to rise, being replaced by colder air in a continuous convection flow.

Radiation
This is the transfer of heat from one body to another by radiant energy through space. Dark surfaces absorb heat, but bright surfaces have a high reflective value. For the latter reason aluminium foil is used as a covering on insulation plasterboards, the bright, reflective foil side assisting in keeping the building cooler in summer and warmer in winter.

Thermal transmittance
The overall transmission rate of heat is known as thermal transmittance, and this is expressed as the heat in watts that will be transferred through 1 m² of the construction when there is a difference of 1°C between the inner and outer temperatures. This is known as the 'U' value, or 'air to air heat transference co-efficient'.

Thermal insulation standards
These are required by the Building Regulations (Part L), and state that the 'U' value of any part of an external wall of a dwelling (except openings) shall not exceed 0.45 w/m² per degree C. The 'U' value for roofs (excluding openings) should not be greater than 0.25 w/m²/per degree C.

The choice of plastering materials used internally has little influence on this low value. A comparison can be made by using the same construction and substituting different materials. Construction:

Brick outer skin (0.84 W/mk), and a lightweight concrete block 100 mm (0.24 W/mk) 'U' values using

(a) 13 mm lightweight plaster 0.45
(b) 9.5 mm plasterboard dry-lined 0.44
(c) 25 mm thermal board dry-lined 0.38.

INSULATION AND ACOUSTIC PLASTERWORK

Sound is transmitted by wave motion and as this reaches a wall or ceiling surface it may be transmitted, reflected or absorbed. Hard surfaces reflect sound waves and porous materials partially absorb sound.

The problem is to get the maximum strength of wanted sound and the minimum of unwanted reflections or reverberation.

Unwanted sounds can be cut down by cladding the wall with a highly absorbent surface. These usually take the form of an *acoustic plaster* which may consist of particles of a lightweight aggregate such as crushed pumice and bound together with a retarded hemi-hydrate plaster. These are sprayed on to form an open textured finish.

Correctly applied acoustic plasters have little strength and are unsuitable for application below a height of 1.8 m, because of the risk of damage. Piecings should be avoided as they are difficult to hide.

Oil bound or similar types of paints should not be used for decoration as these will seal the surface and prevent sound absorption. Sprayed distemper or water colours are suitable.

X-ray plasters

Certain wall and ceiling surfaces require insulation against the penetration of electro-magnetic radiation and X-rays, (hospitals, laboratories, etc). A plaster mix is used containing barium sulphate as the aggregate and retarded hemi-hydrate plaster as the binding agent. This requires only the addition of clean water and the resulting mixture is very heavy. Because of this, plasters of this type should not be applied direct to concrete without a mechanical key.

It is recommended that concrete ceilings be covered by heavy gauge metal lathing securely fixed to a framework or a suspended ceiling. The metal lathing should be rendered and scratched with a deep under-cut key. When this coat has set one, two or three floating coats are applied to bring the thickness to the required specification which may be 25 mm.

Walls are also floated to the desired thickness in two or three coats using the floating grade of X-ray plaster. Good key is essential on the brickwork and between each succeeding coat.

A skimming coat of Finish plaster is used as soon as practicable after the undercoats have set.

PATTERN STAINING

The effect known as pattern staining is the appearance of dark and light patterns on plaster surfaces, usually ceilings. The pattern on a lath and plaster

Figure 72

Section showing heat flow through a plasterboard ceiling (less heat passes through the portion under the joist)

Colder roof space

Convection flow in heated room

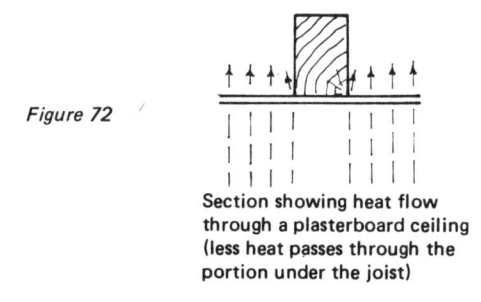

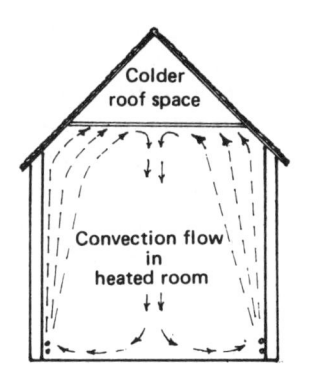

ceiling forms on a more or less complete replica, in light and shade, of the lathing joists, etc. Pattern staining can also occur on the soffits of hollow tile roofs, plasterboard ceilings, partitions and wall linings fixed to battens.

The cause is due to the settling of dust from the air on the plaster surfaces in an uneven deposit. The warmth of the air in the room causes a convection flow. Particles of dust floating in the air are driven on to any surface that is cooler than the air, and they tend to adhere to it. The amount of dust deposited depends upon the dustiness of the air, and the differences in temperature between the warm air of the room and the cooler plaster surfaces. Cooler parts of the plaster surface will have a thicker deposit of dust and after a short time a pattern stain of light and shade will appear (Figure 72).

Remedies

1 To make the plaster surfaces warmer than the air.
2 Reducing the differences in heat flow by adding insulation over the ceiling.
3 Use of a plaster mix having a low thermal conductivity.

For existing ceilings it is best to apply glass silk or slag wool between the joists, and building paper and fibre insulated boards over the joists. Heat losses through the roof could also be checked and minimised.

For new ceilings insulated plasterboards should be used in preference to the normal type. Plastering mixes used should be of the vermiculite type which have good insulating properties.

DAMP-PROOFING

Damp-proofing can be carried out in many ways. One of the simplest methods is to coat the background with a bituminastic type coating such as Synthaprufe which also acts as a bonding adhesive for following plaster applications. The newly applied Synthaprufe should be blinded with dry sand whilst the Synthaprufe is still wet. This will provide good adhesive properties for such plastering applications as gypsum plaster/lightweight aggregate bonding and finish coats. Such a plastering system would provide a damp-proof membrane and also an absorptive finish which would be resistant to condensation.

Another method favoured by some is to thinly coat the background with a waterproofed sand and cement mix. This rendering coat is carefully scratched for key without penetrating through to the background. A plastering floating and skimming system, similar to that described in the previous paragraph, is then applied.

WATERPROOFING

This is the provision of a dense, impervious membrane through which water will not pass. Such a material must be insoluble and the main mixes used are based on combinations of Portland cement, sand and certain waterproofing compounds.

Because sand is a granular material it will contain a proportion of voids in its volume. Sands having a high void percentage are unsuitable, if not impossible, for use in waterproof mixes. Dirty sands with high silt contents are also unsuitable, the silt being soluble in water.

A well graded sand, mixed with sufficient cement to fill the main voids will still have minute voids through which water can percolate after the material has set and dried. Finely grained powders, which also act as lubricants, can be added to the sand and cement mix. The fine grains of the powder helps to fill the smallest voids and lubricate the mix so that it adjusts itself into the densest possible mass. This type of waterproofer is known as a pore-filler.

Other types of waterproofer rely on materials, supplied in liquid or paste form, which combine chemically with Portland cement causing a more complete hydration and crystallisation of the cement grains. The new insoluble compounds produced expand into the smallest voids.

The outer surfaces of waterproofed mixes begin to acquire waterproof qualities after the initial set. This prevents the drying out of the water in the mix which is essential for hydration between cement and water. If the water evaporates before hydration is complete a mass of honeycombed capillary pores will be left behind, previously occupied by water.

Method of application

Extra care is required on the wall surfaces to receive waterproofed rendering, especially if water pressure is behind the wall. Hacking, cleaning, raking out of joints or grouting may be required according to circumstance.

Rendering on walls may require two, three or more coats, each 9 mm thick. Renderings to existing work should extend 600 mm beyond the limits of any apparent dampness. No piecings should be made in rendering coats if possible, but where unavoidable a 125 mm splay should be used.

WATERPROOFING

When waterproofing basements or tanks, subject to water pressure, or severe dampness, the following method can be used. (The mixes quoted are much richer than normally specified but have nevertheless given excellent results when mixed with *chemical* waterproofers.)

1 Corner fillets 50 mm wide are applied in all internal angles with a 2:1 sand/cement mix plus waterproofer.
2 A bonding coat of 1:1 plus waterproofer is spread or thrown on to the wall. The consistency of this mix should be as wet as is possible to apply to a thickness of about 4 mm — 6 mm and left rough for key.
3 A further coat (1½ sand/1 cement plus waterproofer) is applied after a period of 4 to 6 hours. This mix should be mixed to standard consistency and applied with pressure about 6 — 9 mm thick. When this has set sufficiently a spatterdash coat of 2 sand/1 cement plus waterproofer is cast on to form a key for the finishing coat.

If the water pressure expected is likely to be excessive, a second intermediate coat of 2 sand/1 cement plus waterproofer should be applied, followed by a spatterdash coat of 5 sand/2 cement plus waterproofer.

The finishing coat should be approximately 9 mm thick using a mix of 2½ sand/1 cement plus waterproofer. Because of the absence of suction it is an advantage to divide this mix, one half being left dry and the other mixed to standard consistency with water. The wet mix is applied and the excess moisture dried off with the dry mix until the surface can be rubbed up with a float.

Weaker mixes can be used for the finishing coat if condensation is anticipated, or if desired the surface may be scratched for key and then skimmed with a proprietary plaster or strongly gauged lime and plaster.

If water is already percolating through the wall during the period of application the newly applied mix will be washed off the wall surface. To prevent this the following method may be used.

1 Holes are drilled into the wall about 150 mm or 200 mm from the floor. Short lengths of metal tube or conduit about 150 mm long and 12 mm diameter are knocked into the drilled holes and left projecting well clear of the wall surface. The main flow of water will now find it easier to pass through the tubes than through the brickwork joints. This water can now be collected by placing a bucket under each tube outlet or directing the dripping water into a channel or gulley.
2 The base of the wall, where the water pressure is greatest, is coated with the normal mix to which has been added an accelerator. (Proprietary cement accelerators are available for setting times from 1 — 30 minutes.) The size of the mix batch must be decided by the speed of the setting action, the application time and the amount and pressure of water passing through the wall.
3 When the main wall area has been sealed with the accelerated mix the normal waterproofing practice is carried out for the second coat.
4 The second coat is allowed to set thoroughly and the holes are next

sealed in sequence. The tube is taken out and the holes slightly enlarged at the back so that they are wedge shaped, this is to prevent the plug being forced out. A plug is made of neat cement and quick set accelerator, rolling the mixed materials into a cigar shape roughly the size of the hole to be filled. Just before the set occurs the plug is pressed into the hole with pressure and held fast until the set is complete.

5 When each of the holes has been sealed normal waterproofing procedure is carried out.

WATER LEVEL

This is basically a rubber tube with an open-ended glass tube inserted in each end. The two ends of the tube are held togther and filled with water until the water is visible in the glass tubes. If the water has been poured in carefully without entrapping any air as bubbles and the pipe is not kinked, then the surface of the water in each tube will be level with the other. This will be true if the glass tubes are held five, ten or more metres apart.

To use the water level a datum point is usually determined first, at eye level by placing a mark on the wall surface. The tube filled with water and checked for level by holding the ends together as described, needs to be operated by two persons. Each operative covers the open end of the tube with a thumb or finger to prevent any escape of water. One person holds his glass tube to the datum point while the other operative places his vertically held glass tube at the other end of the wall at approximately the height required. At this stage the ends of the tube are re-opened to atmospheric pressure and the surfaces of the water will now adjust themselves until they are level with each other. The two operatives must now raise or lower their glass tubes until one water level is coincident with the datum point. A mark is placed on the wall against the other water level and a chalk line snapped between the datum point and the new mark.

To mark a level line around the room it is usually best, if the length of tube permits, to level every point from the original datum. This method is quicker and less likely to accumulate errors if a great number of levels are to be made.

Purpose made water levels are made with tube lengths of 10 or 20 m. The open ended transparent tubes are of tough plastic and contain a stop valve which is controlled by a screw top cover made of brass, threaded on to the plastic tube. With this type control of handling between the periods of actual levelling is easier, also the tube can be left full of water for long periods. This saves time taken in frequent filling or topping up.

Open ended tubes can be fitted with tight-fitting corks for periods when the water level is not being used or the tube ends suspended vertically above the water tube by a cord attached to a nail in the wall.

COMPLAINTS AND REMEDIES

Many of the plastering defects which occur may be due to causes other than bad workmanship or the use of incorrect materials by the plasterer. Construc-

COMPLAINTS AND REMEDIES

tional faults, including structural movement, weak or badly seasoned furrings, persistent dampness or moisture penetration are some of the many reasons for subsequent failure of plastering systems.

The prevention or correction of defects can best be anticipated or remedied by an understanding of the probable causes.

The following list describes plastering faults with their causes, most of which can be prevented or remedied by the operative taking appropriate counter measures in each case.

Blistering
Caused by local expansion, this may be due to:
(a) A high moisture content and trowelling off before setting action or drying take place.
(b) Delayed hydration caused by later moisture attacks.
(c) Exposure to severe radiant heat.

Bond failure
Due to:
(a) Poor adhesion due to undercoat being gauged too weak, mixed with dirty water or re-tempered when partially set.
(b) Insufficient key.
(c) Finishing coat too strong for the undercoat.

Cracking
Due to:
(a) Structural movements.
(b) Incorrect mixes, including the use of strong cement/sand mixes with subsequent high shrinkage on large panels.
(c) Use of loamy sands with high silt contents resulting in shrinkage on drying.
(d) Plasterboard joints wrongly positioned and/or left unscrimmed.

Crazing
Shows up as hair cracks on finished work and is due to:
(a) Excessive suction or exposure to rapid drying out of skimming mixes, particularly those containing lime putty.
(b) Too much lime putty in lime putty/plaster skimming mixes.
(c) Re-tempering plaster skimming mixes.
(d) The compression of the surface skin of cement mixes by use of a trowel and subsequent shrinkage on setting.

Dry out
Patches of soft, chalky, white dry areas appear which are caused by excessive suction, heating systems, drying winds or too thin application coats.

Patches affected can be lightly sprayed with plain water or a solution of

SPECIALISED TECHNIQUES AND PROBLEM SOLVING

alum in water and allowed to dry. This may be repeated until hardening occurs.

Dusty floors
Caused by:
(a) Too much 'fines' or weak aggregate in the mix.
(b) Insufficient cement in the mix.
(c) Too much water in the mix.
(d) Insufficient curing of drier mixes.

Efflorescence
This is the white powdery deposit on surfaces due to the presence of salt and moisture in the background or plaster mix. The salt is taken up into solution by the moisture. When the solution dries out by evaporation the salt is left as a deposit on the surface.

Simple brushing will often provide a cure unless further deposits occur through continued dampness.

Frost action failures
(a) Do not use frozen materials.
(b) Do not apply on to frozen backgrounds.
(c) Materials applied should have completed the initial set before freezing occurs.
(d) Protect work against freezing whenever possible.

Grinning
This is the pattern of the background showing through on the finished surface. Caused by uneven suction of background materials which may be due to:
(a) Frost in mortar joints (causing excessive moisture in mortar when thawing occurs).
(b) Too thin application coat.

Mould growth on plastered surfaces
Such mildew or fungal growth is due to dampness and poor air circulation, plus the presence of fungal spores.

The mould growth should be scraped off, a fungal mould inhibitor applied and the work dried out by air circulation and heat.

Popping, pitting and blowing
Shown by small blow holes in plastered surfaces and due to unslaked lime particles in the mix.

All observed particles must be removed before patching.

COMPLAINTS AND REMEDIES

Quick setting plaster
Due to the use of:
(a) Old or damp plaster.
(b) Dirty mixing water.
(c) Dirty mixing tools and containers.
(d) Contamination with other plasters or partially set mixes.

Rust staining
Caused by unprotected iron under plastered surface, i.e., EML, electricians' conduits, etc.

It is made worse by plaster mixes, also salt in sand.

Special Metal Lathing Plaster should be used on EML or lime putty added to other plaster mixes.

Salt in sand may prove to be deliquescent, and the moisture attracted will assist corrosion.

Shelling from plasterboards
Due to:
(a) Wrong plaster used.
(b) Too weak plaster/aggregate mix used.
(c) Lime used in the mix.
(d) Mix applied on wet plasterboards.

Sweat out
Dark areas of damp, soft plasterwork, sometimes covered with beads of moisture. Caused by lack of suction, bad air circulation, damp, foggy weather or a hard plaster finish applied over a wet floating coat.

Slightly affected areas can be cured by improving air circulation, providing heat or drying conditions.

CHAPTER SIX

The fixing and plastering of lathing materials

PLASTERBOARDS

Plasterboard is supplied in a variety of sizes to suit various joist centres. The methods used in fixing should be to ensure that the boards are well supported by nails and that continuous joints across the ceiling are avoided if possible to reduce the risk of cracks developing in the finished ceiling.

Each board should be nailed at 150 mm centres with 30 mm galvanized nails for 9.5 mm boards and 38 mm galvanized nails for 12.5 mm boards, care being taken to ensure that the head of the nail just grips the paper of the plasterboard without puncturing it. If the nails are left with a space between the head and the plasterboard then the ceiling will later vibrate and push off small circles of plaster the size of the nail head. If the nail heads have punctured the paper then the plaster behind the nail head will have been crushed and again looseness of the boards can occur and in extreme cases can even fall off. Boards should be nailed not nearer than 13 mm from the board edges.

When fixing plasterboards to a ceiling, the boards should be positioned so that the long bound edges are at right angles to the joists. It is important to cut and fix the first board so that the uncut, unbound edge, rests halfway on the furthest joist that the board will reach. The exposed margin should be parallel to the joist edge. If succeeding boards are also fixed with uncut edges on the two adjacent exposed board edges, the boarding will be kept parallel and square. Succeeding rows are bonded to avoid an unbroken joint across the ceiling as this is a likely cause of cracking (Figure 73).

The plastering of plasterboards is either one or two coat work. One coat work is carried out by using a Board Finish gypsum plaster. Firstly the joints between the plasterboards are filled in and a strip of plaster 80 mm wide laid along all the joints. Before this material stiffens a strip 75 mm wide hessian is pressed into it and tightened as flat as possible. Scrim should not be overlapped at the cross joints because the extra thickness at these points is difficult to hide in one coat work.

Scrim should also be applied in the wall and ceiling angle, but care must be taken to ensure that the scrim left on the wall wing of the angle is as thin and flat as possible.

An alternative arrangement is to leave the walls unfloated until the ceilings are completed, the scrim will then be plastered to the brickwork and easily covered with the floating coat.

PLASTERBOARDS

Figure 73

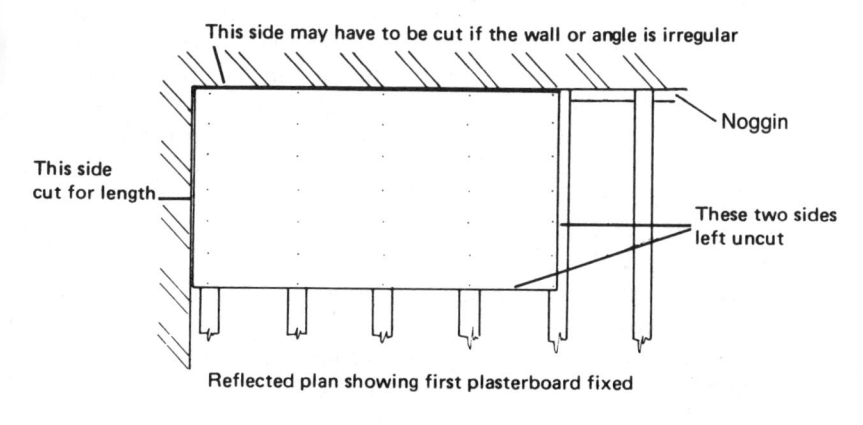

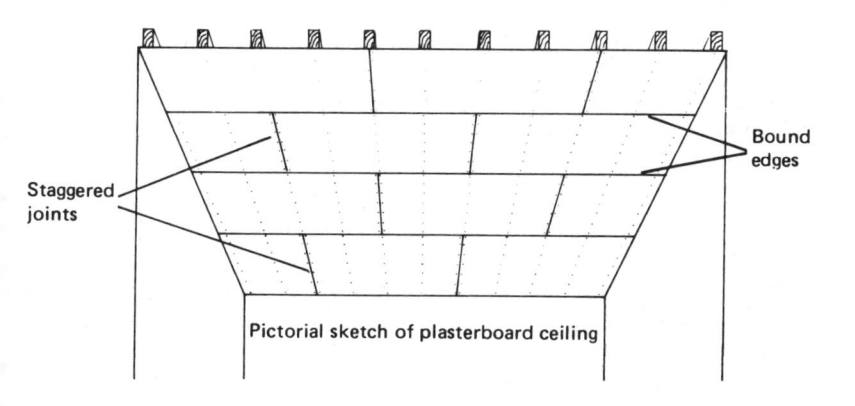

Reflected plan showing first plasterboard fixed, with labels: "This side may have to be cut if the wall or angle is irregular", "Noggin", "This side cut for length", "These two sides left uncut".

Pictorial sketch of plasterboard ceiling, with labels: "Bound edges", "Staggered joints".

Another arrangement (as suggested by some plastering manufacturers, especially for use in replacement ceilings) is to leave the floating coat 25 mm below the level of the plasterboard. Also the plasterboard itself should be left off 12 mm from the brickwork all round the room. Next Board Finish plaster is used to fill the gap and scrim may be trowelled half on the ceiling side of the angle and the rest covered inside the gap (Figure 80). Other plaster manufacturers recommend the latter method, but without the use of scrim. These methods described in Figure 80 are normally confined to repair work.

As soon as the plaster in the scrimmed joints has set, but not dried, the plasterboards should be covered to bring up the level of the ceiling to that of the scrimmed surface. Next an even coat should be laid on over the whole ceiling to a total thickness of 4 mm. Ceiling angles should be wiped out with a feather-edged rule, or straight edge, while the material is still soft and straightened as necessary.

THE FIXING AND PLASTERING OF LATHING MATERIALS

Meanwhile the ceiling material will have stiffened somewhat and at this stage a good method is to give the ceiling a laying down coat with a skimming float. This surface should then be tightened in and laid down with a small amount of softer Board Finish. As the Board Finish sets it should be trowelled to a good finish using as little water as possible.

Two coat work on plasterboards can be carried out in vermiculite/plaster mixes. In each case the plasterboard surface is scrimmed as described previously for one coat work.

The floating coat should be laid on the surface evenly and ruled to a good surface, particular attention being paid to the straightness of the angles. Before the floating coat finally sets it should be lightly scratched to provide a key for the skimming coat.

After the floating has set a skimming coat of Finish plaster should be applied evenly, laid down and then trowelled off to a good finish.

In *lightweight aggregate/plaster* applications on plasterboards the floating material can be bought ready mixed dry and merely requires the addition of water. Vermiculite/plaster mixes are known as lightweight bonding materials and these mixtures obtain an excellent adhesion to plasterboard. (Care should be taken not to confuse these materials with Perlite/plaster mixes which are sold as lightweight aggregate/Browning plasters, and are unsuitable for application to plasterboard.)

Trade names of suitable finishing coat plasters for the undercoats mentioned include Carlite Finish.

Gypsum Plaster Laths are made in widths of 406 mm and lengths of 1200 mm, 1219 mm and 1372 mm. They are nailed at right angles to the run of the joists, and successive rows are bonded to break the joint. A gap of 3 mm should be allowed between each lath. It will be noted that the semi-circular bound edge of the plaster lath runs along the unsupported joint and the square unbound edge rests on the joists. This is a reason for the extra strength of this type of ceiling which does not normally require scrimming to the joints.

Maximum spacing of joists should be 450 mm for 9.5 mm thick lath and 600 mm for 12.5 mm lath.

Skimming plaster laths in one coat work is similar to that of plasterboards except for the following. Because of the bound rounded edges on the long edges of the laths it is possible to omit the use of scrim on all joints except the ceiling angles.

Joints in the laths (3 mm apart) should be filled with neat Board Finish to form a 'welded' or 'rivet' joint as shown in Figure 74. As soon as the material in the joints has stiffened the ceiling can be skimmed as described for plasterboard.

Two coat work on plasterlath, for lightweight aggregate/plaster mixes is also similar to that described for plasterboard, except that no scrim need be applied to the joints.

LATHING

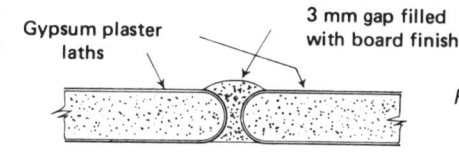

Figure 74: Rivet joint in a gypsum plaster lath

LATHING

Expanded metal lathing

Expanded Metal Lathing is normally supplied in sheets of 2.700 m × 700 mm. It can be fixed to timber joists, or metal runners in the case of fireproof construction.

To fix expanded metal lathing to timber joists the long way of the diamond mesh should be nailed at right angles to the run of the joists. The sheet should be cut to the correct size, held in position and nailed to the centre joist first, using 30 mm galvanised nails or staples at 100 mm centres. The sheet should now be nailed outwards from the centre in each direction, nailing into each joist in turn to keep the lathing as taut as possible. To assist in straining the lath the nails should be driven at an angle away from the centre joist as shown in the sketch (Figure 75).

All joints should be lapped 25 mm over adjoining sheets and wire ties placed every 150 mm along the loose ends using 18G (1.219 mm) galvanized soft iron wire. At joints with the wall the expanded metal lath should be bent down 75 mm flat to the brickwork.

Joist spans recommended are 300 mm centres for 0.5 mm metal and 450 mm for 0.95 mm metal.

Fixing to steel channels, wire ties are used bent like a hairpin. These are pushed up through a mesh, over the metal runner, then down through the mesh again, twisted, cut and bent flat. The sheets should be tensioned and lapped as described for fixing to wood joists.

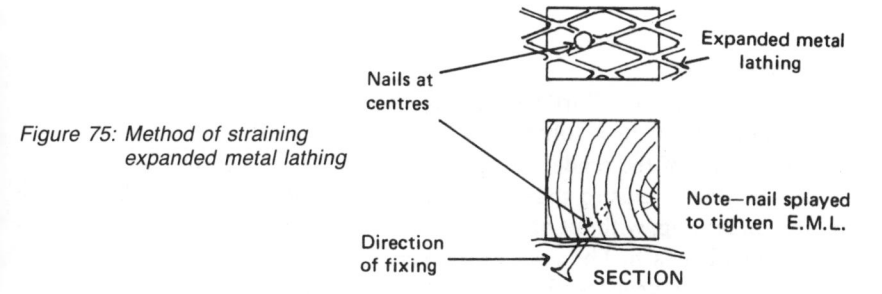

Figure 75: Method of straining expanded metal lathing

When fixing *Rib-lath* the ribs should be fixed with the ribs in contact with and at right angles to the supports. The methods of fixing to timber or steel joists is otherwise as described for ordinary expanding metal lathing, except that the joints should be lapped 50 mm at the ends of sheets on supports. Laps between supports are not advised but if these are unavoidable then the sheets must be overlapped at least 100 mm and wired together with at least two ties to each pair of overlapping ribs. At the joints with the wall a strip of ordinary expanded metal lathing, 150 mm wide, bent to form a right angle with two 75 mm flanges, can be fastened into the angle of the wall and ceiling by nailing or tying with wire.

Hy-rib sheets are used primarily for concrete reinforcement. When used as a lath for ceiling backgrounds it can be used economically where long spans between joists are required. Hy-rib is available in sheet lengths from 2.000 m to 5.000 m in increments of 305 mm, and 455 mm in width. Hy-rib sheets are made in thicknesses of 0.575 mm for spans of up 1200 mm and 0.750 for spans up to 1500 mm between supports.

Hy-rib sheets are fixed with the ribs at angles to, and in contact with, the timber joists or steel runners. The sheets should be fixed with the side ribs lapping over the side ribs of the adjoining sheet and well pressed home. The interlocked ribs are then either clinched together at 525 mm centres using a special Hy-rib punching tool, or the ribs can be wired together using 1.4 mm tying wire at 525 mm intervals.

When Hy-rib sheets are fixed to timber joists they should be nailed with 37 mm staples at 260 mm centres. Special purpose made fixing clips can be used at 260 mm centres when fixing to metal bearers, or alternatively the sheets can be fastened with 2.0 mm tying wire at every rib (87 mm centres).

Plastering on the various types of metal lathwork described can be carried out in raw or gauged lime/sand mixes, or lightweight aggregate/plaster mixes. aggregate/plaster mixes.

Lime plastering mixes are based on haired coarse mortar made from 1 part putty lime to 2 parts of sand with 4 kg of hair per cubic metre of mortar. Raw stuff is rarely used in the present period due to its slow setting time and the raw stuff is either gauged with Portland cement or gypsum plaster.

Rendering coat

When gauged with Portland cement the proportions recommended are 1 volume of cement to 9 volumes of haired coarse stuff.

Application of the rendering coat should be made with sufficient pressure of the trowel to ensure that the material is completely interlocked in the meshes of lathing, without having been broken off or separated from the rendering coat beneath. The average thickness should not exceed 9 mm and the surface must be keyed with a wire scratcher before the setting action takes place.

Floating coat

Mixes similar to that of the rendering coat should be used. Sufficient time should be allowed for drying out before the floating coat is applied. This will vary according to the site and weather conditions, but time should be allowed for initial shrinkage to take place before floating. The floating should be levelled with a straight edge to an average thickness not greater than 9 mm and finished off with a devil float to leave a key for skimming.

The skimming coat can consist of 2 parts lime 1 part sand punched through a fine sieve and the whole mixed with 25 per cent retarded hemi-hydrate plaster.

Lightweight aggregate/plaster mixes for expanded metal lathing are premixed by the manufacturers.

Rendering and floating coats — Lightweight aggregate Metal Lathing Plaster.

Skimming coat — Lightweight aggregate/Finish plaster.

The approximate setting times for lightweight aggregate Metal Lathing and Finish grades is 90 minutes.

Twil-Lath

Twil-Lath metal lathing consists of sheets of steel wire mesh with interwoven slotted backing paper to provide key and reinforcement for plaster or sand and cement applications. The different types available are described on page 29.

G-Lath should be fixed with the long dimension at right angles to the supports. When applied to vertical surfaces the lath should be bent into and around corners to extend at least one stud space on to the adjoining wall. Horizontal joints should lap a minimum of 50 mm (one mesh) and vertical joints 40 mm (one mesh). Horizontal joints should be wire tied or clip tied between supports at 150 mm centres (this is achieved by cutting the vertical wire of the sheet below and bending it over the wire in the sheet above (Figure 76a)).

Vertical joints should be staggered and not occur in line with a framed member of openings (Figure 76b).

In positions where the lathed wall meets a masonry or concrete wall, the lath should be extended by three lath meshes on to the concrete or masonry. Lath applied to ceilings should be bent into the ceiling angle and extended three meshes on to the wall. An alternative is to butt the lath into the internal angle and then fit a reinforcing strip to cover both sides of the angle by a minimum of 75 mm.

G600 is used on internal timber or metal stud walls or partitions, as a backing for tiles or terrazzo, and also for sprayed applications including acoustic plasters. The fixing of G600 is similar to that described for G-Lath except that the vertical and horizontal flaps of the waterproof backing paper should be opened up and extended before fixing. The lath should be fixed at the bottom right hand corner of the wall first, working upwards from right to left with succeeding sheets. To

Figure 76a

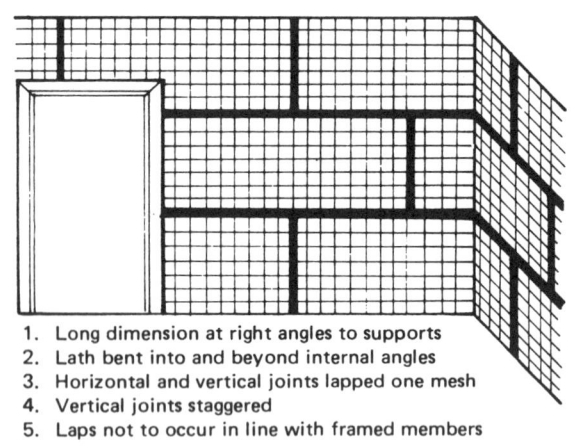

1. Long dimension at right angles to supports
2. Lath bent into and beyond internal angles
3. Horizontal and vertical joints lapped one mesh
4. Vertical joints staggered
5. Laps not to occur in line with framed members

TWIL-LATH

Horizontal joints should be clip-tied or wired every 150 mm

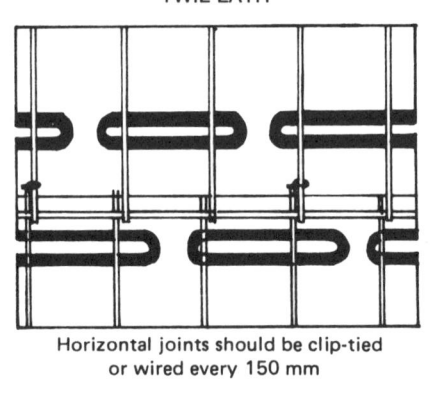

Figure 76b: Sketch showing method of clip-tying of horizontal joints

make sure the lathing is correctly waterproofed, horizontal and vertical joints must be lapped as described for G-Lath.

SS400BP and SS600BP are recommended for external use on timber framed buildings and other external or construction work. The lath provides a waterproof base for the application of a variety of sand and cement renderings and finishes. Stainless steel staples must be used for fixings on external timber frame or similar types of cladding. Other types of rustless fixings are used on internal work.

Wood lath

Wood lath is nailed to timber joists and partitions with 25 mm nails allowing 9 mm space between each lath. The ends should be butt joined and not lapped, and continuous joints across the ceiling should be avoided by breaking the joint at a maximum distance of 900 mm. This is known as *butt and break lathing*. When timber joists are wider than 75 mm, it is essential to

LATHING

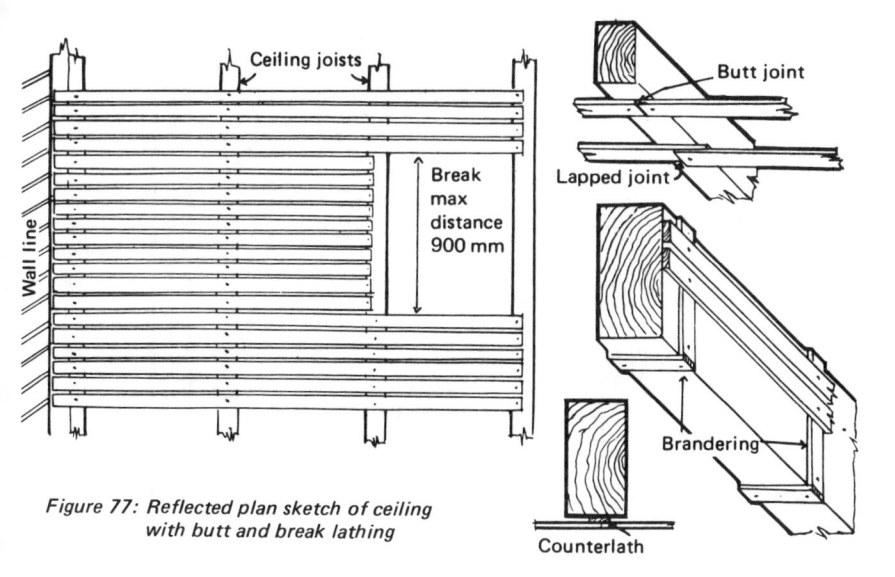

Figure 77: Reflected plan sketch of ceiling with butt and break lathing

provide a mechanical key. This is done by nailing a lath down the centre of wide joists and then *counterlathing* over.

Brandering is used to provide a mechanical key for plastering coats on timber beams and lintels as shown in Figure 77.

Nails used for fixing wood lath need not be galvanized, except when gypsum plaster mixes are used for rendering and floating.

The mixes and methods of application are similar to those described for expanded metal lathing except for the following.

Rendering and floating coats in gypsum plaster/sand and lightweight aggregate/plaster mixes need not be metal lathing grade because there is no risk of corrosion if galvanized nails are used for fixing.

When applying the rendering coat to wood laths the material should be applied diagonally away from the plasterer.

In skimming it is good practice to apply the skimming coat along the run of the laths and not across. This is to avoid the risk of breaking or damaging the key.

Newlath

Newlath is made from high density polythene. It is formed into a pattern of raised studs and these are 8 mm high and face the wall when fixed allowing air to circulate behind the lath. The studs are reinforced with ribs which link them together both horizontally and vertically. On the face of the lath a polythene mesh is formed, this is to ensure an adequate key for the rendering or plastering. Newlath can be fixed both internally and externally to straight or curved backgrounds. The lath is obtained in rolls 1.5 m by 10 m covering an area of 15 square metres.

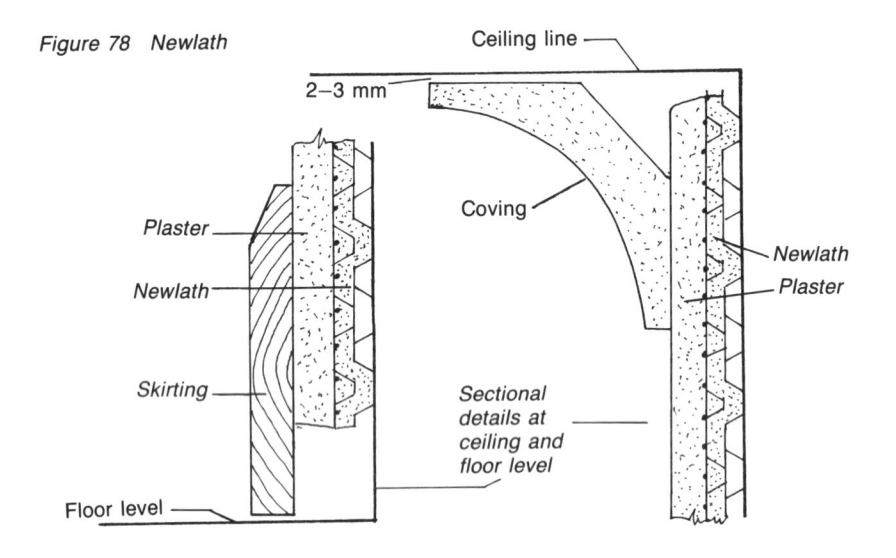

THE FIXING AND PLASTERING OF LATHING MATERIALS

Figure 78 Newlath

Cutting is simply carried out with a sharp knife or shears. Joints between sheets should overlap 100 mm, on one side of the roll there is a plain stud area for lapping which is without mesh. Joints can also be made on the other edges to a minimum of 100 mm and any fixings passed through both sheets.

Fixing is carried out by drilling 8 mm holes 40 mm in depth and 300 mm maximum centres in both directions (16 per m^2). The Newlath is then fixed with large mushroom headed plugs made of polypropylene. Plugs are hammered through the sheet into pre-drilled holes. If Newlath is being used to isolate a damp or stained wall from the plaster any accidental damage could penetrate the water resistant barrier. Damage to the sheet can easily be sealed with a waterproof mastic. When Newlath is being used over a damp wall surface it is better if ventilation behind the sheet is increased by allowing for a free movement of air (see Figure 78). The ventilation can easily be achieved by raising the lath by 20–25 mm from the floor and allowing a gap of 2–3 mm at the ceiling level, (this can be hidden by a feature e.g. a cornice). Special extruded profile is available as an edge treatment similar to stop bead.

Internal plastering to walls is carried out in normal two coat work to a thickness of 13 mm. Materials that can be used are pre-mixed lightweight bonding and finish, renovating backing and finish, also cement undercoats with gypsum plaster finish. One coat universal plasters can also be used. On ceilings the thickness can be reduced to 10 mm, provided the tolerance of the background is good enough. External rendering can be carried out using cement, lime and sand 1:1:6. Because there is no suction in the Newlath the best method of plastering or rendering is to apply one tight coat and then follow up with a second tight coat rather than one straight application. Plasters with a set should not be allowed to stiffen before the follow on and should be carried out with the same gauge.

PLASTERBOARDING FAULTS

Failures in plasterboard ceilings can often be traced to bad practice. Cracks develop due to continuous instead of broken joints across the ceiling, or the omission of scrim.

Staining caused by nails is sometimes due to the failure of the rust prevention treatment, particularly with sherardised nails. Insufficient covering thickness of the one coat skimming is also a contributory cause.

Nails driven at an angle other than 90 degrees will tend to puncture the paper of the board and shatter the plastic sandwich behind (Figure 79). Excessive vibration of the plasterboarded ceiling before plastering could cause the crushed plaster from the plasterboard sandwich to escape leaving a void above the nail. If this is repeated with several nails the board is loosened from the joist and will move freely when the ceiling vibrates. Occasionally this results in the skimming over the nail head being pushed off, because the plasterboard is loose but the nail is fixed in the joist.

A similar situation could occur if the nails are not driven in far enough and the plasterboard can override the shank of the nail. This latter defect occurs but rarely because projecting nails are easily detected and knocked flush.

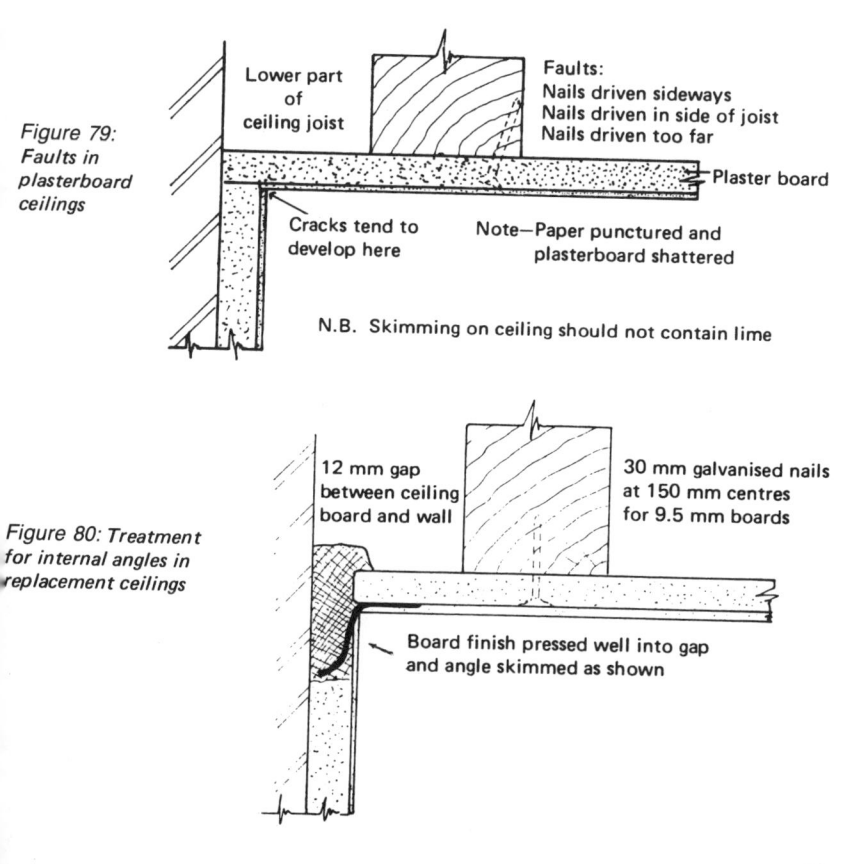

Figure 79: Faults in plasterboard ceilings

Lower part of ceiling joist

Faults:
Nails driven sideways
Nails driven in side of joist
Nails driven too far

Plaster board

Cracks tend to develop here

Note—Paper punctured and plasterboard shattered

N.B. Skimming on ceiling should not contain lime

Figure 80: Treatment for internal angles in replacement ceilings

12 mm gap between ceiling board and wall

30 mm galvanised nails at 150 mm centres for 9.5 mm boards

Board finish pressed well into gap and angle skimmed as shown

Scrimming the internal angle of a ceiling can be awkward if the walls have been previously floated. The extra thickness of scrim on the floated surface is difficult to hide and creates suction problems when skimming.

If the wall has to be floated before the ceiling is skimmed a gap can be left near the ceiling angle for the scrim to be tucked in when the plasterboards are skimmed (Figure 80). On repair work and renewed ceilings this method is particularly suitable.

CHAPTER SEVEN

Dry Construction Methods and Partitions

Newer trends of dry lining walls and the construction of cellular plaster panel partitions have certain advantages. Their present popularity with architects and builders includes time saved in drying out of traditional plastering materials, better thermal insulation and, in many cases, reduced costs.

DRY LINING

This is the application of special tapered edge plasterboard termed Wallboard, to internal wall surfaces. The joints are filled by a special technique but the surface of the boards is left unplastered.

The boards are available in thicknesses of 9.5 mm and 12.5 mm, widths of 600 mm, 900 mm and 1200 mm, and various lengths between 1.8 m and 4.8 m. A width of 900 mm and thickness of 12.5 mm is recommended as suitable for most jobs.

Boards are fixed vertically and should be approximately 25 mm less in length than the floor to ceiling height.

Ceilings should be completed before the walls are boarded.

Method

The walls are straightened by applying a series of linable dots at pre-determined spaces to form a grid base for bedding the subsequent plasterboard to a perfect alignment. The dots used are pieces of thin bitumen impregnated fibreboard 75 mm long by 50 mm wide.

Firstly the walls are marked vertically at 450 mm centres. The high spots of the wall are then found by testing with a straight edge or line to determine the minimum thickness which can be used.

Two dots are placed in each internal angle and plumbed as for normal plumb dot and screed method. The upper dot should be 230 mm from the ceiling and the lower dot 100 mm above floor level.

Intermediate dots should be lined in vertically at 1.07 m centres.

Dots should now be fixed at 1.8 m centres horizontally, and when set all intermediate dots bedded and lined in from existing dots (Figure 81).

Fixing the plasterboards

The dots should be allowed to set hard and this usually takes about an hour from fixing.

Figure 81: Grid base for dry lining

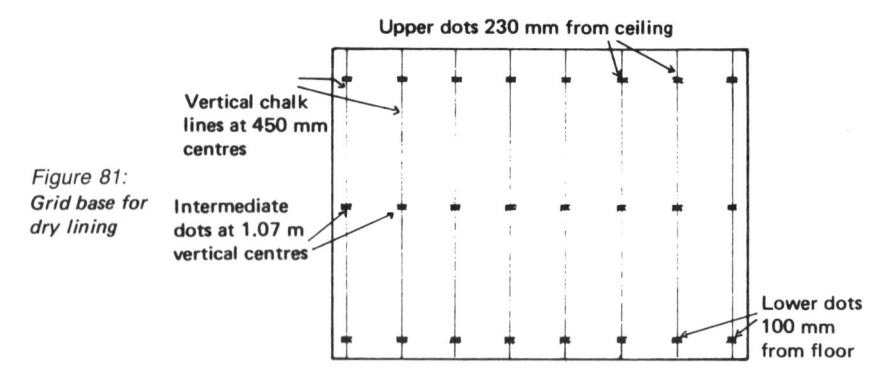

- Upper dots 230 mm from ceiling
- Vertical chalk lines at 450 mm centres
- Intermediate dots at 1.07 m vertical centres
- Lower dots 100 mm from floor

Dabs of neat Gyproc Bonding Compound plaster, should now be applied vertically to the wall surface between the dots. The dabs of plaster should be the full length of the plasterers' trowel and thick enough to stand proud of the dots. Gaps of 50 mm to 75 mm should be allowed between each vertical plaster dab.

Sufficient plaster dabs for one board only should be applied and the dabs adjacent to joints should be kept back about 25 mm.

The plasterboard should be placed in position with the bottom end resting on a rocker board or foot lifter (Figure 82). The board should firstly be tapped back tight to the dots with a straight edge and then lifted with the foot lifter until it is tight to the ceiling.

A check should be made to make sure that the leading edge is plumb and central down the line of the vertical dots, and then the boards are nailed into the dots at the edges of the boards. Special double headed nails should be used and they should be driven home until the first head has slightly dimpled the plasterboard paper surface (Figure 83).

When the plaster has set (in about 1 hour) the nails can be removed by twisting and pulling out with pliers. The nails may be re-used many times.

This procedure is repeated around the room. Cut edges should be placed to the internal angles and masked by a full board when re-starting on the adjoining wall.

Window reveals, soffits and narrow widths of up to 450 mm wide are bedded without dots. Bound edges of plasterboard should always be fixed to master the cut edge of the adjoining board.

Another system is now being accepted by the manufacturers and the trade which removes the use of pads/dots. The most important requirement of the new system is to ensure that a continuous ribbon of bonding compound proceeds around the perimeter of the wall and at all junctions such as doors/windows. Cavity ventilation behind the boards is not now considered necessary. The boards are simply bedded on to the wall surface with dabs of material not using either pads or nails. The whole operation relies on the skill of the operative and has increased the speed of application. It is now possible to use both 900 mm and 1200 mm boards.

Figure 82: Rocker board or foot lifter

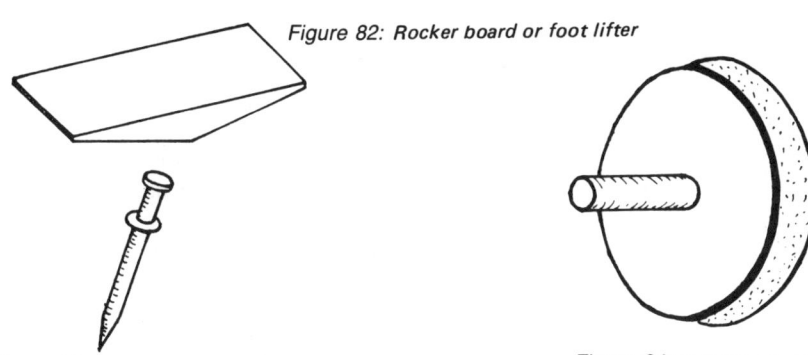

Figure 83: Double headed nail

Figure 84: Jointing sponge

Treatment of joints

Special materials for filling and finishing of joints have been manufactured by British Gypsum Ltd. They are Joint Filler, Joint Finish, tape and corner beads.

Joint Filler sets in about 45 minutes but Joint Finish hardens by evaporation only and can be retempered if required.

The joints are filled by coating with the joint filler mix, inserting the joint tape and filling up the joint trough flush with a further application of the filler mix.

When this has set a broad band of Joint Finish mix is applied over the joint about 200 mm to 250 mm wide. The edges should be feathered out with a jointing sponge. When partially dry (not less than 1 hour in warm weather and longer in winter) a further similar application should be made.

After a further period of drying the jointing sponge (Figure 84) should be dipped in a very thin slurry of the joint finishing mix and the slurry distributed over the whole of the plasterboard surface. This can best be achieved by using light circular strokes, the object being to even up the texture between the plasterboard and the jointed area.

External angles can be strengthened by the application of special jointing tapes some of which are strengthened with metal strips. Metal corner beads can be used for external angles where maximum strength is desired.

The use of mechanical aids for the joint filling and taping process can increase the speed of operations once the techniques have been mastered. These mechanical jointing tools include a taper for straight joints, a roller for internal angles, different finishers for straight joints and internal angles, a nail spotter and a loading pump.

DRY PARTITIONS

These are made of two plasterboards, separated by and bonded to a core of cellular construction. The complete panel is used as a lightweight, non-load-bearing wall unit.

The surfaces are made for immediate decoration or as a base for gypsum plaster applications. Plastic covered surfaces are also manufactured.

DRY CONSTRUCTION METHODS AND PARTITIONS

The sizes vary in thicknesses of 50 mm, 57 mm, and 63.5 mm in widths of 600 mm, 900 mm and 1200 mm. (The 50 mm thick panels are in 900 mm widths only.) Lengths of panel vary from 1800 mm up to 3700 mm except for the 50 mm thickness which is made only in lengths of 2350 mm.

Dry partition panels can be cut to any required size by use of a fine tooth panel saw. Cut edges should, if possible, be placed in an internal angle. Where the cut edges have to be left shown, they should be cut back with a Surfoam tool, at an angle of 45° for about two thirds the thickness of the plasterboard. The paper burrs should be removed with fine sandpaper.

If the panels are to be plastered they should have grey outer surfaces. Normal plastering practice for work on plasterboards can be used, including scrimming of joints, one coat work in Board Finish and two coat work in approved lightweight aggregate/gypsum plaster bonding mixes.

Partitions which are not to be plastered should have ivory or plastic covered surfaces. Ivory faced panels with tapered edges should be jointed and finished as described in the section on dry-lining. Square edged panels are available if the joints are to be covered with special cover strips.

Fixing the panels

The panels are normally fixed without horizontal joints, individual panels being the full floor to ceiling height.

Lengths of timber batten 37 mm x 19 mm are nailed or screwed to ceilings and walls. (For 50 mm panels the batten thickness should be 30 mm.) A sole plate of 19 mm thickness and the width of the panel thickness is nailed to the floor.

The first panel is placed over the ceiling batten and on top of the sole plate and then slid along to engage over the wall batten. 300 mm lengths of 37 mm x 19 mm timber batten should be inserted half its length into the base of the panel on the sole plate. A length of 37 mm square section timber batten is then fitted vertically half way into the exposed edge of the fixed panel. 30 mm x 2 mm galvanized nails are then driven through the panel edges into the battens at 230 mm centres.

The second panel can now be positioned by sliding the new panel over the exposed half of the joint batten. This is repeated until the partition is completed.

It is usual to work away from each wall junction, terminating in a doorway if such is contained. The correct doorway width, allowing for frame thickness, is then treated as before except that 37 mm square battens are driven flush with the panel edges before nailing. Short lengths of 37 mm x 19 mm batten should be nailed above the door height to the upper edge of each panel in the opening. A short piece of panel is cut to size and slid over the exposed short battens and nailed. The door frame can then be nailed to the flush side battens of the opening.

When erecting panels between fixed points without a doorway opening the panels should be fixed from each end of the room as before described. Three slots are cut into the edges of each exposed end of the panels and each side of the opening. 37 mm x 37 mm battens are now driven in flush into each exposed edge of the panels. Nails are driven into the battens through each of the slots, leaving the nails protruding about 12 mm. The final panel is then placed in the opening and the exposed nails are tapped sideways until the battens are in the correct fixing positions. The panel edges can now be nailed in the normal way and the temporary nails removed (Figure 85).

Details of joints for external angles and T junctions are shown in Figures 86, 87.

Metal furring wall lining systems

This is a method of fixing plasterboards to solid backgrounds. The plasterboards are power screwed to metal furring channels previously bedded to the wall background.

The system provides a dry lining and thermal insulation in one operation. It is not suitable for use in continuously damp or humid conditions.

Materials used include special metal furring channels supplied in 2260 mm lengths, metal furring stops in 150 mm lengths, drywall adhesive, special self drilling and tapping screws with countersunk Phillips heads. Plasterboards used include tapered edge wallboard, with or without a vapour check backing, tapered edge insulating wallboard with a backing of aluminium foil, tapered edge wallboard with a backing of expanded polystyrene plus an additional vapour check if required.

When setting out, the following thickness allowances should be made from the face of the background as follows:

(a) 25 mm for 12.5 mm wallboard.
(b) 33 mm for 12.5 insulating wallboard.
(c) 38 mm for 25 mm thermal board.
(d) 44 mm for 32 mm thermal board.

Electric or other service pipes can be accommodated behind the plasterboard lining provided the cavity is sufficiently wide, otherwise partial chasing may be essential or a greater clearing allowance made by a thicker bed for the metal furring.

The metal furring system should not be commenced before the ceiling has been completed.

In the initial setting out, the high spots on the walls should be determined, and their positions plumbed and marked on to the ceiling and floor. An allowance of 12 mm is made for the depth of the furring channels and adhesive, and marks should be lined in on the ceiling and floor the full length of the walls. These lines provide guide lines when bedding channels in position. An allowance of 20 mm should be made for insulating grade wallboards.

If a parallel straight edge is used, it is an advantage to add the width of the straight edge to the previous allowance, and this margin marked as the guide line. When bedding the metal furring channels, the exposed edge of the straight edge can be tapped back on to the line (Figure 88).

DRY CONSTRUCTION METHODS AND PARTITIONS

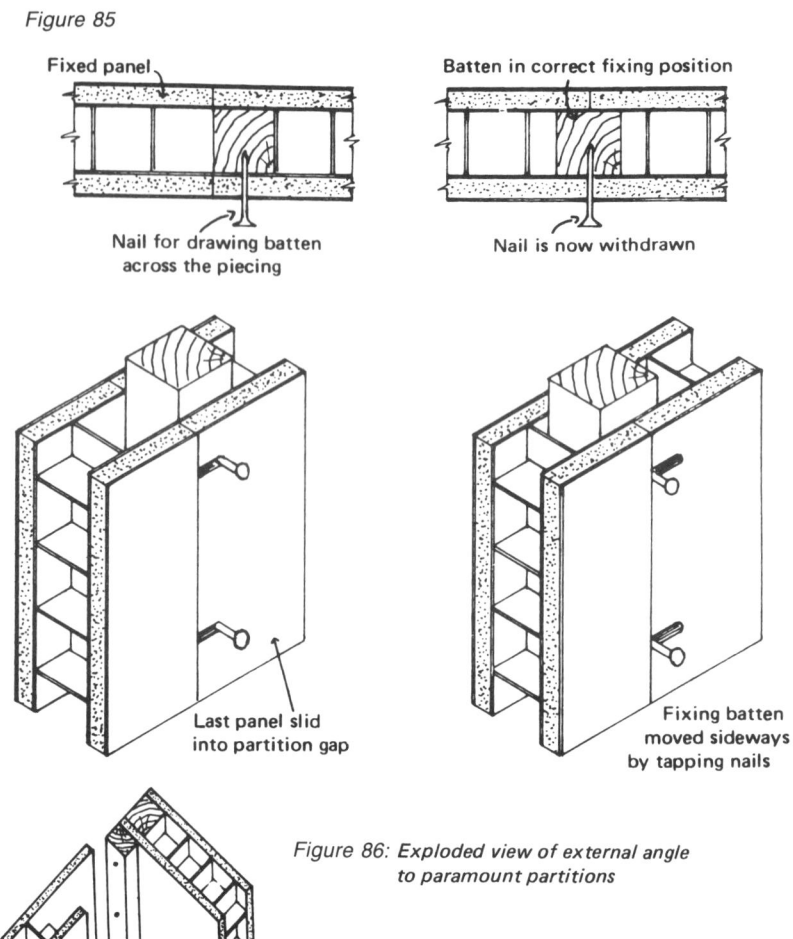

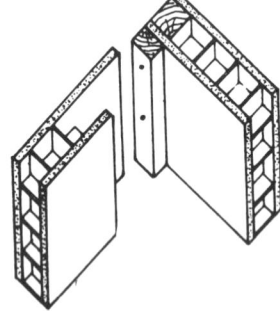

Figure 85

Fixed panel
Nail for drawing batten across the piecing

Batten in correct fixing position
Nail is now withdrawn

Last panel slid into partition gap

Fixing batten moved sideways by tapping nails

Figure 86: Exploded view of external angle to paramount partitions

Figure 87: T-junction assembly

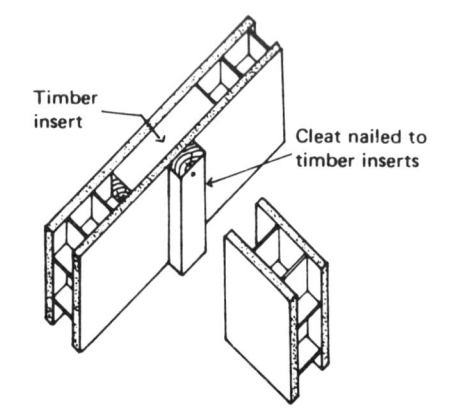

Timber insert

Cleat nailed to timber inserts

The walls are marked vertically at 600 mm centres. Channels should be fixed as close as possible to internal and external angles, allowance being made for the extra projection of the facing board at external angles.

The drywall adhesive should be mixed in a bucket with water to the consistency of soft putty. It has a setting time of about two hours.

Dabs of the adhesive 200 mm long should be applied on the vertical lines at intervals of about 450 mm centres, always ending with dabs at the bottom of the wall. A length of metal furring channel is placed over each line of dabs while the adhesive is still soft and the channel tapped back with the parallel straight edge to the guide lines on floor and ceiling. The channels should be firmly embedded with no projecting adhesive.

When all the vertical furring channels have been fixed, the short metal furring stops are bedded horizontally at the top and bottom of the wall. These are lined in with a straight edge from the vertical channels (Figure 89).

In positions where horizontal services have to be accommodated behind the wall lining, the vertical channels can be cut and fixed to allow a suitable gap for the conduits to pass horizontally. Gaps of up to 300 mm are permissible.

The plasterboards are fixed by screwing to the metal furring channels after the adhesive has set. The first board should be positioned so that the leading edge is on the centre line of a furring channel. Using a power screwdriver and appropriate length screws, the boards are screwed to the channels at 300 mm centres, also to the metal furring stops at top and bottom of the wall. The screws should be driven in so that the heads are slightly below the paper surface of the board without fracturing it.

Holes for electric switch or fuse boxes should be cut out of the plasterboard before fixing. Window reveals and narrow widths may be bedded directly to the background if ordinary wallboards are used. Other types of wallboard will require the metal furring fixtures.

Methods of making allowances in the fitting of boards at reveals or external angles are shown in Figures 90, 91.

Cut edges of boards should be masked by bound edges.

The tapered joints, external angles and screw head holes are then treated as described in the jointing procedure for dry lining.

LAMINATED PLASTERBOARD PARTITIONS

These consist of three layers of plasterboard bonded *in situ* with a special bonding compound and supported by a surrounding timber frame.

They can be constructed in 50 mm and 65 mm single leaf partitions, also 150 mm and 200 mm double leaf partitions or separating walls.

The 50 mm partition is composed of a core of 19 mm plasterboard plank with a layer of 12.5 mm wallboard on each side. The 65 mm partition is similar except that the outer layers are 19 mm plasterboard plank.

Services, such as electricians' conduits, can be positioned in the core along the centre line of the partition.

Timber battens 38 mm x 25 mm should be fixed to the partition boundaries

DRY CONSTRUCTION METHODS AND PARTITIONS

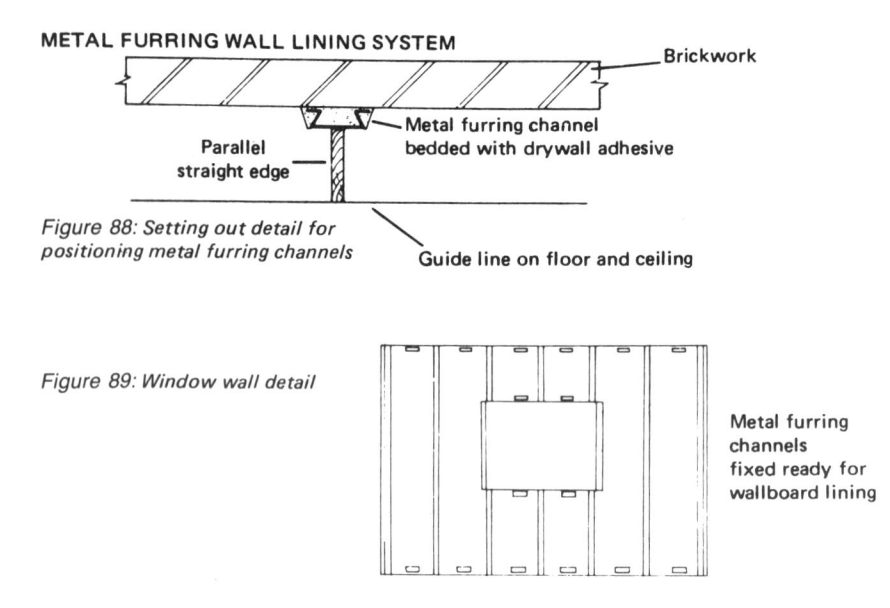

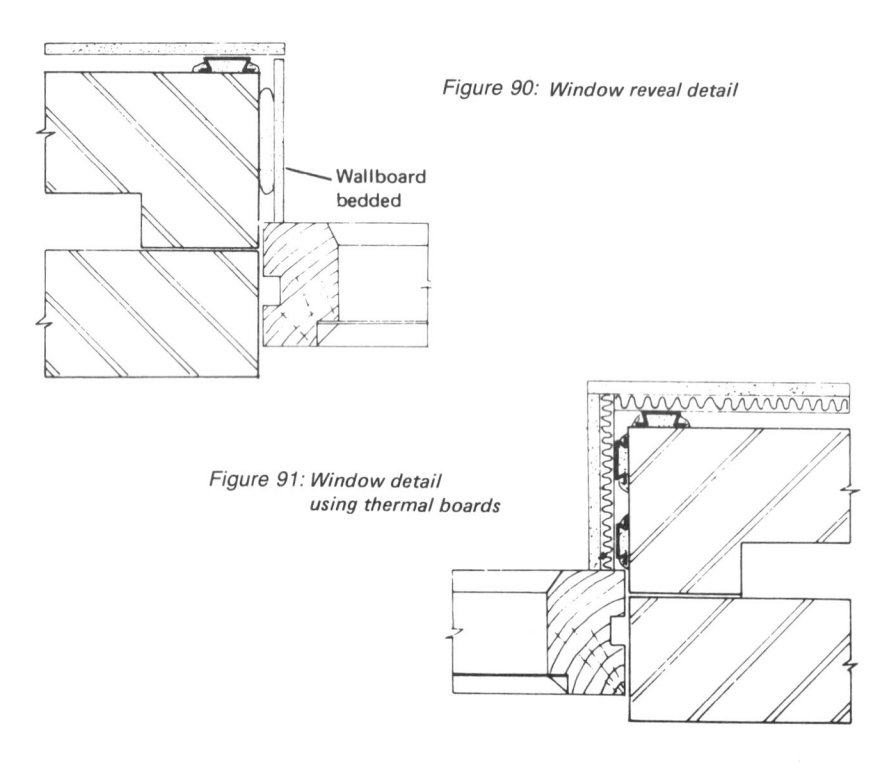

METAL FURRING WALL LINING SYSTEM

Brickwork

Metal furring channel bedded with drywall adhesive

Parallel straight edge

Guide line on floor and ceiling

Figure 88: Setting out detail for positioning metal furring channels

Figure 89: Window wall detail

Metal furring channels fixed ready for wallboard lining

Figure 90: Window reveal detail

Wallboard bedded

Figure 91: Window detail using thermal boards

at wall, floor, ceiling and at door frame openings. The battens are fixed to one of the 25 mm sides. (On concrete surfaces 25 mm square section battens are recommended, see Figure 92.)

For 50 mm partitions, the outer layer of one side is cut to a floor to ceiling height and then nailed to the battens at top and bottom with three equally spaced 30 mm x 2 mm galvanized nails. When using tapered edge wallboards, the ivory side is placed on the outer face. The first layer should be completed along the length of the partition. To prevent the wallboards getting out of alignment, adjoining wallboards should be skew-nailed together with three equally spaced 38 mm panel pins on the inner faces of the boards.

Bands of bonding compound mixed to a thick creamy consistency should be applied to a minimum of 5 mm thickness in vertical bands on the inner face of the fixed wallboard at 300 mm centres. Every joint must be covered by a band. The 19 mm plank core is then bedded to the outer wallboard, previously fixed, starting with a 150 mm width. Succeeding 19 mm plank core lengths are bedded, full width, to ensure that the joints in the core and outer layers do not coincide (Figure 93).

The third layer of boards is bedded to the middle layer in a similar manner, again staggering the joints. This layer, with ivory face outwards, should be nailed to top and bottom battens as described for the first layer.

The joints and nail head holes should be filled and jointed as described for dry lining.

The fixing method for 65 mm laminated partitions is similar to that described for 50 mm partitions except 19 mm planks are used for the outer layers, and 300 mm wide starting strips are used for the core instead of 150 mm. To achieve a two-hour fire rating for the 65 mm partition, the outer layers should be skew-nailed along the edges of the planks with 60 mm x 2.6 mm galvanized nails at 300 mm centres. They should be spaced to alternate nails on the adjoining plank so that the maximum spacing along the joints is 150 mm.

Sketch details for corner and door jamb are shown in Figures 94, 95.

METAL-STUD PARTITIONS

This is a lightweight partition system which consists of a steel frame with one, two or three layers of plasterboard screwed to each side. The partition is non-loadbearing but gives a high degree of sound insulation and fire protection. It is made up from 48 mm, 70 mm or 146 mm metal stud framing plus the plasterboard thicknesses.

The framing consists of galvanized rolled mild steel lightweight stud and channel sections, the plasterboards being screwed to the framework with 32 mm or 36 mm self drilling and tapping drywall type screws with Phillips heads. A suitable powered screwdriver with magnetic bit and adjustable depth control is recommended for fixing the boards.

Services can be accommodated in the centre of the partition if required. Extra insulation in the form of glass or mineral wool quilts can be inserted in the framework between the cladding if desired.

DRY CONSTRUCTION METHODS AND PARTITIONS

LAMINATED PLASTERBOARD PARTITIONS

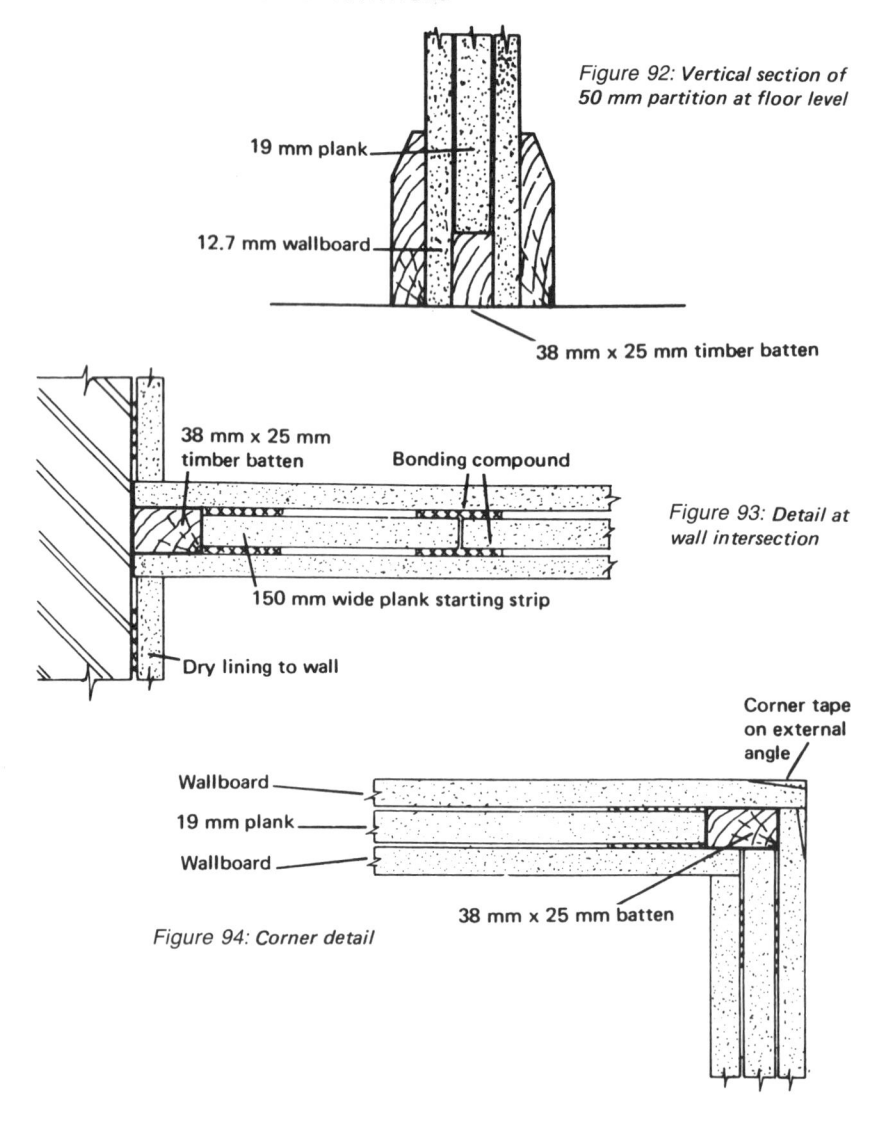

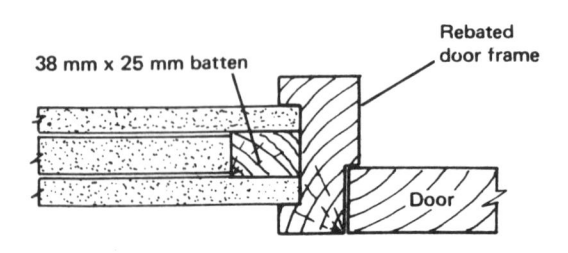

Figure 92: Vertical section of 50 mm partition at floor level

19 mm plank
12.7 mm wallboard
38 mm x 25 mm timber batten

38 mm x 25 mm timber batten
Bonding compound
150 mm wide plank starting strip
Dry lining to wall

Figure 93: Detail at wall intersection

Wallboard
19 mm plank
Wallboard
38 mm x 25 mm batten
Corner tape on external angle

Figure 94: Corner detail

38 mm x 25 mm batten
Rebated door frame
Door

Figure 95: Door jamb detail

The maximum height of the partition for 75 mm finished thickness is 2.6 m; 3.7 m for 100 mm width; 4.2 m for partitions with 146 mm studs.

The metal studs and channels can be cut with hacksaw or tinsnips. The erection sequence is to fix channels to the floor and ceiling at 600 mm centres, butt joining at angles and T junctions. On concrete floors a timber sole plate may have to be fixed first or a damproof membrane inserted under the channels. Studs are then cut to fit from web to web of the ceiling and floor channels, and twisted into position vertically. The spacing of studs is normally at 600 mm centres but adjustments have to be made to set back the studs at each side of openings to receive timber grounds for door and window frames. Timber grounds are fixed by screwing through from the channels. Services can be passed through cutouts in the metal studs and are inserted at the same time as the framework erection.

When fixing the 12.5 mm wallboards, 32 mm drywall screws are used at 300 mm centres for single layers of boards and 36 mm screws for two layers. The inner layer of boards is screwed to the studs and channels on all four edges. Intermediate studs need not be screwed. At external angles the screws should be placed at 200 mm centres. Joints between boards of separate layers should be staggered.

Jointing finishing of the board lining is as described previously for dry-lining.

Sectional details of framework construction are shown in Figures 96, 97, 98.

PLASTERBOARD COVES

These consist of lengths of plasterboard with cove sections. They are available in the following sizes: 100 mm girth in 3000 mm lengths; 127 mm girth in lengths of 3000 mm, 3600 mm and 4200 mm. The terminal members at wall and ceiling for both types are 6 mm. Full wall and ceiling projections are 67 mm for the 100 mm girth cove; 83 mm for the 127 mm girth cove.

The plasterboard coves are intended for direct decoration and are used to improve the appearance of, and to hide any cracks in, the wall/ceiling angle.

Before fixing, the area of wall and ceiling adjacent to the internal angles must be cleaned and any old or painted plaster surfaces keyed by scratching.

Guide lines should be snapped around the room 81 mm away from the wall/ceiling angle.

The lengths of cove should now be measured on the wall line and the mitres cut in a mitre box if available. Alternatively the first length may be butted into the angle and a scribed joint formed against this. Joints of 3 mm should be allowed between straight lengths and it is an advantage to cut the joints at an angle instead of being square butted.

When the lengths have been cut (preferably with a fine toothed saw) the sawn edges should have the burrs smoothed off with fine sandpaper. The cut lengths should be tried in position for accuracy before fixing.

If the cove is to be nailed the position of the joists should be found by probing and these points marked on the ceiling.

METAL STUD PARTITIONS

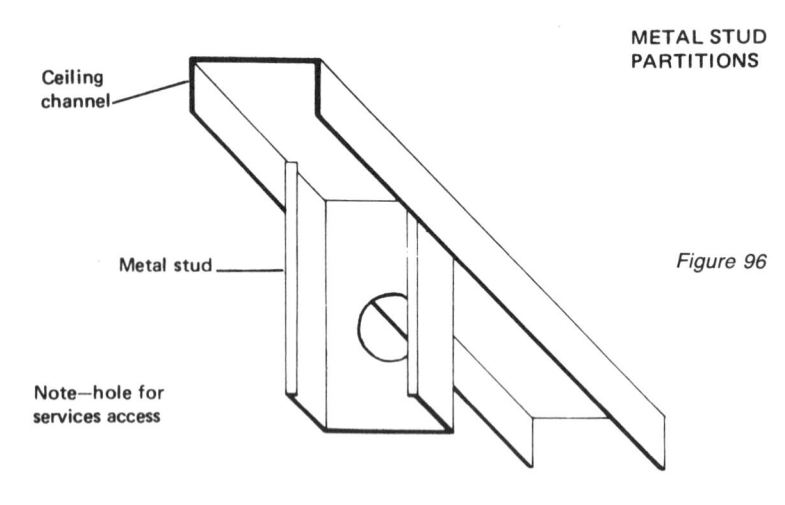

Ceiling channel

Metal stud

Note—hole for services access

Figure 96

Figure 97: Ceiling and floor details single skin construction

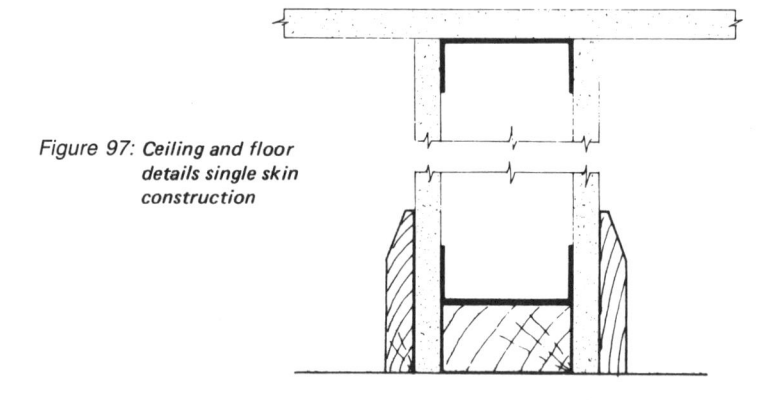

Figure 98: Corner detail double skin construction

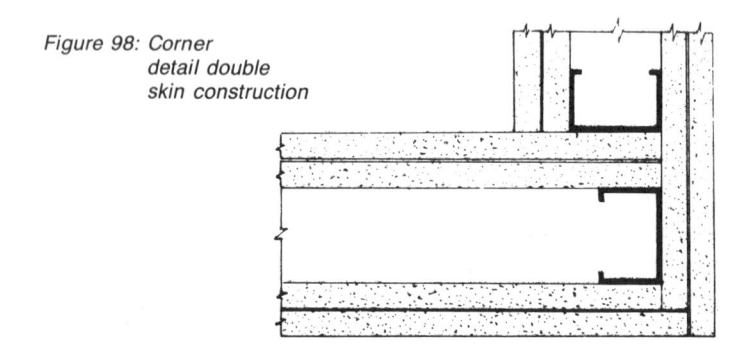

PLASTERBOARD COVES

Special bedding materials are available from manufacturers and some are recommended for mixture with hot water. The bedding mix should be made to a fairly stiff consistency and spread about 3 mm thick along each bedding edge on the reverse side of the cove, which can now be pressed into position with a sliding action until solid. Galvanised nails may be inserted at each end of the cove length and also in the middle if the adhesion or key is suspect. The nails should be driven tightly enough to make a slight impression in the paper, but not to penetrate through the paper into the core.

Joints should be butted and made good as the work proceeds. Excess joint filler should also be cleaned off whilst still soft and nail holes spotted with this mix. When dry any set material left on the cove surface can be smoothed off with sandpaper or wiped clean with a damp cloth.

CHAPTER EIGHT

External Plastering

PEBBLE DASHING

This is one of the most popular forms of external rendered finishes, and can briefly be described as a layer of pebbles thrown on to a freshly applied coat of cement mortar.

The various points concerned with pebble dashing can be considered under the following headings:
 Scaffolding
 Preparation of background
 Rendering
 Methods of applying the pebbles

Scaffolding

The best type of scaffolding for use in pebble dashing is undoubtedly an independent tubular type, well braced in itself and/or tied in to window openings with extenders. Putlogs fastened into the walls should not be used as they slow down the essential speed of dismantling the scaffold during the progress of tall pebble dashing jobs. If the scaffold is to be left erected until the main area of dashing has been completed, then the holes caused by the putlogs in the wall can cause awkward dry piecings. These may be noticeable if not patched carefully.

Regarding safety precautions the working platform should be at least three planks wide with a toe board and guard rail on the outer edge. The guard rail must be at least 900 mm high and the distance between toe rail and guard rail not more than 657 mm.

If the planks used are 50 mm thick then the maximum span between putlogs must not be greater than 2.55 mm.

Ladders must be secured top and bottom and must protrude above the landing platform by at least 1.05 m or have a good handhold provided.

It is a good plan when arranging the height of the scaffold 'lifts' to measure a reasonable distance from the floor to a comfortable working height. This may be about 1.8 m to 2.1 m depending upon the height and reach of the plasterers. In certain cases, especially gable ends, it may be wiser to count down from the ridge the number of courses and identify the position of each staging down to ground level. It must be remembered of course, that the lowest staging must be at least 1.9 m high on a public site or thoroughfare to allow passage underneath (Figure 99).

Figure 99

PEBBLE DASHING

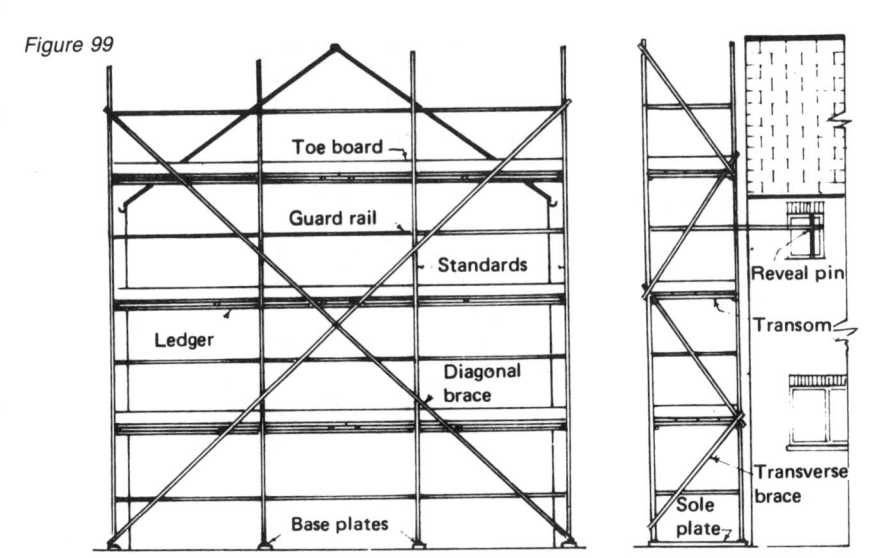

Preparation of background

If the background is clean new brickwork, then no preparation is required beyond damping down excessive suction with a hosepipe or water bucket.

Old and dirty brickwork may require the joints to be raked out, and smooth, hard bricks should be hacked and roughened.

Timber lintels, steel beams, etc. must be covered with expanded metal lathing and rendered about a week before the main rendering is carried out.

Rendering

The materials used are clean, well graded sand gauged with cement, three to one by volume and waterproofer added. An even coat of approximately 9 mm thick should be applied over the background, straightened and then keyed immediately with a wire scratcher. This rendering coat is best left for a minimum of three or four days before proceeding with the pebble dashing coat.

Application of the pebbles

The type of pebbles used may be limestone spar, Dorset spar (brown pea shingle), marble chippings or other crushed stone. Sizes used vary from 6 mm — 25 mm depending upon effect desired.

The dashing coat should be fairly fatty and can be sand/cement gauged at three to one with a plasticiser, but a more usual mix is sand/lime cement gauged at 6/1/2. A slight increase in lime content may be allowed if the wall is protected from severe weather.

Application is best carried out by two plasterers, one applying the dashing coat, the other casting the pebbles. The rendering at this stage must be damped to kill the suction. A tarpaulin or similar sheet can be placed at the

EXTERNAL PLASTERING

base of the wall to catch surplus pebbles falling from the paddle. All gullies, rainwater outlets, etc, must be covered to prevent blockage.

The dashing coat can now be applied about 9 mm thick and while this is still soft the pebbles are thrown into it (Figure 100).

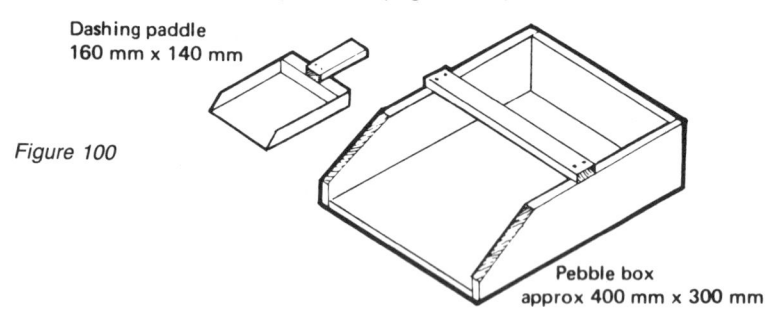

Dashing paddle
160 mm x 140 mm

Figure 100

Pebble box
approx 400 mm x 300 mm

When casting the pebbles a single layer covering the bottom of the paddle is thrown on to the soft dashing coat at right angles to the wall, and with sufficient force to embed the pebbles to half their depths. Pebbles thrown 'sideways' will tend to push the dashing coat into uneven bumps and hollows and must be avoided. With careful practice, skill will soon be acquired at throwing the pebbles at right angles to the wall from varying positions, above the head and even at foot level.

It is best not to attempt to pat in with a trowel or float after throwing on the pebbles, because this often disturbs rather than strengthens the bond between the pebbles and the dashing coat. Also, tapping back with a trowel often leaves marks of the trowel edge and in any case spoils the natural look of well cast pebbles.

It is wasteful in time and materials to have two or three layers of pebbles on the paddle when throwing on to the dashing coat. Only the first layer can be embedded, the rest will fall and be wasted unless they are collected and cleaned.

Piecings are difficult to hide and should be avoided if possible. On tall buildings dry piecings need not be necessary if separate gangs are used on each scaffolding 'lift', the lower gang in each case following a few feet behind the gang above and keeping the piecing 'green'.

Sometimes piecings can be positioned behind down spouts, or shortened by having a joining between a door head and window sill or similar (Figure 101).

If dry piecings have to be used they are best cut back with a saw tooth pattern, as this tends to make the joint less obvious. The individual saw tooth cuts must be greater than 90 degrees to enable the job to be carried out with a normal plastering trowel instead of a gauger or pointing trowel (Figure 102).

When pebble dashing up to a dry piecing the dashing coat is applied about 2 mm behind the original dashing coat. This ensures that when the pebbles are thrown into the new coat they will be pushed forward in line with the old. Great care is required to make sure that the new dashing coat is adjusted

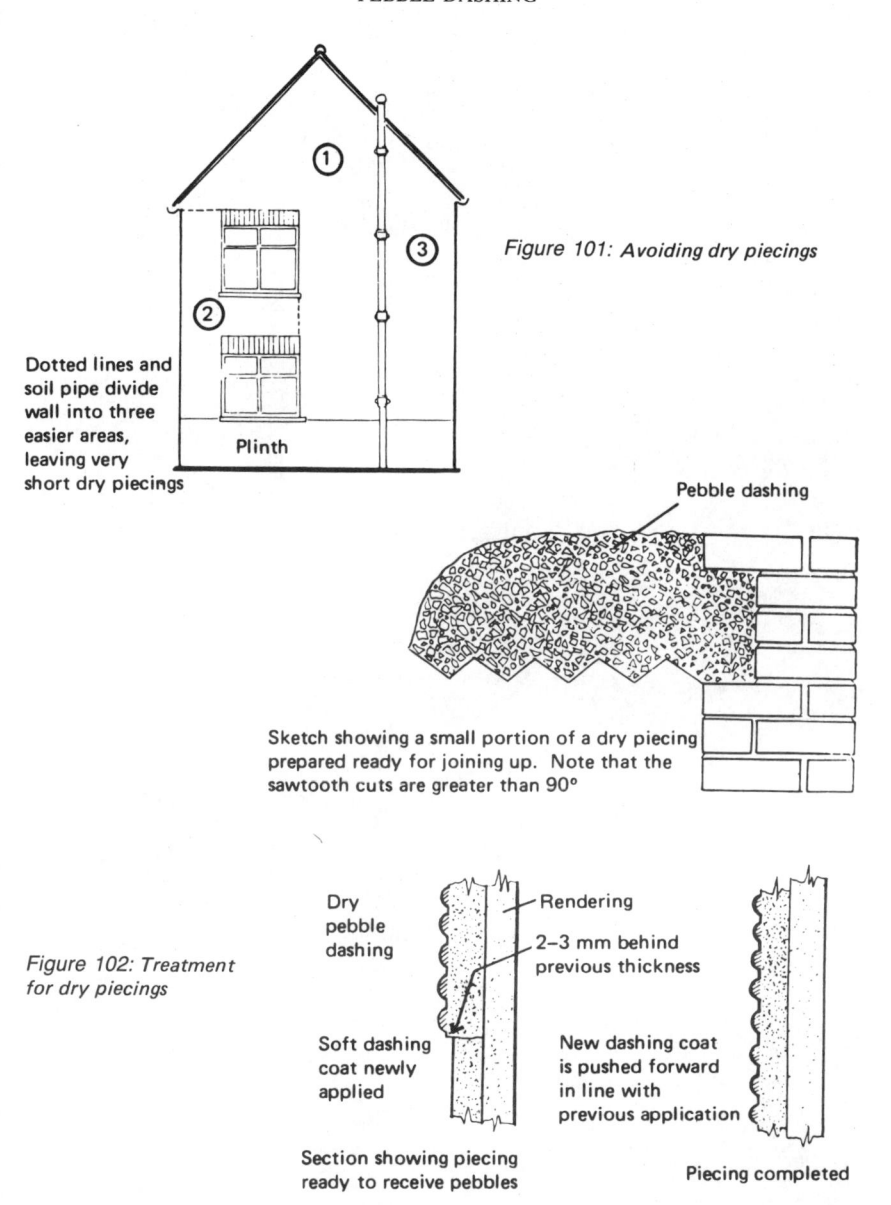

PEBBLE DASHING

Figure 101: Avoiding dry piecings

Dotted lines and soil pipe divide wall into three easier areas, leaving very short dry piecings

Plinth

Pebble dashing

Sketch showing a small portion of a dry piecing prepared ready for joining up. Note that the sawtooth cuts are greater than 90°

Figure 102: Treatment for dry piecings

Dry pebble dashing

Soft dashing coat newly applied

Section showing piecing ready to receive pebbles

Rendering

2–3 mm behind previous thickness

New dashing coat is pushed forward in line with previous application

Piecing completed

in thickness until no apparent difference can be noted. Otherwise a collection of dirt over a period of time will show if the lower piecing protrudes above the original.

Putlog holes often show bad piecings for a considerable period because a thicker dashing coat is often used. This causes a difference in colour, and if this is coupled with untidy patching, then they stand out for a long period like so many 'sore fingers' in an otherwise unbroken expanse of pebble dashing.

117

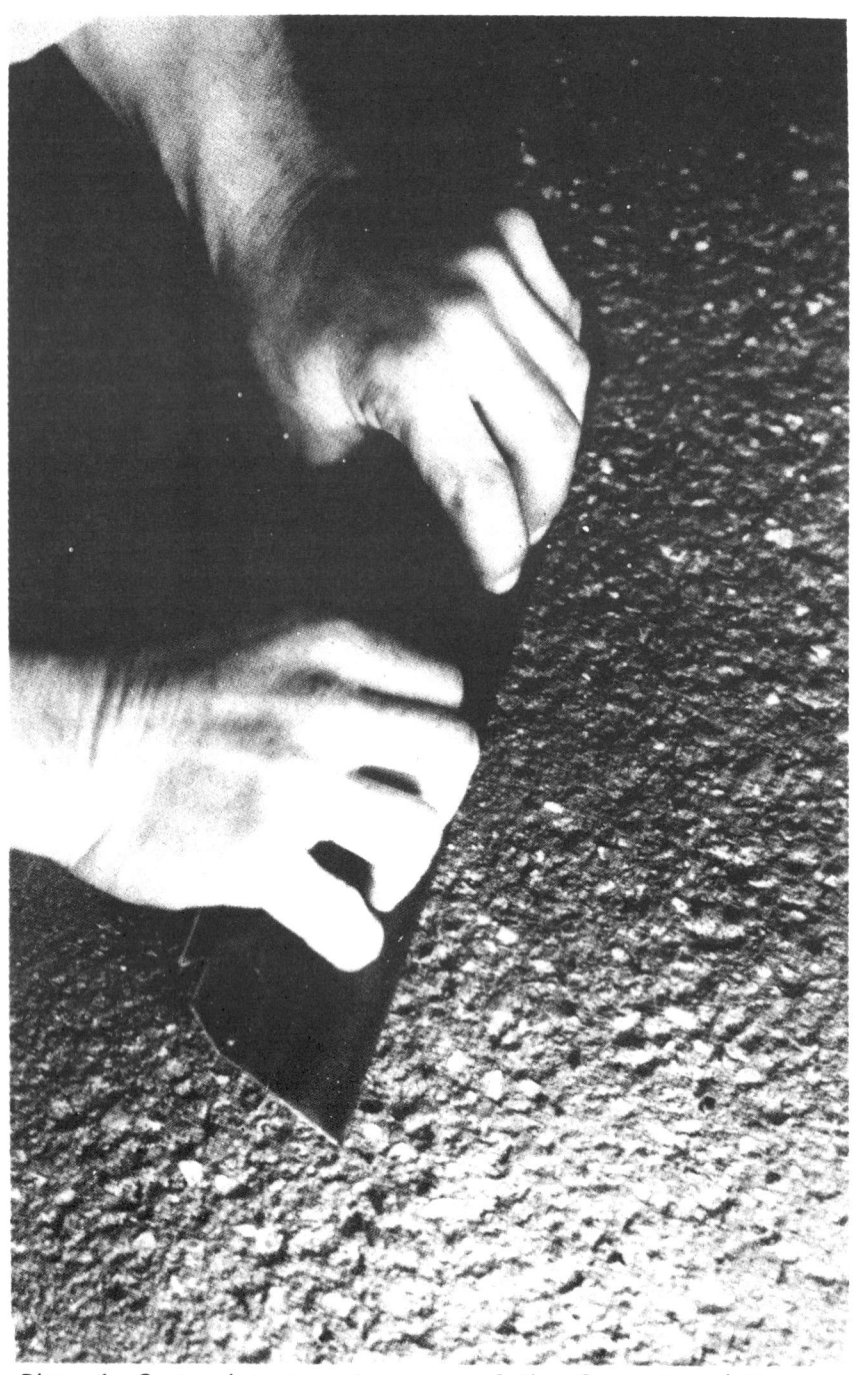

Plate 1: Scraped texture (courtesy of the Cement and Concrete Association)

Plate 2: Alpine finish (courtesy of the Blue Circle Group)

Plate 3: Cullaplast finish (courtesy of the Blue Circle Group)

Dirty pebbles can be washed easily in a sieve with a hosepipe and water. The pebbles are left to dry and mixed in with other pebbles before re-use.

External angles are best carried out by holding a darby or straight edge to one side of the angle and applying the dashing coat straight up to it. The rule is now held to the side of the angle just coated and the dry side is now covered. Before removing this rule the angle is dashed, after which the rule can be slid off carefully. Next the straight edge is washed and held over the newly dashed side of the angle and the remaining side is pebble dashed. With this method there is no fear of an irregular shaped external angle (Figure 103).

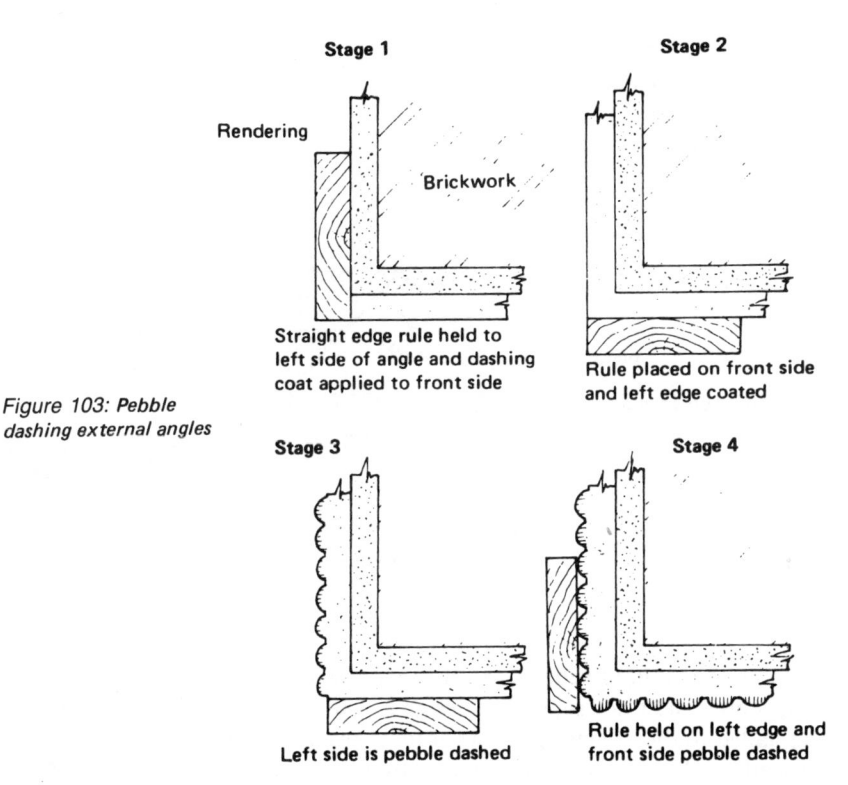

Stage 1

Rendering

Brickwork

Straight edge rule held to left side of angle and dashing coat applied to front side

Stage 2

Rule placed on front side and left edge coated

Figure 103: Pebble dashing external angles

Stage 3

Left side is pebble dashed

Stage 4

Rule held on left edge and front side pebble dashed

PATTERNS AND DECORATIVE EFFECTS

Pebble dashing can be used in conjunction with cement window linings, verge bands, quoin stones and plinths, etc. In some cases walls are divided into simple geometric panels by cement bands and the spaces filled in with pebble dashing.

Contrasting colours of pebbles can be used effectively to break up the monotony of large areas. The coloured bands of pebbles can be first applied between parallel rules which have been nailed to the wall. When the rules have been removed the band of pebble dashing can be treated as a dry piecing (Figure 104).

121

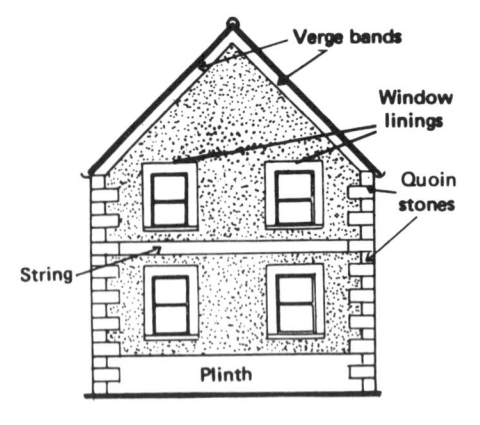

EXTERNAL PLASTERING

Figure 104: Pebble dashing used in conjunction with cement bands

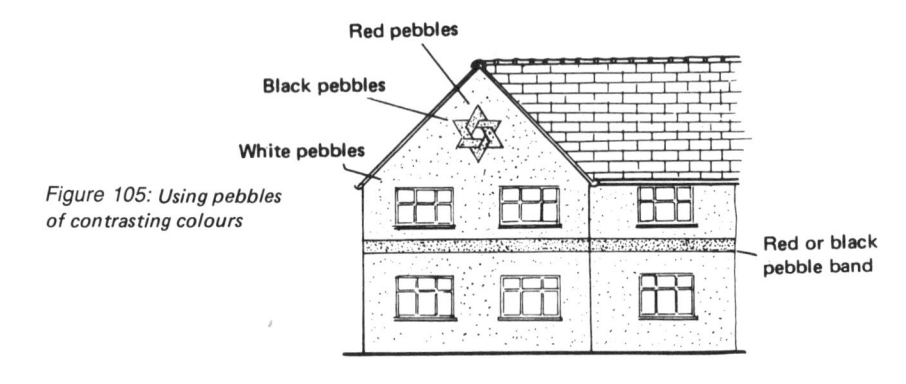

Figure 105: Using pebbles of contrasting colours

Different coloured pebbles can be used in various designs, geometric patterns, motifs and trade signs. The design is usually cut out as a wood template and nailed to the rendering in one or more pieces. After this the main area of the wall is pebble dashed. To complete the design the template is removed and the exposed rendering is pebble dashed, great care of course being taken with the dry piecing to ensure that the new pebbles make a perfect joint with the old (Figure 105).

Bell castings are made at window and doorway lintels and at the base of bay windows and pediments, etc. The purpose of bell castings is to direct rainwater clear of the wall or opening, and also as a decorative feature.

To carry out this job a core can be made by first nailing expanded metal lathing over the lintel and rendering this in sand and cement. It is easier to form if a length of timber can be nailed or wedged under the bell cast and the core completed by pressing strong sand and cement mortar on to the timber and coving back to the rendering above.

ROUGH CASTING

The following day the bell casting can be pebble dashed with the main wall area and the timber form removed. If the under surface is found to be unsatisfactory it can be touched up with sand and cement. The underside can be pebble dashed if required, provided that first of all the soffit is hacked for key and then pebble dashed in the normal way (Figure 106). If a drip groove is required this can be formed by nailing a length of half round wood on top of the timber form before fixing.

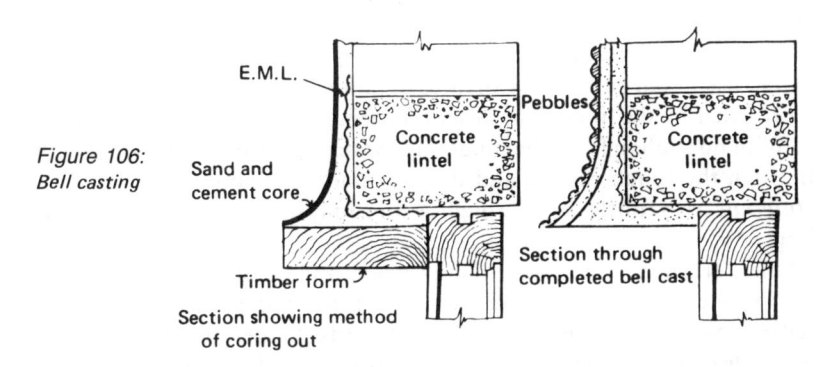

Figure 106: Bell casting

An alternative to the above method is to fix an external render stop bead (also termed a *bell cast stop bead*) at the front lower edge of the windows or doorway head. The bell cast formed to this edge will not be so bold as the previously described method but is preferred by some because of its neatness and ease of application.

External render stop beads are ideal for use as a finish at the base of walls especially when it is important that the damp proof course is not spanned by the cement rendering. The bell cast bead is fixed immediately above the damp course level and the cement rendering finished down to its edge for a neat bell cast finish (Figure 42).

ROUGH CASTING

This is a type of external textured finish in which pebbles or crushed stones of suitable size are mixed with a cement mortar and thrown on to the wall as a wet dash. In the north of England rough casting is also known as Scottish Harling.

The distinction between pebble dashing and rough casting is that in the former dry pebbles are thrown on to a soft dashing coat, whilst in rough casting the wet mix is thrown on to a dry rendering.

The latter method ensures a much stronger job which will outlast pebble dashing by many years. This is due partly to the fact that the pebbles are completely enveloped by the matrix, and also that maximum adhesion is ensured by throwing on to the wall instead of application by trowel.

The texture achieved by rough casting is generally considered to be inferior to that of pebble dashing, though many prefer its more rugged effect.

Preparation

The scaffolding and background preparations are the same as that previously described for pebble dashing.

Rendering

When considering the rendering coat it must be borne in mind that some suction from this coat is not only permissible but advisable. So therefore it is not normal practice to include a waterproofer in the rendering coat. The rendering mix should not be stronger that 3: 1 sand and cement nor weaker than 2: 1: 9 cement/lime/sand.

In the case of tall buildings with three or more lifts of rough casting, the suction provided by the lime in the latter rendering mix would be a great advantage. With no suction from the rendering, water drains down the wall behind, or through the rough casting and this can cause sliding in the lower 'lifts'.

Materials used for the actual rough casting may include any or all of the following:

pebbles, crushed stone, sand, lime, cement and colouring pigments.

The pebbles used vary in size from 6 mm to 18 mm and crushed stone, including granite chippings, from 3 mm upwards. Larger, rougher stones give a coarser texture more suitable for rural districts, and the smaller, smoother pebbles are better suited to industrial areas.

It is important that clean, sharp, well graded sand is used because this not only strengthens the finished job, but helps considerably in making an even spread when casting.

Proportions for the rough casting coat should be three of aggregate to one of matrix, and can be any of the following variations depending upon the finished texture required.

	3 PARTS AGGREGATES			1 PART MATRICES	
Coarse pebbles	Medium granite chippings	Fine sand		Lime	Cement
	3				1
6	1	2		1	2
4		2			2
6				1	1
3	1	2			2

Permutations of the above are infinite, provided they agree with the proportion of three to one aggregate to matrix, and that the cement content is at *least* 50 per cent of the matrix used.

Before starting the rough casting application, it may be necessary to cover up stock brick facings or glass windows with a sheet or similar cover. A temporary mask of hardboard is ideal for protecting facing brick in-bands and out-bands at external angles.

The rough cast mix should be sloppy enough to ensure an even spread on the rendering. The paddle and box used in pebble dashing are also suitable for rough casting.

Usually the mix will have to be thrown on with a fair amount of force, and it is easier to allow it to slide off from the end of the paddle at speed, rather than a direct throw. If the rough cast mix is cast correctly it will hit the wall at right angles to the throw, also this gives a better texture and prevents bunching and sliding. It must also be thrown with sufficient force to propel the softer part of the mix forward as it hits the wall. The force of impact then bursts the soft stuff sideways to envelop the pebbles completely and fill the keys in the rendering. This also has the effect of exposing the maximum amount of pebbles, though of course these will be covered with a film of matrix.

With regard to piecings, these are best kept short or avoided if possible. Because of the roughness of the texture piecings are easier to hide compared with pebble dashing. One of the obvious faults to guard against is the accidental formation of the herring bone effect in a rough cast piecing. This is caused by throwing in one direction only on the upper side of the piecing, and at this stage it is unnoticeable. However, if the rough cast mix is thrown up to the piecing from below in the *opposite* direction only, the herring bone effect at the piecing is strikingly obvious.

The direction of the casting should be consistently altered. Also no attempt should ever be made to touch up slack or bald places with a trowel as the fault will be made even more obvious. The correct method is to avoid bare places, but accidental omission can be made good by flicking on small quantities of soft stuff from the end of a flexible gauger. Another method that can be used is by dipping a bristle brush in the soft mix, and rapping it smartly against a piece of timber when held just in front of the bare spot.

SPATTERDASH

This is a method of forming an adhesion to smooth, dense backgrounds such as concrete and hard facing bricks.

A slurry mix of cement and coarse sand or fine chippings gauged at between 1:1 and 1:2 (cement/aggregate) is thrown on to the background and allowed to set before applying the next coat.

Due to the introduction of cement and plaster bonding adhesives, spatterdash is not used as much as previously but it is most efficent if carried out correctly on *clean* surfaces.

In parts of Ireland spatterdashing is known as scudding, and the paddle used is called a scudder.

TYROLEAN FINISHES

There are two types of Tyrolean finish — normal and rubbed. These are applied to provide a protective rendering and decorative finish to wall surfaces. The method of application for both normal and rubbed texture are the same except for the finishing of the rubbed texture, this is described later.

Firstly, the background is prepared as for normal cement renderings by hacking, brushing, damping, etc, if these are necessary. Next the wall surface is rendered with a mixture of Portland cement, lime and sand in the proportions 1: 1: 6. This is applied to a minimum thickness of 9 mm, straightened with a rule, or darby, and lightly rubbed up with a plain float without nails. No key is necessary. Slack places and hollows are undesirable and difficult to hide with the finishing coat.

The rendering coat is allowed to set and dry until it gives adequate suction for the Tyrolean finish. The finishing coat should be one volume of water and two to two and a half volumes of Cullamix. (Cullamix is a mixture of Portland cement, silver sand and colouring pigment.) This mixture is applied to the surface of the floating coat by use of the Tyrolean hand spray machine in a minimum of three layers to a total thickness of 6 mm to 8 mm.

The Tyrolean hand spray machine consists of an open ended galvanized sheet metal container in which is suspended a horizontal rotating cylinder (Figure 107). A large number of spring steel tines are fastened radially to the surface of the cylinder, which can be rotated by means of a cranked handle on the outside of the cylinder.

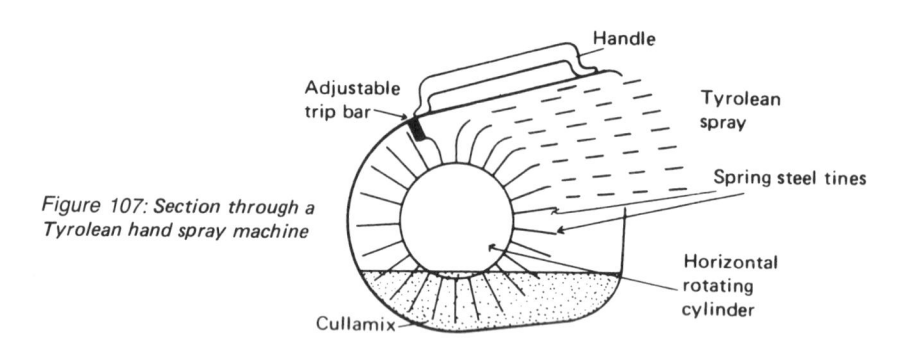

Figure 107: Section through a Tyrolean hand spray machine

At the rear upper end of the container is situated a trip bar which can be adjusted to bend the spring steel tines to any desired deflection as the cylinder is rotated.

The wet mixture of Cullamix is poured into the base of the container and as the cylinder is turned so the tips of the tines become loaded with a small pat of the mixture. When the tines reach the trip bar they are bent backwards and on passing the bar spring suddenly forward, flicking the pat of mixed cement out of the mouth of the machine chamber on to the wall.

When applying the Tyrolean finish the operative should ensure that the finished work is built up to a honeycomb texture by carefully avoiding dense applications in any area of the wall.

This can best be achieved by altering the direction of application of each layer. Also by keeping the machine constantly moving to avoid too great a concentration of the flicked mixture on any part of the wall, until each sparse layer has hardened by suction.

The first layer can be flicked on at right angles to the wall with the mouth of the machine about 450 mm from the wall surface. The second can be applied at an angle of 45° from the left and the third layer applied at 45° from the right. A further one or even two layers may be necessary in building up to the full thickness. The build up of the right honeycomb texture and thickness will depend upon suction and correct adjustment of the Tyrolean machine.

It is advisable to mix small batches of the Cullamix and water, because if the mixture is allowed to stand without stirring the heavier sand and cement will settle out at the bottom of the container. For the same reason the machine should not be overloaded with material and the mixture well stirred immediately previous to loading.

The trip or flicker bar adjuster should not be set beyond the second notch when the machine is new. As the bar wears after a period of time, then it can be adjusted to give the correct deflection by setting it to a lower notch.

Piecings must be avoided. If the area to be covered is too large to permit of three or more layer applications in one day, then one or two full layers may be applied and the final layer or layers on the following day.

The flicked particles of the Cullamix should just touch each neighbour thus creating a cellular or honeycomb texture. This type of texture can be damaged easily by hard usage, so it is advisable therefore to provide a protective plinth not less than 300 mm high to the base of walls.

Power operated machines are also available for the application of this finish and are particularly useful for application to soffits.

Rubbed texture

The normal texture described may be altered by rubbing off the high points with a carborundum or similar type of stone. This can best be achieved when the normal texture has been allowed to set for a minimum of twenty four hours, or up to three days in winter.

The stone should be lightly rubbed over the surface with a circular motion, rubbing off the high points but leaving the indents untouched. After rubbing, all the dust must be well brushed out.

A rubbing block can be made by casting a block of the Cullamix similar to that used for the Tyrolean application. A suitable size for the rubbing block would be 150 mm x 50 mm x 35 mm.

PLAIN FLOATED CEMENT WORK

This type of cement work is usually carried out in two coats, rendering and finishing, though for special work a floating coat may be applied additionally. In three coat work the floating coat would be plumbed and screeded as described for plain interior plastering.

Rendering coats should be applied 9 mm to 15 mm thick, straightened and combed with a wire scratcher to give key for the following coat or coats. The materials used are sand and cement in the proportion 3:1. It is an advantage to incorporate a waterproofer in with this mix to protect the background and to regulate the suction for the finishing coat.

Floating coats, if used, are done with a similar type of mix but as stated are screeded and ruled in to give a straight surface for the finishing coat.

The finishing coat mix will vary between 1:3 cement/sand and 1:1:6 cement/lime/sand, depending upon the severity of climatic exposure to be expected. Completely sheltered positions could even have a 1:2:9 cement/lime/sand mix, though this is too weak for normal usage.

The addition of even a small amount of putty lime or hydrated lime will reduce the risk of crazing which is one of the biggest risks with plain floated cementing. These additions however reduce the strength of the finished work so a compromise must be reached to balance the requirements of the finished work and the type of exposure, wear and tear expected. For northern areas of the British Isles or other exposed areas a straight 3:1 sand/cement mix would give longest life, or up to one quarter putty lime added to give a mix 3:1:0-¼ sand/cement/lime.

The finishing coat is applied about 9 mm thick, working from left to right. This is straightened with a rule, rubbed over with a large straight grained float filling in all slack places. As this coat begins to dry it should be fined down with a smaller float approximately 150 mm x 75 mm until a stone like textured surface is obtained over all the wall. This can only be accomplished if the final rubbing up or fining down is done at the right stage. If the finishing coat is too wet streaks will be left and later shrinkage will occur causing crazing. When the finishing coat is too dry the final rubbing up will cause scratch marks on the finished work and loose sand particles over the surface.

For these reasons it is best for two or even more plasterers to work on large areas of finishing work to ensure that the piecings are kept 'green'. Dry piecings (piecings left overnight) should be avoided or kept as short as possible. Continuity of work throughout the finishing period should be maintained by arranging staggered breaks to avoid certain areas becoming too dry for finishing after meal breaks.

An excellent finish can be obtained by following the final rubbing up with a sponge or bunched up piece of soft cloth.

COLOURED CEMENT WORK

If the finished surface is to be coloured this can be accomplished by (*a*) adding a colouring pigment to the normal sand/cement mix or (*b*) using Cullamix.

Whatever the method used it is important that the same proportions of sand, cement, colouring *and* water are used in every mix. With method (*a*) it is best to mix all the dry materials necessary to complete the job on a clean banker and then fill it into dry sacks until required for mixing with water.

It is essential when mixing colouring pigments with cement and sand that all the dry ingredients are passed through a 3 mm sieve to ensure thorough dispersion of the colour throughout the mix.

ASHLAR JOINTING

Unless the precautions mentioned about mixing are carried out, different densities of the colour will be obvious in the finished work.

Other causes of defects in the uniformity of colour are due to (i) different thicknesses, (ii) uneven suction and (iii) scouring of hardened areas.

Differences in thickness will give a difference in density, hence the importance of straightening the undercoat and finishing coat.

Uneven suction due to different types of background materials such as concrete lintels, brickwork or coke breeze will affect the rate of water absorption from the finishing coat and so alter the texture and colour of the finished work.

If certain areas have hardened before the final rubbing up then excessive scouring will bring sand to the surface and colouring will be lost. To add water to soften back the sand may be even more noticeable.

Proprietary brands of colouring pigments are available but the following is a list of mineral colouring agents:

Colour	*Pigment*
GREYS AND BLACK	LAMPBLACK
	CARBON BLACK
Blue	Ultramarine Blue
Dark red	Iron oxide
Bright red	Mineral Turkey red
Purplish red	Indian red
Brown	Metallic brown oxide
Yellow	Yellow ochre
Green	{ Chromium Oxide
	{ Greenish Blue Ultramarine

All the colours mentioned require between 2 kg and 4 kg of colour pigment for each 50 kg of cement used except for greys and black when up to 0.5 is usually sufficient. It is not possible to make light colours such as cream, blue and pale green from ordinary Portland cement and therefore white cement must be used. For darker colours ordinary Portland cement is quite satisfactory.

ASHLAR JOINTING

This is often termed 'blocking out' and is a method of lining in joints on plain floated finishes in imitation of ashlar stonework.

Jointers of various shapes and thicknesses are used to mark out the joint lines but 6 mm wide is an average thickness. Two are shown in Figure 108.

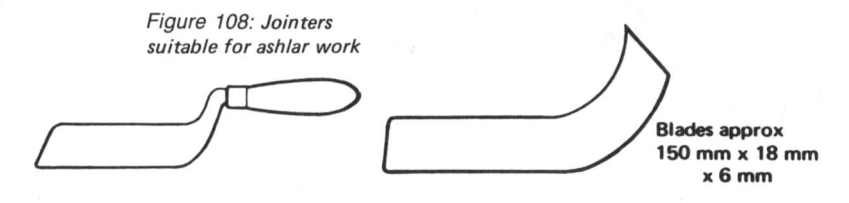

Figure 108: Jointers suitable for ashlar work

Blades approx 150 mm x 18 mm x 6 mm

EXTERNAL PLASTERING

The joints should be lined in as soon as possible after the rubbing up of the finishing coat is completed. Firstly the horizontal joints are marked on the right and left hand vertical edges of the wall surface from a gauge staff, and chalk lines snapped across the full width of the wall. These can now be cut out and lined in by holding a straight edge to the chalk line and cutting a shallow groove about 3 mm deep with the edge of the jointer, and then smoothing down the joint with the smooth outside edge of the jointer.

The positions of the vertical joints may have to be adjusted to finish up with a series of full size blocks. These are now marked from a gauge staff across the top and bottom joint lines of each lift in turn, and the alternate vertical joints cut and lined in as before with the aid of a straight edge.

Because of the extra labour involved in this type of work dry piecings are often unavoidable. The best method of dry piecing is to cut away all surplus material 6 mm below the last horizontal joint. On the following day the 6 mm band can be used as a finishing member and the next day's application rubbed up right to the joint. Any material in the joint can easily be cleaned out and the piecing will not be apparent when the work dries out.

Vee joints can be cut into plain floated cement finishes to line out the wall surface into imitation masonry with a much bolder effect that is also longer lasting than ashlar jointing.

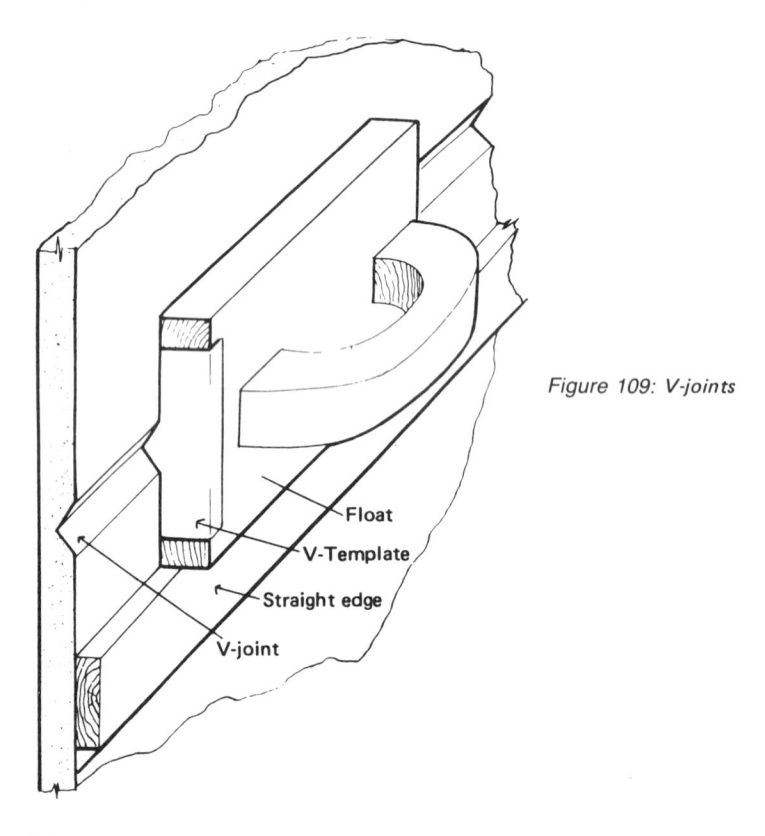

Figure 109: V-joints

ARTIFICIAL MASONRY

The vee joints can be formed by nailing a V shaped template to the front edge of a float, and cut out as shown in Figure 109 after marking out as described for ashlar jointwork.

Because of the depth to be cut for the vee joints it is important that the rendering coat is reasonably straight and that the finishing coat is thick enough without being cut right through to the backing coat. Any small stones, pebbles and even very coarse grit in the mix will prove a nuisance when cutting the vee joints, so the dry mix should be passed through a 3 mm sieve. Ragged arrisses can be touched up with a small chamfered float.

RECESSED JOINTS

These are rectangular joints cut into plain faced cement work and can be formed in two or more ways. One method is to straighten and line out the rendering coat. Strips of planed timber laths, of the section required for the size of the desired joints, are nailed to the lines on the wall. The edges of the laths are slightly splayed to assist easy withdrawal later (Figure 110).

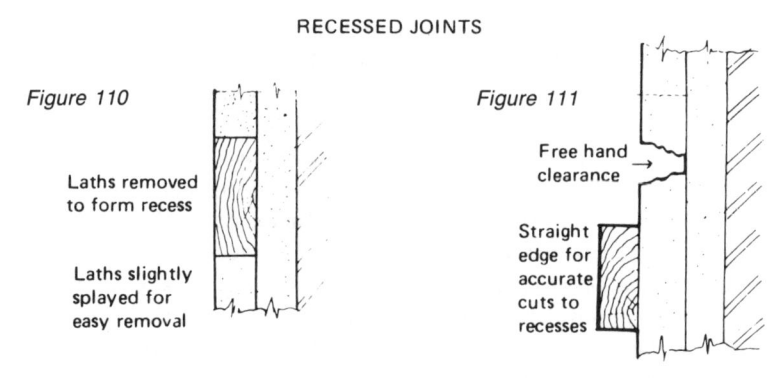

RECESSED JOINTS

Figure 110

Laths removed to form recess

Laths slightly splayed for easy removal

Figure 111

Free hand clearance

Straight edge for accurate cuts to recesses

The spaces between the laths are filled in with the finishing coat, which is ruled off, tightened in, rubbed up and fined down. The laths are withdrawn and the recesses are now rubbed up with a very thin coat of sand and cement using a narrow float.

A better method is to put the finishing coat on and line in the position of the recessed joints. Between the lines, in the space to be removed, a slice of the finished sand and cement can be removed free hand to give about 12 mm clearance back to the rendering coat. A straight edge can now be held to the lines and the surplus material cut off with a trowel in a similar manner to cement skirtings. The arrisses can be rubbed up as the work proceeds and the recesses rubbed up later as previously described (Figure 111).

ARTIFICIAL MASONRY

Imitation square snecked rubble stonework can be carried out by first marking out the desired joint lines on the rendering coat in chalk. This can be

Figure 112: Artificial masonry

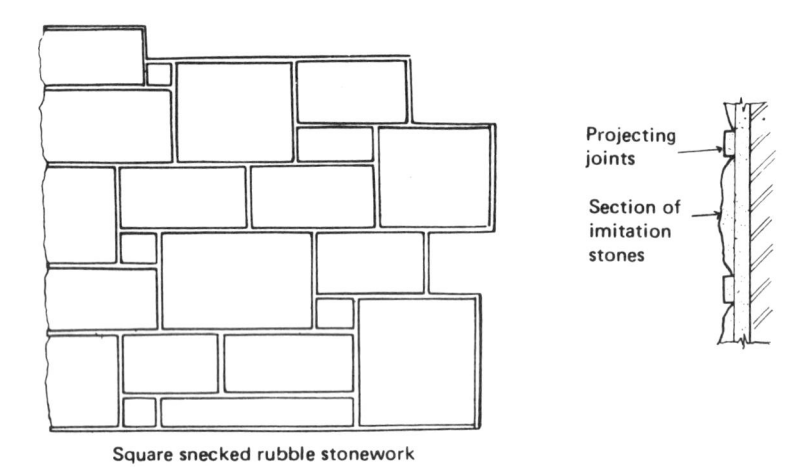

Projecting joints

Section of imitation stones

Square snecked rubble stonework

done from a scale drawing for accuracy of detail or free hand (Figure 112).

A thin band of sand and cement, previously passed through a sieve, is laid over each marked joint about 18 mm wide. This is rubbed up and cut off while still reasonably soft with a 12 mm thick rule held narrow edge on, so that both edges are cut at the same time before the rule is removed.

After the joints have been finished and allowed to harden, preferably overnight, the stones are filled in to roughly the shape of axed stonework. The edges of each 'stone' must be tucked well in below the level of the joint to give boldness and neatness to the finished appearance. The thickness of individual 'stones' will vary considerably as in normal stonework and as the cement dries it can be 'rubbed up' with a dry tool brush after first roughening slightly with screwed up newspaper, wire brush or similar.

The work is made much more effective if 'stones' of several different colours are used. Particularly suitable for this type of work is coloured Stonite, but other coloured cements are all right providing that the colours are not too strong or glaringly unnatural.

CEMENT LETTERING

Cement bands, geometrical designs and lettering are done in various ways. If the bands or designs are simple they are merely marked out on plain faced cement work and cut out with a trowel or small tool to the required shape.

Cement lettering is carried out by firstly cutting out all the individual letters full size in cartridge paper, or similar, and coating both sides with three coats of shellac.

The area to be lettered is finished with plain floated sand and cement, then the upper and lower lines of the lettering are snapped on the cement face.

Now the individual letters are placed between the lines, centrally in the panel and to the correct spacing, and the lines drawn around the outline of each letter.

Inside the lettered outlines the cement work can now be deeply keyed and the position of each letter marked above the left hand edge of each. The positions of the upper and lower edges of the letters are also marked outside the area to be lettered.

A layer of sand and cement can now be applied about 6 mm to 9 mm thick over all the area to be lettered but taking care not to obliterate the top and side markers. As this material dries it is rubbed up to a good finish. The horizontal upper and lower lines for the lettering can now be snapped back on to the new surface, and each letter template in turn can be placed between the lines in its appropriate position as indicated by the left hand marker above the recently applied mix. A small tool can now be used to cut away the material around the letter back to the original finishing coat. With care, all that is usually necessary is to cut cleanly through the top coat of cement and the surplus outside the template will peel away when lifted with the small tool.

It is much easier to apply the letters after the finishing coat has hardened, but this method risks the possibility of the letters being detached through frost or other physical damage. If the whole work is completed on the same day the lettering and finishing coat will set and harden together and no separation can occur through poor bond.

If the cement lettering is required to be much bolder than 15 mm thick then precast letters are best. Assuming that the letters are to project say 20 mm then two battens 25 mm thick are placed slightly wider apart than the depth of the letters on a level base without suction. Semi-dry sand and cement is now consolidated and ruled in with a straight edge bearing on the two battens, the surface is then rubbed up with a wood float. The letter templates can now be placed on the surface in turn and the shape of each letter cut through to the base removing all surplus material.

Because of the semi-dry mix it is essential to cure the cement carefully, so the letters can be covered with damp sacks held from the surface of the letters by laths or similar. After a period of 36 hours the letters can be removed from the base and immersed in water for 7 days.

At the end of this period the letters can be bedded in their correct position, and spacing, to the rendering coat with a strong mix of sand and cement. The wall can now be given the finishing coat working carefully around each letter, taking care to cover the bedding thickness to ensure maximum weatherproofing.

SCRAPED TEXTURE

Plain floated cement surfaces have certain disadvantages and probably the greatest of these is crazing caused by shrinkage of the surface skin. The greater the density of this skin, due to trowelling or consolidation by rubbing up, will result in a proportionately increased final shrinkage.

EXTERNAL PLASTERING

Risk of crazing due to surface shrinkage can be virtually eliminated by scraping the surface skin away.

The methods employed will vary according to the materials used in the finishing coat. In all cases the finishing coat is straightened and all hollows filled in with a float, but excessive rubbing up is unnecessary and should be avoided.

After a period of from 4 to 24 hours, depending upon the degree of hardness of the set due to climatic conditions or type of materials used, the surface skin should be scraped away to a depth of 2 mm — 3 mm. The tools used for scraping include a hacksaw blade or blades mounted in a frame, a float with several dozen nails protruding from the base as a scratcher, a float with the base covered with a piece of expanded metal lath or a float with the base covered with bottle-top caps having serrated edges (Figure 113).

Whatever tool is used to scrape off the surface skin it is important to carry out the scraping before it is too hard because of the extra labour involved. If it is not yet ready for scraping the scraping tools become clogged with soft residue. As the surface is being scraped any dust remaining should be brushed away. To avoid the risk of scratches due to stones in the mix, the dry ingredients used should be passed through a fine sieve. An excellent material for finishing with scraped texture is Cemrend which is a ready mixed material obtainable in various colours and requiring only the addition of water. It has a slightly weaker strength than some Portland cement mixes and is easier to scrape, but a stronger mix may be required for a plinth if this provision is necessary.

ENGLISH COTTAGE TEXTURE

An unusual feature in the application of this texture is the fact that the material is spread and textured firstly at the base of the wall, working upwards and from left to right. Suitable mixes for the finishing are as described for plain floated finishes but colouring is more often added to accentuate the effect of light and shade.

Figure 113: Scraping tools for scraped texture

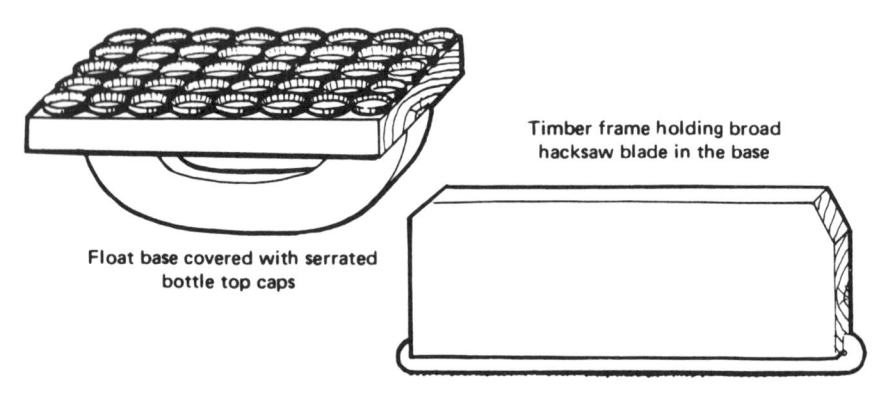

Float base covered with serrated bottle top caps

Timber frame holding broad hacksaw blade in the base

STIPPLE

The material is applied the full width of the trowel with plenty of edge to give a torn effect. Each trowelfull slightly overlaps the last, altering the angle of each successive torn spread and taking care to avoid flat or bald areas.

OLD ENGLISH COTTAGE TEXTURE

The material used for this texture requires to be fairly fatty so a mix of 1: 1: 6 cement/lime/sand can be used or 3: 1 sand/cement plus a plasticiser. Colouring again increases the effectiveness of this fairly heavy texturing.

Application is by means of a broad nosed gauging trowel commencing at the top left hand corner of the wall. The trowel is loaded at its tip and pressed to the wall surface so that the material just squeezes over the tip and upper sides of the gauger as it is pressed forwards, downwards and withdrawn. This is repeated across the wall panel with each trowel application just touching its neighbour. The next row commences *between* and immediately below each trowelfull above, to give a staggered or bonded effect. This is repeated in successive rows until the panel is completed. Large areas are unsuited to this type of texture and its use is best confined to bay window or other smaller panelled areas.

FAN TEXTURE

Again a fatty mix is essential for fan texture so a mix of 3: 1 sand and cement plus a plasticiser and colour is suitable. Small panels are also most suited for its application.

Suction must be controlled so a rendering mix incorporating a waterproofer is essential.

The finishing coat is applied by spreading the mix on 12 mm thick and immediately impressing upon this the shape of a series of fans with the toe of the plastering trowel. This is accomplished by pressing the toe of the trowel into the soft finishing at an angle of approximately 45 degrees to the wall surface. The first indentation should be horizontal and the trowel nose raised, twisted slightly to the right and pivoted from the right hand corner, then pressed forward again. These indentations are repeated until the nose of the trowel is horizontal on the right hand side of the pivot point by which time a semi circle of radial indents will have been completed. The next fan shape should just touch its neighbour without leaving any bald spots. Succeeding rows of fans should commence immediately below and between the fans above, taking care to ensure that the new indents slightly overlap the previous row to avoid bare spaces and ensure continuity of texture.

STIPPLE

This can be achieved in many ways but the essential conditions for a good finish are that the background or rendering coat should be straight and that there should be no suction.

A suitable mix is 2:1 sand/cement plus a plasticiser, and colouring material if required. Proprietary mixes such as Cullamix plus Cemprover are excellent.

Methods of application and finishing are various but an easy method is to use a banister type hand brush and scrub on to the wall a liberal coat of a fairly soft mix. As the application takes up slightly on the wall it can be stippled by dabbing with the brush, plucking up the surface to the desired height of stipple required. If a softer, more uniform, stipple is necessary then the original stipple can be softened down as the work stiffens on the wall by dabbing again with a clean brush.

Stippling floats can be bought or made and used on trowel applications if a deeper texture is required. Floats of this type have small pieces of rubber attached to the base.

Plain floats can be used to dab against softer surfaces to form a stipple, but the texture is difficult to maintain for consistency.

FINGER WHORL TEXTURE

This consists of a series of 'figure of eight' whorls or curved indentations on plain floated finishes. The severity of bare flat surfaces is relieved by scoring over the surface with the finger tips protected by rubber finger stalls.

DRAPERY TEXTURE

Achieved by combining the surface of the newly applied finishing coat with horizontal wavy lines. A stiff bass broom about 25–50 mm thick and 150–200 mm long is ideal for applying the texture. The undercoat must be waterproofed to eliminate suction and a sand and cement mix plus a plasticiser should be used for the finishing coat to ensure the soft, fatty consistency essential for this type of texturing.

DEPETER

A form of pebble dashing except that the pebbles are pressed into the soft finishing coat by hand. Large sea pebbles are most often used, but shells, flints or other types of flat stones may be used.

SGRAFFITO

This means scratched work and consists of two or more coats of coloured plaster or cement work. Contrasting colours are applied in sequence over each other as soon as each coat has stiffened. The finishing coat is brought to a good finish and a cartoon or design is held to its surface and the outline pounced or pricked through with a sharp compass point or similar. The design is now cut through to expose a different colour below. As stated two or more contrasting colours can be exposed to show up the design details which may be a motif, advertising sign, recessed lettering, coat of arms or similar.

EXTERNAL CEMENT FEATURES

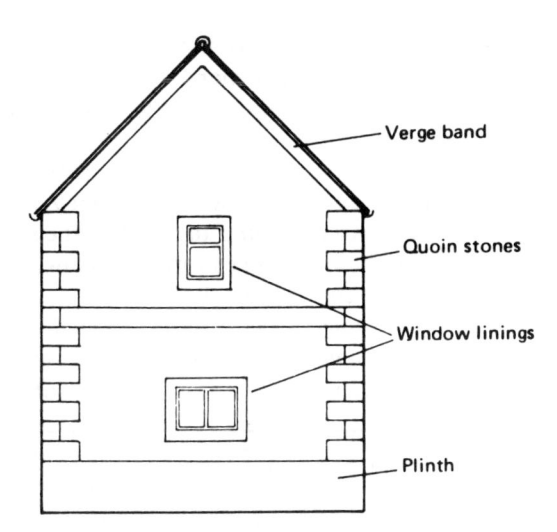

Figure 114: Cement features to a gabled end

Verge band
Quoin stones
Window linings
Plinth

Quoin stones

These are imitation corner stones at the external wall angles. They are normally formed after the rendering coat has been finished. A rule can be nailed to the external angle, allowing 18 mm thickness, and a band of sand and cement applied to the rule edge and the width of the largest quoin stone plus 25 mm. This band of material can now be straightened, rubbed up and fined down. The outline of the quoin stones is marked from a gauge staff, chalk lines snapped and the horizontal joints cut out with a V joint template nailed to a float.

The vertical joints are cut next and the surplus material outside the stones is cut away with a trowel.

Plinth

This is a tall skirting used on external walls as a protection for the base of the wall against damage or deterioration of weaker mixes, and in certain cases to improve the appearance and provide a good line of finish for such textures as pebble dash or Tyrolean, etc.

In certain cases a rebated rule can be traversed from the face of the rendering coat to give a uniform thickness of 18 mm. This can be tightened in, rubbed up, fined down and the top edge lined through with a snapped chalk line. The top can now be cut off with a trowel, either square or chamfered, and finished off by rubbing up with a float.

A suitable mix for a plinth is 3:1 sand/cement plus a waterproofer.

Door and window linings, verges and other bands are applied in a manner similar to that described for plinths.

Vermiculated and *reticulated* work is usually confined to quoin or key stones, though occasionally plinth stones may be covered with this type of finish. A finishing coat of sand and cement is applied to the area to be covered, rubbed up and fined down in the normal manner. The shape of the vermiculations or reticulations is marked on the surface either free hand or pounced through from a traced drawing with a nail. In the case of reticulations, the larger spaces between the narrow band lines is cut back with a small tool and left pocked or plain according to the texture required about 6 mm deep. The width of the raised lines should be about 9 mm to 12 mm with a bolder raised or rebated margin around the edge of the stone. Vermiculations are similar except that the raised lines form a narrow continuous worm-like ridge over the complete area (Figure 115).

If bolder projections for the reticulated work are required they can be carried out in a manner similar to cement lettering. In this way when the surplus cement is cut away it will leave a clean rubbed up finish below.

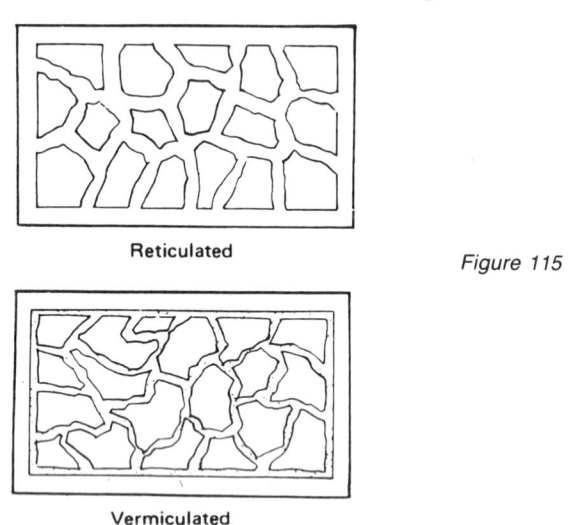

Reticulated

Vermiculated

Figure 115

MODERN PROPRIETARY FINISHES

Mineralite

This is a type of external rendering using a premixed cement and aggregate of extremely hard coloured minerals some of which resemble crushed glass in appearance.

The rendering coat should be a 2:1:8 cement/lime sand mix applied about 9 mm thick and scratched with a fine comb, or an Aquacrete/sand mix gauged at 1:4 applied as above.

A minimum of 72 hours should be allowed between the rendering and floating coat. The mix for the floating coat is as described for the rendering and the application straightened and rubbed up to a good surface and then keyed in the same manner as the rendering coat. The floating coat may be omitted if the brickwork or background is sufficiently straight and true.

Mineralite application

Mix the Mineralite in a watertight container, adding the Mineralite to the water and stirring for small quantities, larger gaugings can be mixed mechanically. Approximately 9 litres of water are required to 50 kg of Mineralite. The mix should stand for 10—15 minutes before use.

Apply the Mineralite about 6 mm thick and level with a float and then tighten in with a steel trowel. The surface is allowed to stiffen up slightly and it is then traversed with a foam sponge rubber roller. The roller must be damp and only sufficient pressure used to ensure the blotting of the excess cement fat to expose the maximum amount of aggregate. If the surface is roughened rolling must be discontinued until the Mineralite has stiffened sufficiently. The roller must be washed frequently.

The surface is next patted with a clean damp felt float to expose the maximum amount of aggregate by dabbing off any remaining surface cement. The pad should be scraped and washed as necessary by dousing the felt float in a bucket of water, and scraping and drying it against a piece of timber batten held with one end resting in the bucket. Any loose aggregate can be lightly patted in with a steel trowel and the work should be protected from frost, rain, and strong sunshine.

After a period of 48 hours the Mineralite surface should be washed with a diluted solution of hydrochloric acid and water to remove any traces of surface laitance and leave the true colour of the Mineralite aggregates exposed. Before applying the acid, any timber or other building materials likely to be attacked by the acid should be protected by covering with wax or hosed with water as the acid is applied. The operative should wear protective clothing and the use of gloves and goggles is recommended.

The hydrochloric acid should be mixed with four parts of water in a rubber or plastic pail. Before applying the acid solution the wall surface should be damped with clean water to prevent the Mineralite absorbing the acid. The acid solution should be brushed on to the wall surface whilst the Mineralite is still wet. This process should be done fairly quickly without missing any patches and care should also be taken by the plasterer to avoid inhaling any fumes or gas given off by the acid. Scrubbing must be avoided or the surface aggregate will be loosened.

Immediately the surface has been treated with the acid solution it should be washed down with clean water.

Coverage

Mineralite covers approximately 3.5 square metres per 50 kgs.

Glamorock

This material is manufactured for both trowel and spray applications.

The trowel applied Glamorock consists of natural coloured rock particles bonded with a transparent vinyl resin, and as no pigments are added fading cannot occur. It provides a textured finish which is waterproof yet permits normal breathing of the backing. The vinyl resin bonding agent gives excellent adhesive properties to most sound building backgrounds yet is flexible enough to allow for normal expansion and contraction. There is also no shrinkage on setting and as the colours are fadeless and the surface is reasonably self cleansing little or no maintenance should be required.

Trowel applied Glamorock is supplied in coarse and medium grades in sixteen different colours which can be intermixed if desired to give any desired blend. Special blends can be specially manufactured to order. Both coarse and medium grades are supplied, with undercoat, ready mixed for application.

Backgrounds

Glamorock coarse and medium can be applied to fair faced concrete, asbestos and many other clean, sound building backgrounds. Brickwork, concrete and breeze blocks must be rendered with a suitable sand and cement mix, straightened and rubbed up to a good finish. No key is necessary or advisable. Backgrounds should be free from salts, grease or oil and not subjected to constant dampness.

Application: undercoat

This consists of a finely powdered basic rock and clear setting plastic cement, having a consistency similar to that of a thick paste. It should be emptied on to a clean spot board, sprinkled sparingly with clean water and then mixed thoroughly to a good working consistency.

The surface to be covered should now be damped slightly by brushing with a wet brush, after which the undercoat is applied with a tight even coat. Only sufficient thickness to completely obliterate the background or rendering coat need be applied. The undercoat must be allowed to set before the finishing coat is applied.

Finish coat

If an area requiring the contents of two containers is to be covered, then it is advisable to premix a sufficient quantity to complete the job. Any material not required for immediate use can then be returned to the containers and resealed to avoid drying and setting.

The material to be used should be placed on a clean spot board, sprinkled sparingly with clean water and mixed thoroughly until a slight foam appears. The wall surface should be damped slightly with water immediately prior to the application of the finish.

MODERN PROPRIETARY FINISHES

Both Glamorock coarse or medium finish coats should be applied tightly with the trowel held at an angle. The plasterer should aim at applying only sufficient thickness to ensure complete coverage by a single layer of the crushed stone particles in the mix. The thickness should not be greater than 3 millimetres, but if extra thicknesses are necessary they can be built up in two or more coats.

As the surface of the finish coat starts to set, and loses its tackiness, it should be tightened in by lightly trowelling over with a clean damp trowel held almost flat to the surface.

No water should be applied on to the finished surface until it has hardened and protection should be provided from rain until setting is complete.

The materials should not be applied in conditions of frost, though low temperatures will have no effect once the set has taken place.

Glamorock should be stored in air-tight containers in conditions which are not subject to excessive heat, cold or dampness.

Coverage

Undercoat approximately 0.75 kg — 1 kg per square metre.
Medium finish approximately 2 — 3 kg per square metre.
Coarse finish approximately 3 — 5 kg per square metre.

Spray-applied Glamorock known as Sparklon is a specialist material and its application is carried out through the firm's own contracts department. It has many similarities to the coarse and medium finishes described, but the finished thickness is 2 mm only and the texture is much finer.

Alpine finish

This is a pre-mixed decorative white finishing material which only requires the addition of water before use. It consists mainly of white Portland cement with specially graded white aggregates and is a durable, weatherproof finish suitable for exterior or interior use. The finished texture is similar to other continental cement work.

A suitable backing coat for Alpine finish is 1:3 cement/sand, straightened and brought to an even plain float finish. The surface should then be scratched lightly with straight horizontal lines for key. Weaker floating coats such as 1:1:6 cement/lime/sand can also be used, but any suction from the floating coat should be stopped to ease the application and texturing of the finish coat. A special stabilising solution is recommended for application when the floating coat has dried out after a period of at least four days, and at least twelve hours before applying the Alpine finish. The provision of a plain plinth at ground level is also recommended to avoid discoloration of the white finish.

Application

The Alpine finish coat is applied by trowel as 'tightly' as possible so that the large aggregates in the mix tend to 'chatter' under the trowel.

The texturing is achieved by drawing a float across the recently applied material before it has had time to stiffen. The larger aggregate will be dragged along the wall surface leaving a series of narrow grooves to form the 'dragged' texture. The float strokes should be either horizontal or vertical only for the whole area, although window or door reveals can be dragged vertically as a contrast to a horizontally 'dragged' main area.

Coverage

The coverage of Alpine Finish is 5m^2 — 6m^2 per 25 kg.

Sandtex Cullaplast

Cullaplast is a ready mixed decorative rendering material composed of selected aggregates with a synthetic resin binder, plus suitable high quality pigments to provide a selection of coloured finishes. It is claimed to have a weather resistant finish, able to withstand cracking and crazing.

Backgrounds suitable for application of Cullaplast include most fair faced building surfaces which are clean, sound and dry. These include cement; (lime): sand floating coats, asbestos cement panels, smooth *in situ* concrete and, for interior use only, plasterboard, gypsum finishing plaster, plywood and similar boards.

For external work the construction design should ensure that no moisture can penetrate behind the Cullaplast. A plinth should be provided at ground level.

Backings should be smooth, without key, dry and of good alignment. Cullaplast primer can be applied by brush or roller to the backing and allowed to dry for three to five hours. A similar colour of primer is used to match that of the Cullaplast rendering finish.

Application

Cullaplast is applied by trowel to an even coating of the thickness of the aggregate only. A suitable texture can be achieved by rubbing up with a plastic float while the material is still workable. A plastic trowel can also be used flat to the surface to produce an alternative texture if desired.

Coverage

The coverage of Cullaplast is 8 m^2 — 10 m^2 per 25 kg bag; primer coverage is 30 m^2 — 35 m^2 per 5 litres.

Cemrend

Cemrend is designed for external and internal rendering to be hand applied in one coat. The 40 kg bags contain cement, additives and specially selected aggregates in dry powder form. Cemrend can be obtained in white and a range of colours.

MODERN PROPRIETARY FINISHES

Mixing

The bags should be rolled to ensure that the ingredients are evenly distributed before clean water is added. Mixing can be carried out with a tumble mixer or with a suitable drill and whisk attachment. The material when applied in one coat does not experience the shrinkage cracking that is often associated with strong cement renderings. It is advisable to use only white during the winter months, or damp weather, as lime bloom will form temporary discoloration.

Application

All surfaces should be prepared as for normal renderings. After mixing the Cemrend it is applied to the surface in one coat, this can be formed in two applications to build up to the thickness required of between 15 to 25 mm. The Cemrend is ruled out and any slacks or hollows filled in. The coat is allowed to harden for between 5 and 16 hours depending on the weather conditions and then scraped to form an open textured surface. Cemrend should be applied continuously and joints avoided.

Coverage

The coverage of Cemrend is 1.6 to 1.8 m^2 per 40 kg bag or 40 to 45 m^2 per tonne.

Cullarend

Cullarend is a self coloured projection render based on sand/cement. It can be finished with different textures including light honeycomb, heavy rough cast and a scraped finish.

Mixing

Cullarend can be mixed in a suitable batch mixing machine, many of the spray machines mixing is carried out within the spray machine, (see section on mechanical plastering).

Application

The method of application will depend on the finish selected. For a scraped finish, the material is sprayed on in ribbons to build up a thickness of 20 mm in one application, it is then ruled off and finished as Cemrend. For a textured finish one application is made to a thickness of 10 mm, it is sprayed, ruled out and allowed to stand for up to 16 hours. The surface is allowed to steady up sufficient for the second textured application to be sprayed. If 16 hours is exceeded it may be necessary to damp and scratch the surface before the textured coat is applied.

Coverage

Cullarend applied 15 mm thick will cover 1.6 to 1.8 m^2 per 40 kg bag or 40 to 45 m^2 per tonne.

CHAPTER NINE

Moulded work *in situ*

Mouldings vary from very simple to quite intricate shapes. The various moulding members, or individual parts of the moulding are mainly Roman or Greek in origin. Roman mouldings are formed from parts of the circle (Figure 117), whereas Greek mouldings are parts of an ellipse or other conic sections such as hyperbolas and parabolas (Figure 116). Other styles of moulding members, including Gothic, are often made up of free-hand outlines to give curved shapes which throw effective shadows or show high relief (Figure 118).

In the sketch of the Roman moulding members it should be noted that the terminal members are usually in line with the centre from which the curve was struck. Mouldings are made up of one or more members separated by their terminal members or fillets (straight members) (Figure 119).

To form these mouldings for solid or fibrous plasterwork, *running moulds* are made. These are templates cut to the outline required and horsed up, or strengthened, in different ways to enable the necessary shape to be formed in plaster or cement.

The simplest form of running mould is a thumb mould, and this merely requires the outline of the moulding cut to shape in sheet metal and strengthened with a timber horse or background.

It will be seen from the sketch of a thumb mould shown in Figure 120 that the horse or frame is cut in the shape of a right angle at one side only. When a thumb mould is used to form a curved front moulded surface to a plaster truss, or bracket, the right angled end is used as a guide to keep the thumb mould moving in a correct line.

A box or frame is made to the correct size and profile required for the plaster bracket. Plaster is fed into the box and the thumb mould traversed over the curved front part of the frame, keeping the right angled end tight to one side of the frame. In this way the profile of the thumb mould cuts out the moulded front of the bracket without wavy or twisted lines.

To make the profile for a thumb mould a tracing can first be taken of the full size moulding detail. This shape can be transferred on to a piece of sheet metal with the aid of carbon paper and cut out to within 1 mm–2 mm of the required outline with a pair of tinsnips. The metal profile should now be filed accurately to the marked outline. This is best accomplished by fastening the metal profile in a metal vice so that the portion of the profile to be filed is

MOULDED WORK *IN SITU*

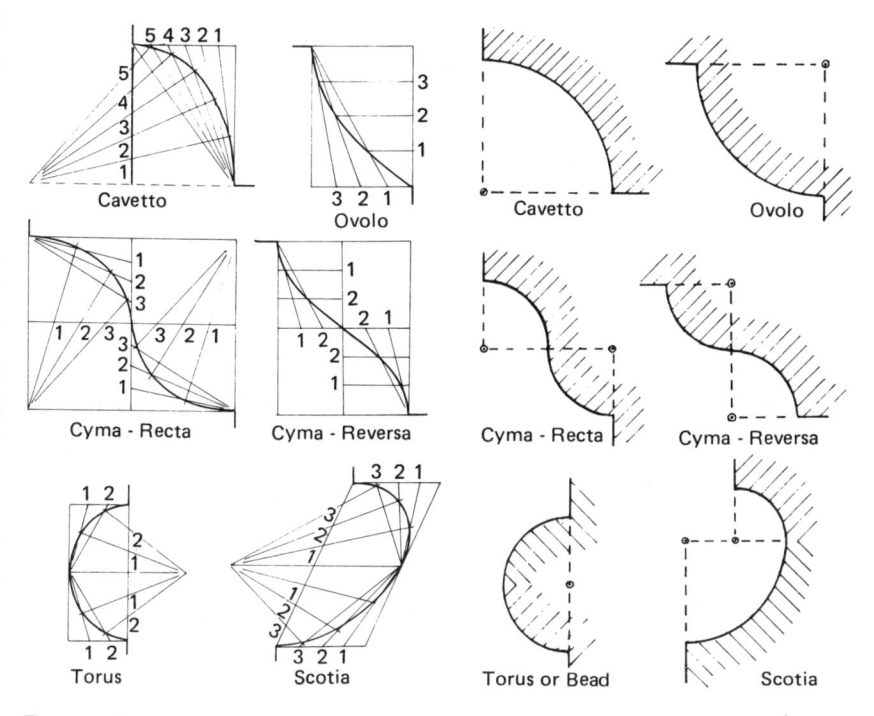

Figure 116: Grecian moulding members

Figure 117: Roman moulding members

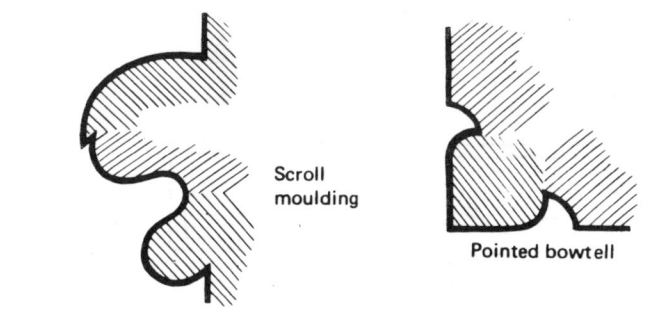

Figure 118: Gothic mouldings

within 3 mm of the vice jaws. Suitable files for use are 175 mm or less and square, half round and rat tail in section. They give best results if used with a forward cutting action only. When the carbon outlined filed shape has been reached it should be checked back with the original profile and approved or adjusted as necessary. If the file has been used in one direction only the back outline edge of the profile will be bent over slightly to form a burred edge. This should now be filed off flat, with the straight edge of a file held flat to the back of the sheet. Finally any scratches or imperfections in the outline of the profile can be removed by rubbing carefully over the moulded edge with a nail or emery cloth.

145

MOULDED WORK *IN SITU*

Figure 119: Composite mouldings

- Weathering
- Cyma Recta
- Cornice section for external cement work
- Drip member
- Ovolo
- Cavetto

- Cavetto
- Cyma Reversa
- Internal cornice section

- Torus
- Section for base mould of column

- Abacus
- Ovolo
- Astragal
- Section for capital of column

- Cavetto
- Cyma Reversa
- Fillet
- Archivolt section

- Ceiling panel mould
- Ovolo

THUMB MOULD
The dotted line indicates the cut away portion of the horse

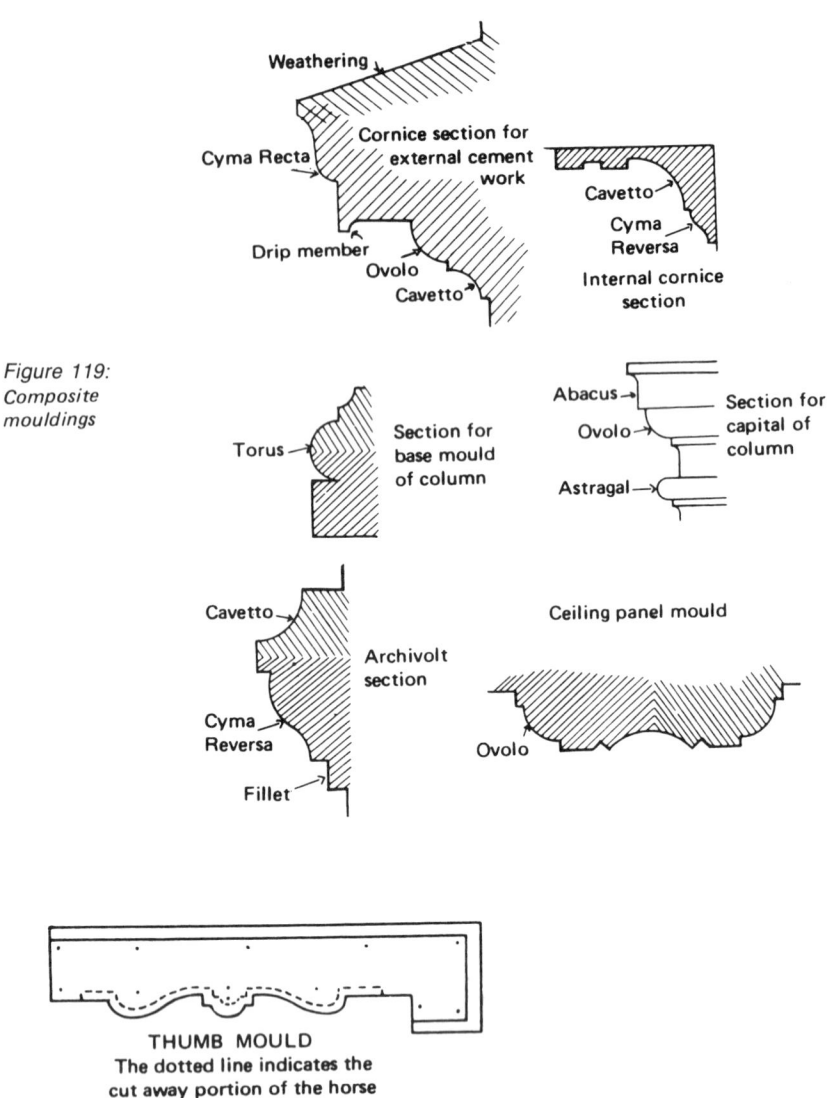

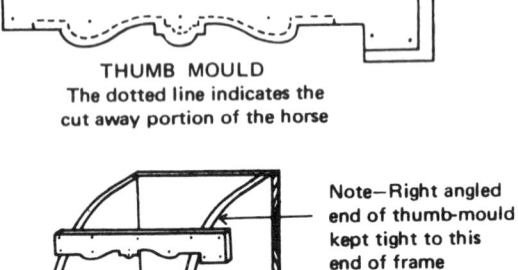

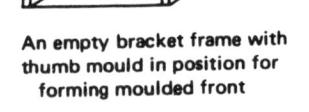

Note—Right angled end of thumb-mould kept tight to this end of frame

Figure 120

An empty bracket frame with thumb mould in position for forming moulded front

MOULDED WORK *IN SITU*

The horse or back frame of a thumb mould can be made from a piece of timber about 12 mm – 15 mm thick. The profile should be placed over the timber and its outline marked out in pencil. A new outline can now be marked 3 mm above the moulded edge as shown by the dotted lines in the sketch. The reason for this being that the timber may swell on being wetted and so protrude beyond the metal profile during the actual running or forming of the moulded surface.

The new reduced outline of the horse can now be cut to shape with coping saw or chisel and rasped to a good finish. Holes should be punched through the metal profile, which can then be nailed to the horse to complete the mould.

A panel running mould is shown in Figure 121. This requires an extra piece of timber nailed at right angles and braced as indicated. The duplication of names for parts of the mould is due to local terminology in various parts of the country. Straight lengths of panel moulding can be formed by running the mould against a rule on a bench or wall surface. The gauged plaster is fed into position by trowel or gauger and the mould is passed over, the metal profile cutting the setting plaster to the correct section required. Two, three or more gaugings may be required to build up the moulding until it is filled out to the full section along its whole length free of imperfections.

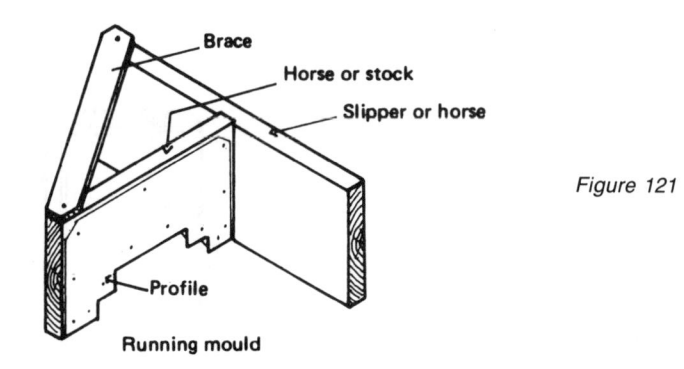

Figure 121

Curved lengths of panel moulding are formed by using a gig stick with the running mould. The gig stick is a radius arm and its use is shown in Figure 122. The most accurate method of running an arc to the correct radius is to place the pivot point so that it is in line, or radial, with the metal profile. This is also shown in Figure 123. With this type of pivot point it is important that the hole in the metal pivot is a good fit with the headless round headed nail used, otherwise a different cut will be obtained each time the mould is passed over the curved moulding.

A flush bead dado mould is shown in pictorial view, Figure 124. The method of running this *in situ* is shown in the sectional detail (Figure 125).

When used on ramps or staircases the mould may be horsed up as in Figure 126.

MOULDED WORK *IN SITU*

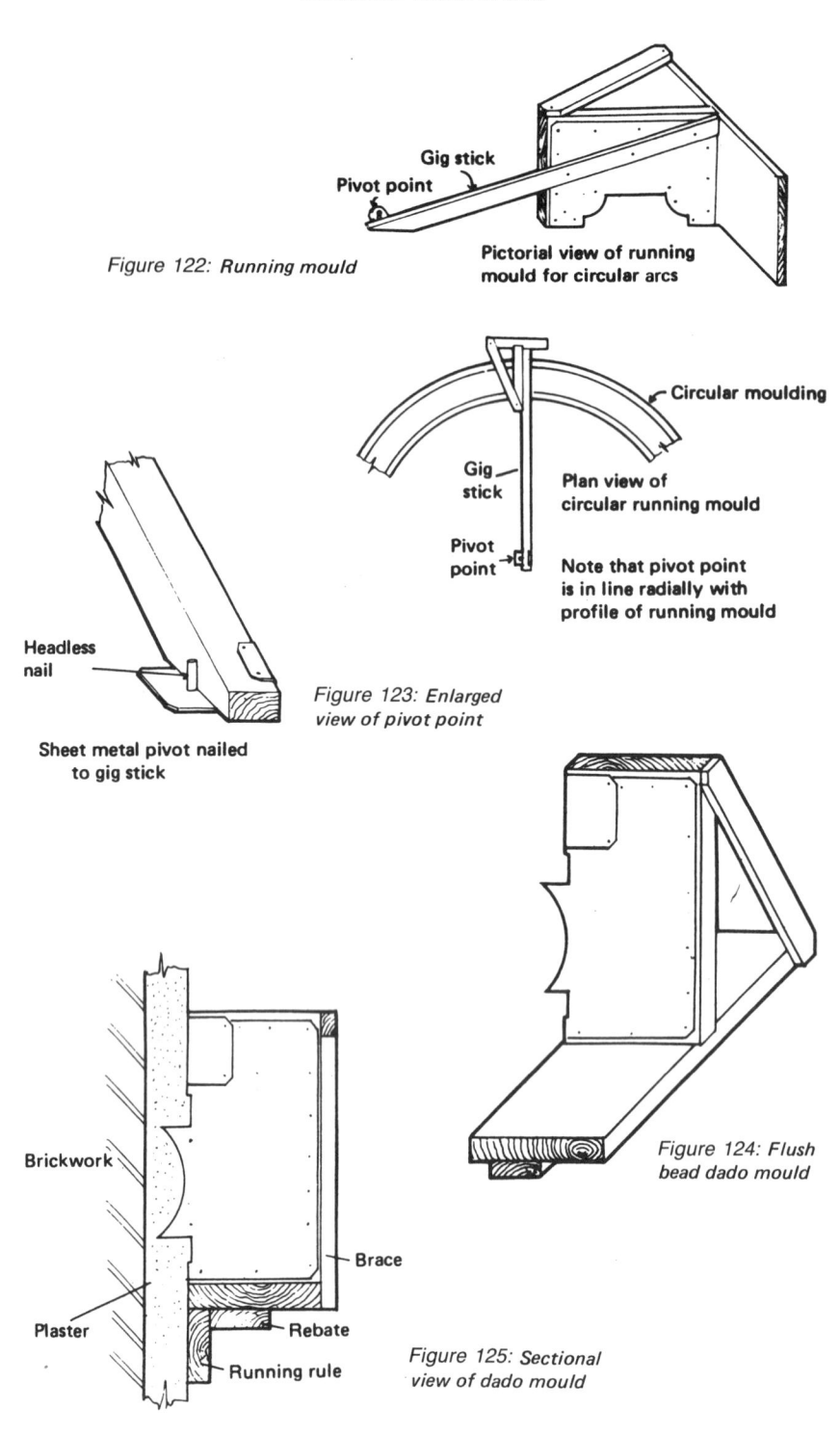

Figure 122: Running mould

Gig stick
Pivot point

Pictorial view of running mould for circular arcs

Circular moulding

Gig stick

Plan view of circular running mould

Pivot point

Note that pivot point is in line radially with profile of running mould

Headless nail

Figure 123: Enlarged view of pivot point

Sheet metal pivot nailed to gig stick

Figure 124: Flush bead dado mould

Brickwork

Plaster

Brace

Rebate

Running rule

Figure 125: Sectional view of dado mould

148

MOULDED WORK *IN SITU*

Figure 126: Ramp or bare faced mould

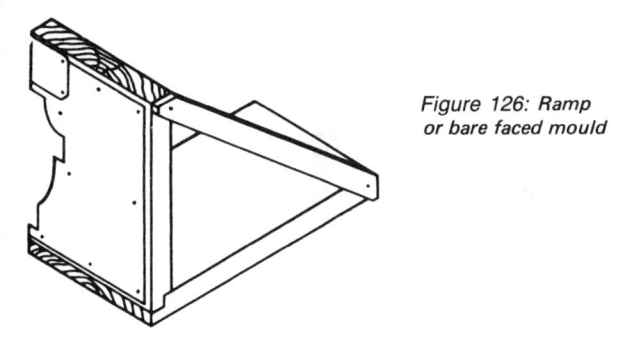

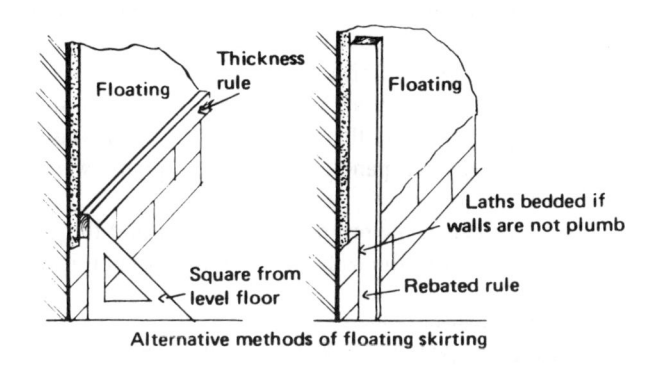

Alternative methods of floating skirting

- Floating
- Thickness rule
- Square from level floor
- Laths bedded if walls are not plumb
- Rebated rule

Figure 127: Moulded skirtings

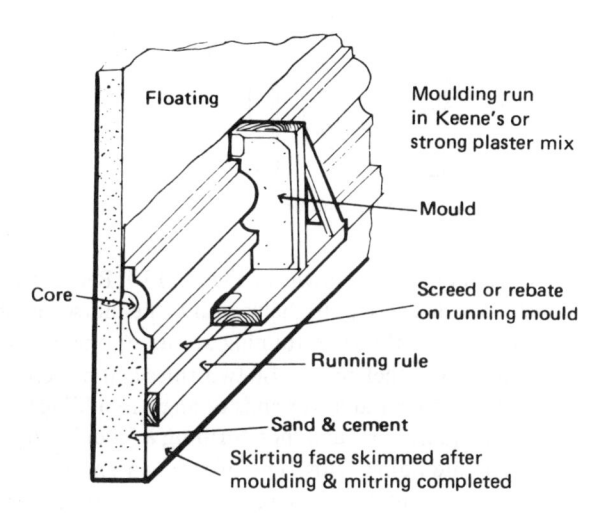

- Floating
- Core
- Moulding run in Keene's or strong plaster mix
- Mould
- Screed or rebate on running mould
- Running rule
- Sand & cement
- Skirting face skimmed after moulding & mitring completed

149

MOULDED WORK *IN SITU*

Figure 128: Panel moulds

Sunken panel ruled in from stiles

Stiles floated plumb and linable

Rebated floating rule for forming sunken panels

Sunken panel mould
Nib runs in sunken panel

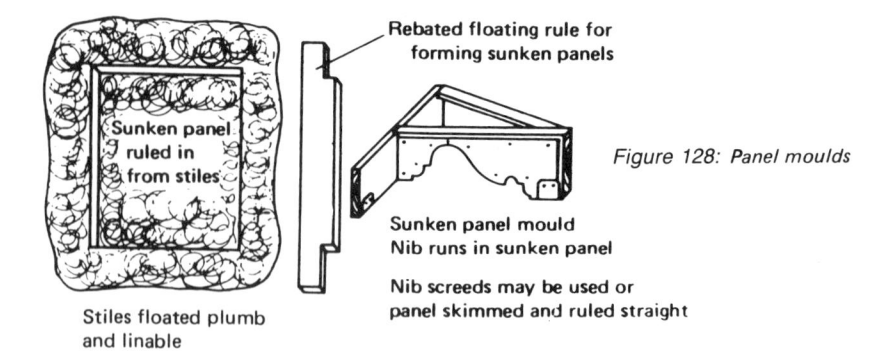

Nib screeds may be used or panel skimmed and ruled straight

Figure 127 shows methods of forming a moulded skirting. The nib must project beyond the line of the slipper the exact width of the base plinth. Note also the skimming member formed when running to enable the plain plinth to be skimmed to a good finish as the last operation.

Another type of running mould which has a nib projecting beyond the line of the slipper is given in the sketch, Figure 128. This is necessary when running the sides of a sunken panel, because the slipper always runs on the outside of the panel.

For raised panels the reverse is required in that the nib is recessed behind the line of the slipper to the width of the raised panel thickness (Figure 129).

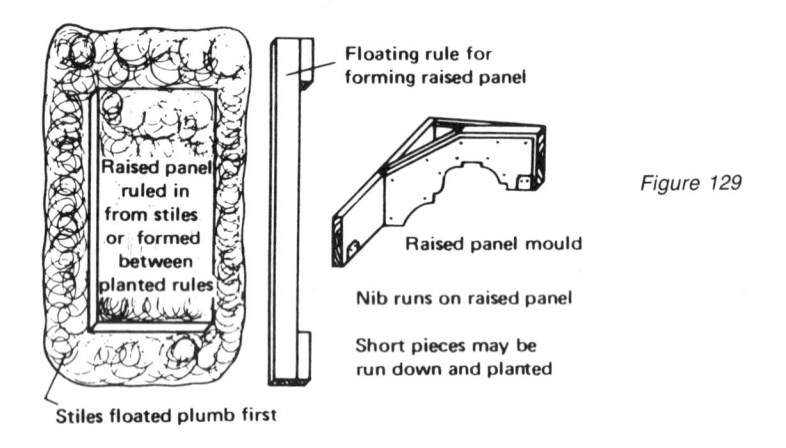

Raised panel ruled in from stiles or formed between planted rules

Stiles floated plumb first

Floating rule for forming raised panel

Raised panel mould

Nib runs on raised panel

Short pieces may be run down and planted

Figure 129

Angle moulds are shown in Figure 130. Because of their shape it is usually considered necessary to halve joint the horse as shown to strengthen the mould. Two methods of running are used, one against two screeds and a running rule, the other by use of two rules and no screed. Stopped ends are often formed at upper and lower ends of angle mouldings by bringing the last 75 mm or so to a square arris and then forming a 'cock's breast' or curved sweep to the run angle moulding. This type of stopped end is also known as a 'bird's beak'. Splayed, segmental and cyma-recta curved sheet metal templates are held to

MOULDED WORK *IN SITU*

Figure 130: Angle moulds

Figure labels: Nails for skimming finish; Pictorial views of suitable angle moulds for methods shown; Brickwork; Floating; Rule; Nib; Running rule; Rabbet; Profile; Slipper; Stay; Sectional views showing setting out; Note—Screeds are essential for lower method; Brickwork; Screed; Nib; Screed; Running rule; Slipper; Stay; Note—Skimming member

Segmental   Splayed   Bird's beak or Cyma-recta

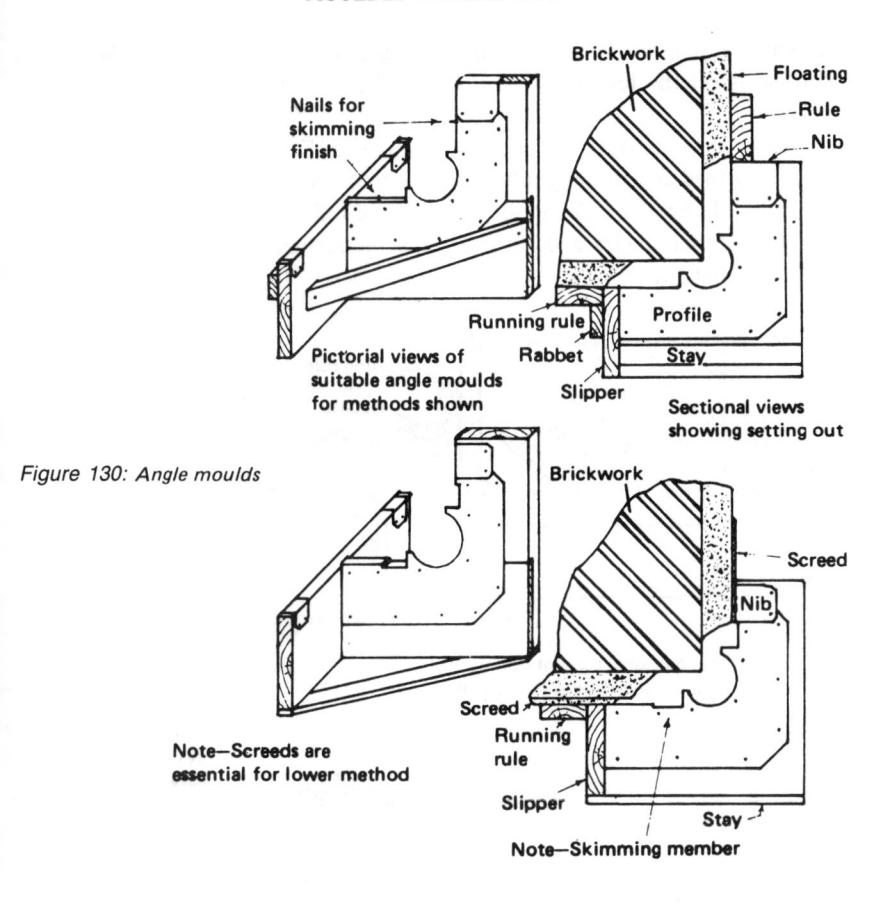

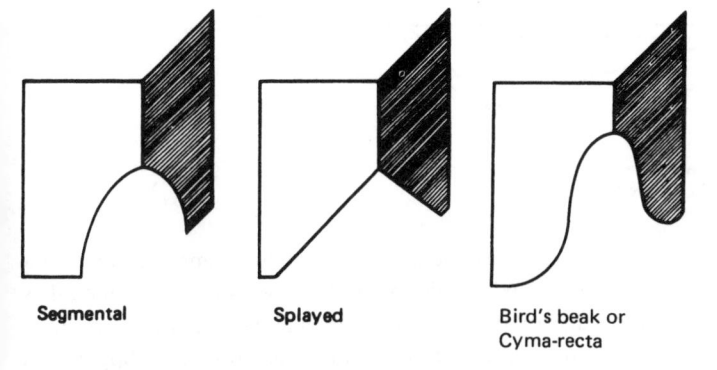

Figure 131: Stopped end angle templates

MOULDED WORK *IN SITU*

the angle and filled in with strongly gauged plaster (Figure 131). When this material is set the template is removed and the arris touched up with a small tool.

Double angle moulds can be made as shown in Figure 132. The slipper bears against a screed and running rule and the nib forward against another rule. With this method there is a saving in time because only one screed is required and the two external angles will automatically be parallel without the necessity of careful setting out.

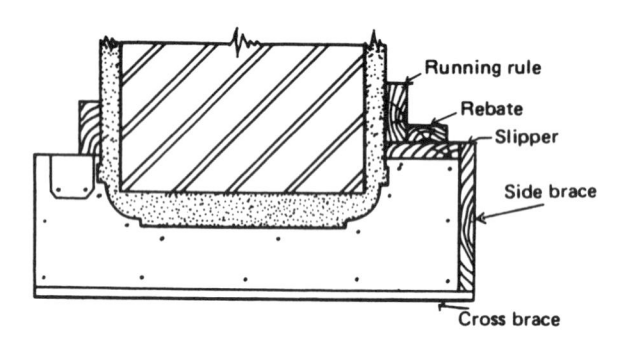

Figure 132: Double angle running mould

Horizontal sectional view of double angle running mould in position for running

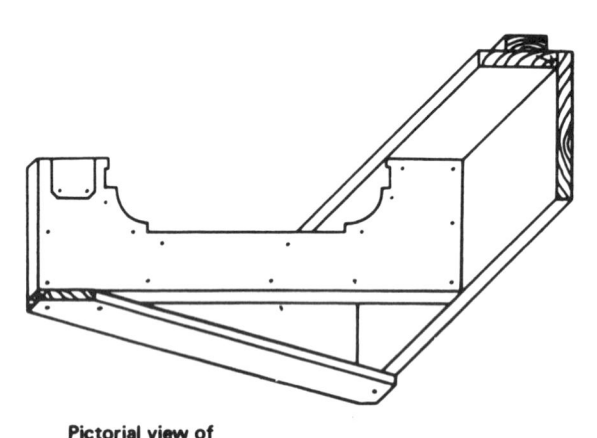

Pictorial view of double angle running mould

When running large single or double angle moulds to beams it is an advantage to use a hanging mould. With this method the weight of the mould is transferred to the top edge of one running rule, making the task of running much easier. Because of the extra stresses downwards from the slipper it is advisable to construct the mould much stronger than normal, using screws to fasten the slipper to the horse to prevent withdrawal or loosening during the running (Figure 133).

MOULDED WORK *IN SITU*

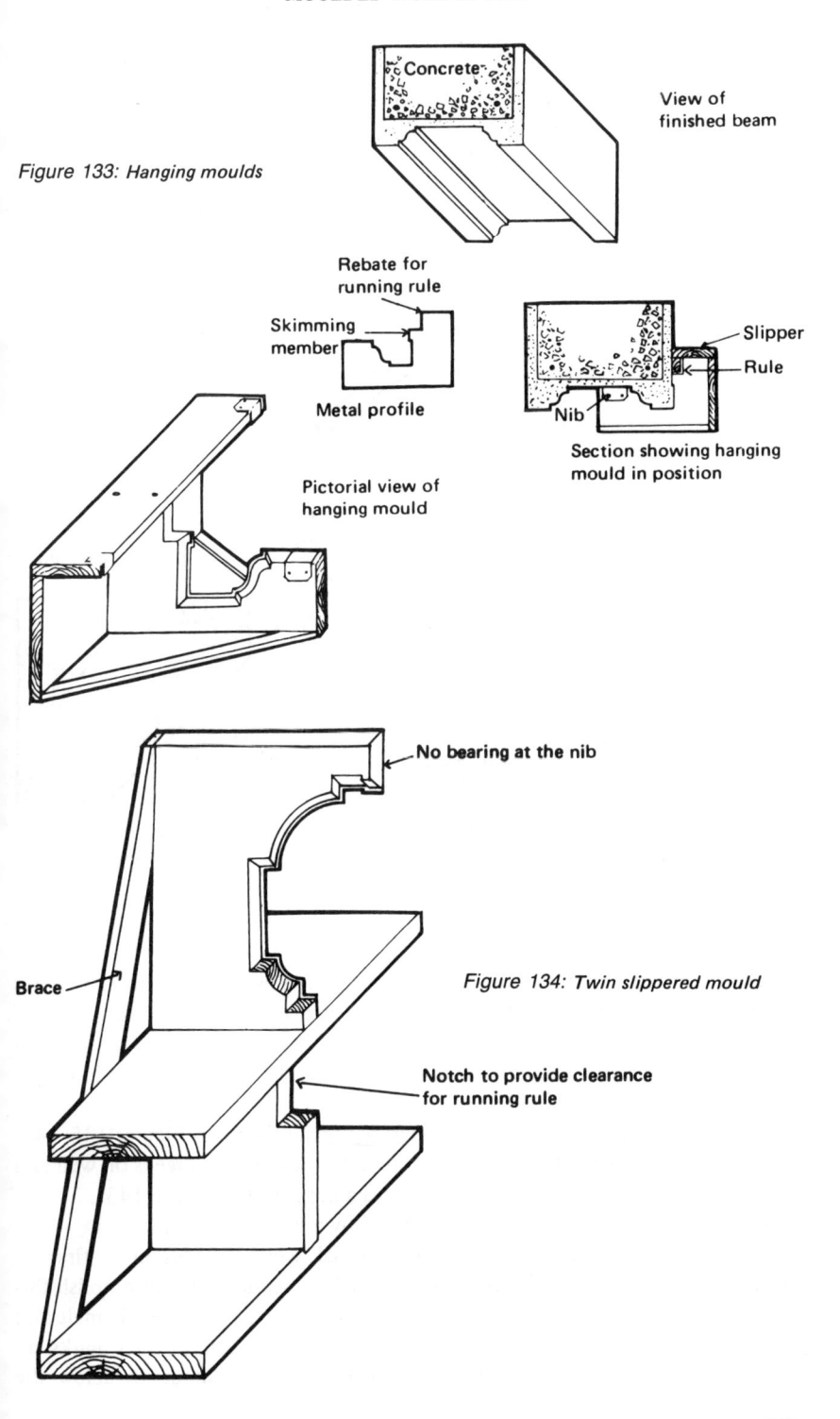

Figure 133: Hanging moulds

View of finished beam

Rebate for running rule

Skimming member

Metal profile

Slipper

Rule

Nib

Section showing hanging mould in position

Pictorial view of hanging mould

No bearing at the nib

Brace

Figure 134: Twin slippered mould

Notch to provide clearance for running rule

MOULDED WORK *IN SITU*

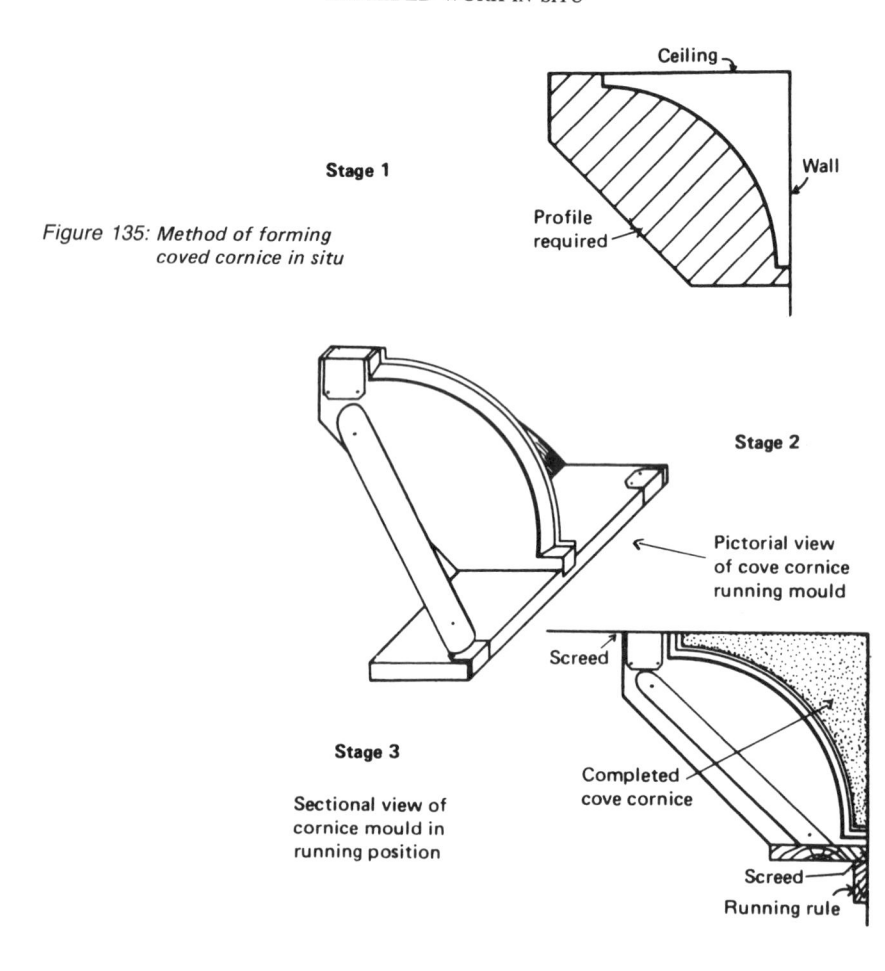

Figure 135: Method of forming coved cornice in situ

Stage 1

Ceiling, Wall, Profile required

Stage 2

Pictorial view of cove cornice running mould

Stage 3

Sectional view of cornice mould in running position

Screed, Completed cove cornice, Screed, Running rule

Twin slippered moulds (or double-horsed moulds as they are sometimes called) are used when there is no bearing for a nib slipper. In this type of mould a broad screed is carefully made and the running profile is cantilevered from the two parallel slippers, often with the rule between as shown in Figure 134.

Cornice moulds for internal and external use are shown in Figures 135, 143. The main difference is that internal cornice moulds run on screeds on wall and ceiling, whereas an external cornice requires a nib rule (see Figure 143).

A combined cornice and angle mould is shown in Figure 136.

When running cornice moulds for internal work, screeds must first be formed as running bands for the wall and ceiling bearings. To accomplish this the running mould is held to the correct position in the ceiling/wall angle and the ceiling projection and wall depth marked with a pencil. Further marks are placed 50 mm outside the ceiling projection and a similar depth below the

MOULDED WORK *IN SITU*

Figure 136: Combined cornice and angle mould

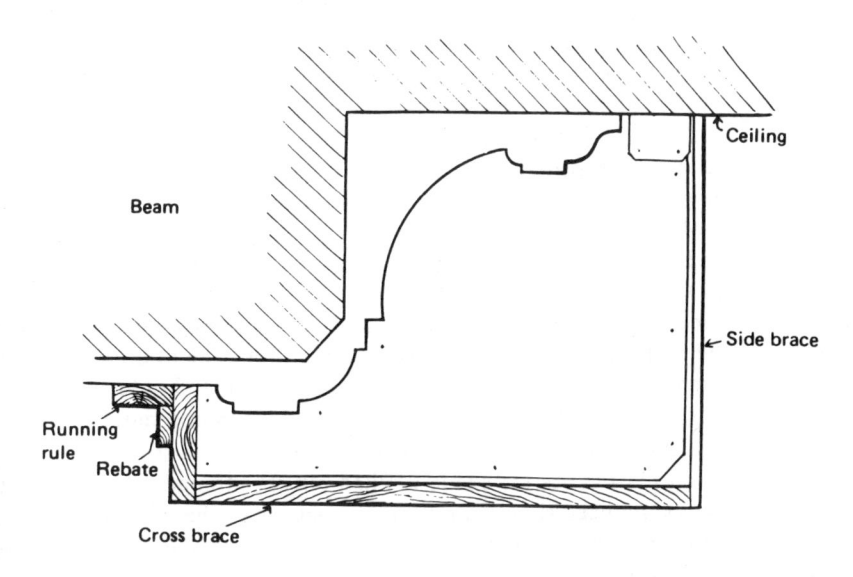

first wall mark. Coloured chalk lines can now be snapped on the wall and ceiling to give 50 mm wide parallel lines all round the room.

Strongly gauged putty lime and plaster can now be applied between the parallel lines, straightened and 'sweetened in' by dry trowelling after filling in and tightening. The screeds when finished should be as thin as possible, but straight in length and square in the angle.

The running mould should now be held in each corner in turn and the position of the nib projection marked carefully. Snapped chalk lines are now struck on each ceiling screed.

The running rules, usually of 50 mm x 12 mm planed timber, are nailed to the wall commencing at the right hand end. The cornice running mould is held in the ceiling angle with the nib touching the chalk line, and the running rule is now nailed so that it touches the slipper. This is repeated every 375 mm or 450 mm along the wall, raising or lowering the running rule before nailing to ensure a perfectly straight line for the ceiling member.

If the screeds are true the wall member will also be perfectly straight and linable with the ceiling member. It is advisable to 'trap' the running rules, or fix them more firmly, by placing small dabs of strongly gauged putty lime and plaster at intervals of 450 mm or so along their lengths.

Cornice moulds having a shallow section which hug the wall and ceiling can be run in putty lime and plaster gauged strongly. Larger sections are usually cored out first with a coarser type of mix, to give added strength and save the use of more expensive finishing mixes.

The running mould must first be muffled by covering the original profile with an extra thickness projecting approximately 3 mm beyond each member. One method is to cut and file a completely new larger profile to the size required and nail this over the original. Another method is to cover the original profile with 3 mm thickness of neat plaster.

This latter method is best achieved by driving small nails into the timber horse of the mould immediately behind the profile, leaving the heads protruding. Gauged plaster can now be fed around the nail heads and built up beyond the outline of the profile. As the plaster sets it can be cut back with a small tool leaving 3 mm projection and smoothed up. When hard the muffle can be coated with shellac and greased if possible.

The cornice can now be cored out with sand/lime mortar gauged with plaster using the muffled mould which should be run along the rules tight to the rule, wall and ceiling screeds. When the core has set it should be deeply scratched for key and the muffle removed from the running mould.

The finishing coat for the cornice can now be applied with strongly gauged lime putty and plaster. As the shape is filled out the strength of succeeding gaugings can be reduced slightly and the consistency of the mix slightly softened. The final polishing coat can be fairly weak and very soft and the running mould passed over finally with a splash of water for a clean smooth finish.

Cornice moulds should be run to the maximum possible lengths on each wall to keep the mitres as short as possible.

Internal mitres have to be cleaned out after the cornice lengths have been run, by trimming off surplus material until a joint rule can be traversed along each member clear inside each angle.

Strongly gauged mixes, similar to those previously used for running the lengths of cornice, are then placed in the mitres with a gauging trowel roughly to the shape of the extended members. A suitable joint rule, usually about 300 mm in length, is traversed over from the existing moulding members to cut the mitre to the correct shape. When the members have been extended from each side of the mitre until they meet, they are smoothed over with softer, weaker gaugings until a good finish is obtained.

Joint rules must be perfectly clean and held lightly but firmly in contact with the existing moulding members. It must be held parallel with all moulding members, particularly curved shapes. The joint rule should always be traversed away from the external edges of members otherwise tearing of the arris will occur.

External mitres can often be formed by the cornice running mould. Exceptions to this are when the ceiling members recede, or are recessed, in which case the recesses will require to be filled in and finished by joint rule later.

To run the external mitres it is necessary to extend the running rules beyond the external wall angle for the distance of the extra ceiling projection. This can be achieved by nailing a piece of rule behind the projecting piece of

MOULDED WORK *IN SITU*

running rule to form a rebate. The running mould can thus run beyond the length of the wall and past the external angle (Figure 137).

Another method is to nail a piece of rule to the underside of the mould slipper forming a rebate equal to the thickness of the running rule. Again the running mould can now be used to run beyond the normal length of the wall past the external angle.

It is best to work from right to left by running the right hand side of the external angle first, taking care to build up and finish the cornice at least a projection width beyond the external wall angle face.

The running rule can now be removed and a new one fixed on the remaining side of the external angle, taking care to adjust it to the exact height so the members will intersect perfectly. The running mould is now used to cut away the surplus set material by passing it from right to left along the rule, aided by cutting to within the last 3 mm or so with a chisel or gauger. By this means the shape of the external mitre can be run without the use of a joint rule, except in the case of recessed ceiling members as previously described.

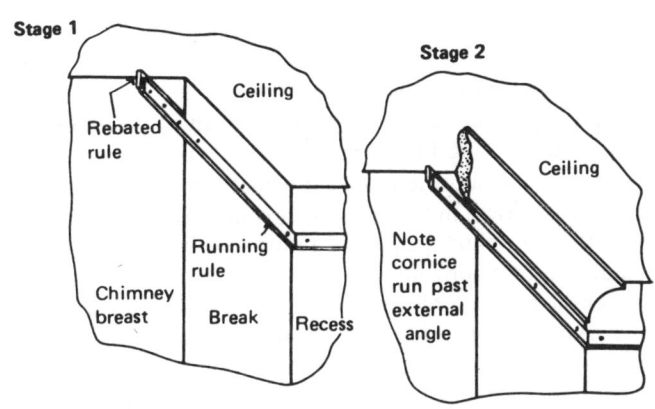

Figure 137: External mitres

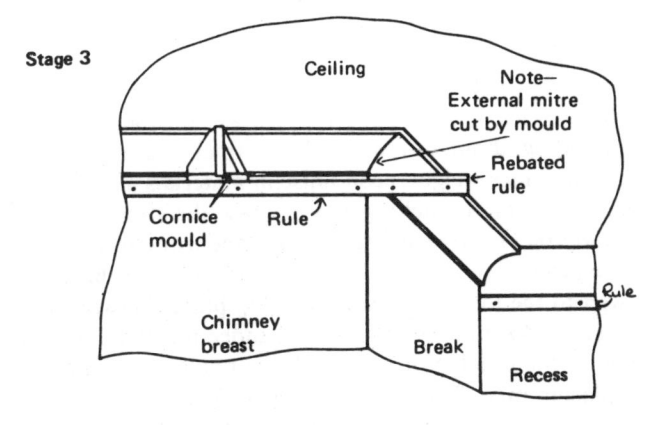

157

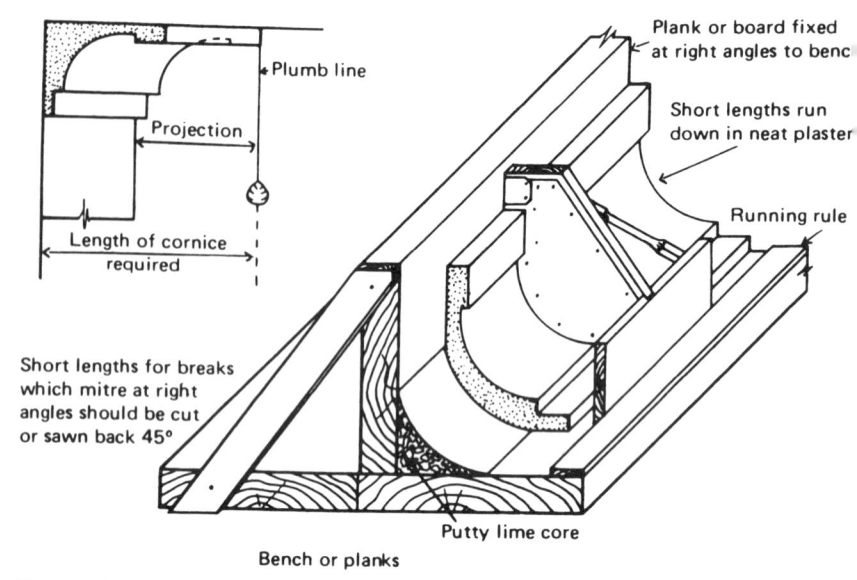

Figure 138: Method of forming short breaks and returns

Short breaks and returned ends

These can be formed by running down short pieces of cornice on a bench, cutting and planting into position or by scribing and cutting the cornice *in situ*. On site the short lengths are usually run down on an improvised bench of planks as shown in Figure 138. It is important to remember when marking and cutting the short lengths or returns for bedding that the ceiling line of the moulding is kept horizontal.

After the piece has been cut, it should be held in position first to ensure sufficient clearance for bedding. A strong putty lime/plaster mix is used for bedding, care being taken to ensure correct intersection of members, and in the case of returns, bedded square and level. Before bedding, the back edge of the pieces should be scraped clean of any putty lime left on from the core and keyed with a deep undercut key.

Bracketed cornices

When cornice moulds have a very heavy or thick section it is usual to run the moulding about 25 mm — 35 mm thick, leaving a hollow space behind. This is to reduce weight, economise in materials and often labour as well, to reduce the risks of cracking and increase the strength (Figure 139).

Wood brackets are constructed of timber roughly cut to the desired profile allowing 25 mm — 35 mm of cornice thickness. The brackets can be nailed or screwed to the joist sides or to nogging pieces between the joists at normal joist centres 400 mm to 425 mm. Expanded metal lathing can be nailed to the outline of the bracket profiles and this rendered with a strong gauged sand/lime/plaster mix. The rendering coat should be deeply undercut for key after which the cornice can be cored out and finished as described.

MOULDED WORK *IN SITU*

Scotch bracketing is a simple method of bracketing using short lengths of wood lath cut to size and bedded across the wall/ceiling angle. Bands of strong gauged stuff are applied along the ceiling and wall about 50 mm inside the cornice profile. The bands need to be about 50 mm wide and 25 mm thick. Whilst this material is still soft the laths are pressed well into its depth to ensure complete coverage of the lath ends. The laths should be spaced about 9 mm apart along the length of the wall. Clearance from the profile of the running mould to the face of the laths should not be less than 18 mm.

Rendering can be carried out with strongly gauged haired coarse stuff and then cored out in the usual way.

Metal brackets — These are used in fireproof construction work or other positions where the previous two methods are unsuitable. Mild steel or wrought iron bars of various sections up to 25 mm x 6 mm can be used. They can be bent to match the profile but are usually left straight to span the angle as shown in Figure 139.

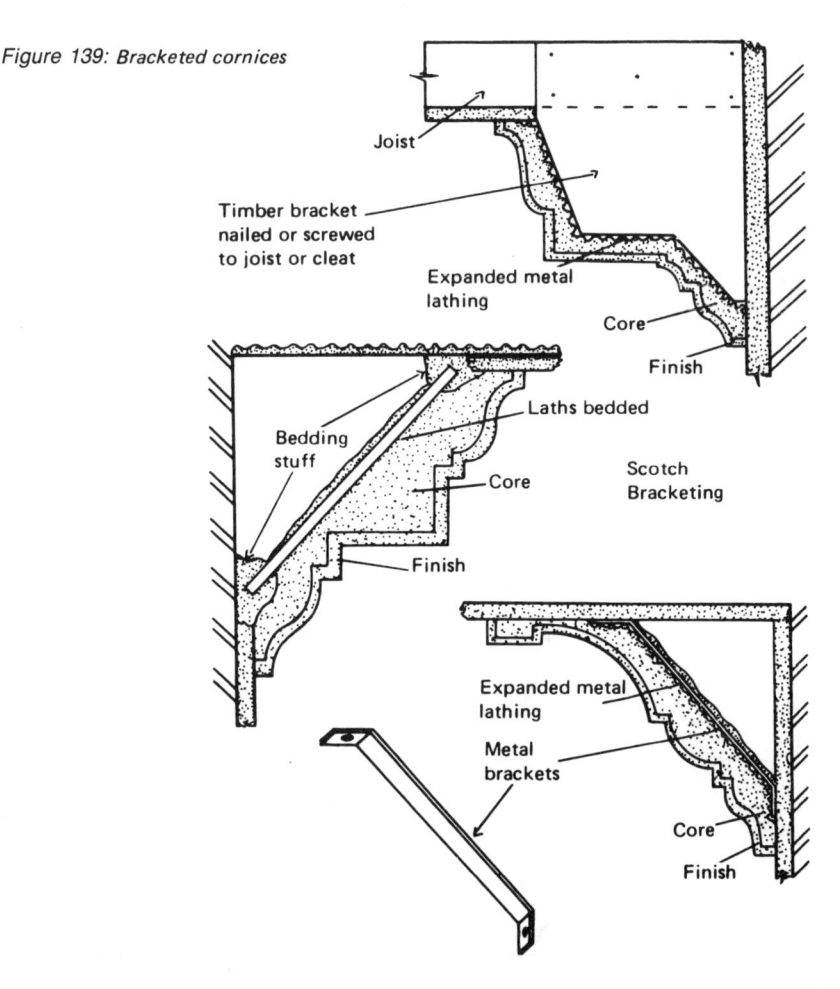

Figure 139: Bracketed cornices

MOULDED WORK *IN SITU*

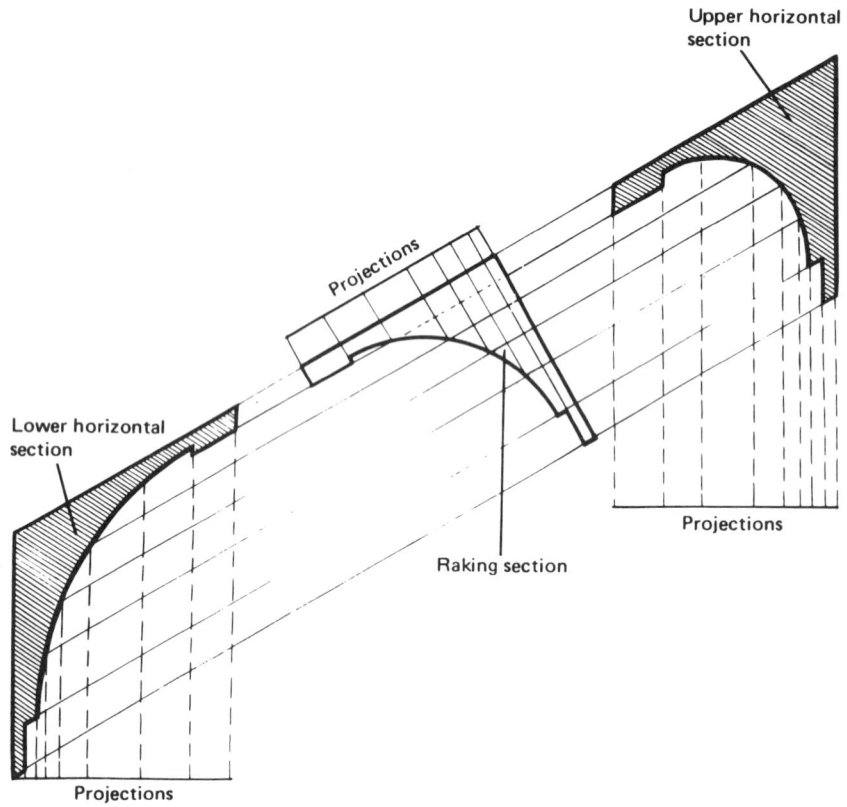

Figure 140: Cornice raking mould

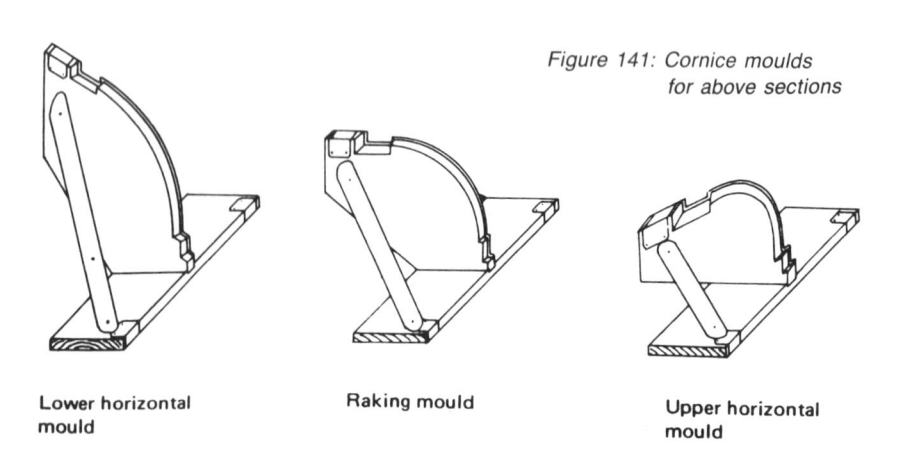

Figure 141: Cornice moulds for above sections

Lower horizontal mould

Raking mould

Upper horizontal mould

In ordinary construction work the brackets are nailed or screwed to joists in the ceiling and plugs in the wall. With fireproof construction they may be wired, welded, or merely bedded in position with cement.

Expanded metal lathing is wired to the brackets by looping soft galvanised wire through the mesh then behind the bars and out through to the front again. The ends are twisted, cut and pressed flat.

Rendering, coring and finishing as before described.

Raking moulds are used for angles other than horizontal or vertical, such as cornice moulds for internal angles on raking ceilings, dado moulds on staircases containing attached piers, cornices for triangular or segmental pediments, etc.

Staircases having raking ceilings surrounded by beams will require three different cornice moulds each one having a different profile (Figure 140). When developing the correct sections it is usual to draw the 90° internal angled cornice section between lines of the correct ceiling rake as shown. The upper and lower sections are developed by projecting lines vertically from the horizontal projections shown, to intersect the raking moulding lines.

Differences in the three cornice moulds required are shown at the base of Figure 141.

Figure 142 shows a method of obtaining upper and lower horizontal sections for a dado moulding on a staircase containing an attached pier.

EXTERNAL CORNICE

Cornice moulds used for outside work have a functional as well as decorative purpose. They are designed and positioned to ensure that the lower part of the building is protected from excessive rainwater running down the wall face.

To do this effectively an external cornice must have a reasonably large projection, contain a throating which includes a drip member and also have a weathered top. The weathering is to make sure that the rainwater is fed quickly towards the crown member of the cornice. Any rainwater which does not fall clear of the wall from the projection will run down the face of the cornice until it reaches the drip member. At this stage the water will be unable to travel up into the throated recess and will therefore drip clear of the wall face.

External cornices are occasionally formed on brackets but are more usually cored out in brickwork or concrete. Special preventative care against deterioration through rust or water penetration must be taken when the cornice is formed on brackets. This should include the use of heavy gauge bitumen coated or galvanized expanded metal lathing, and also the provision of a lead flashing for the whole of the broad flat upper surface of the weathering.

The expanded metal lathing should be cored out with haired coarse stuff strongly gauged with Portland cement or if no putty lime is available a mixture of sand and cement 5: 2 plus a plasticizer is suitable.

Concrete cores should be spatterdashed or treated with a bonding adhesive

MOULDED WORK *IN SITU*

Figure 142: Staircase dado raking mould

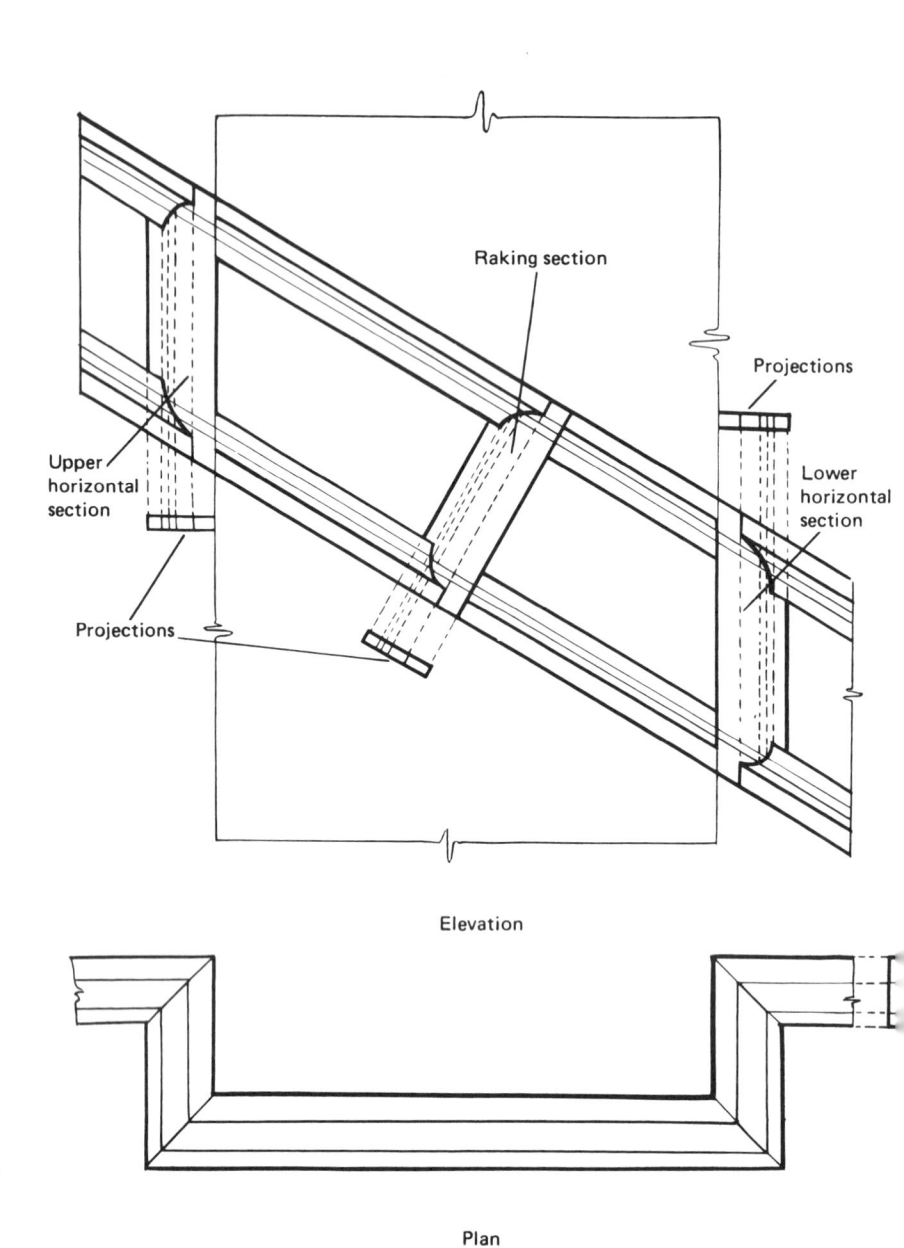

Elevation

Plan

EXTERNAL CORNICE

before use. Brickwork cores may contain irregular shapes requiring excessive thicknesses especially if the ends are not axed roughly to the shape of the cornice profile. Dubbing out coats should be applied on the day previous to running the cornice.

The mould is run on a rule nailed to the wall and against a nib rule bedded on the weathering core as shown in Figure 143. Positions for the two rules are determined by holding the cornice mould at each end of the wall in turn, plumbing the parallel back edge or measuring its projection from a plumb line. The position of the wall rule can be marked on the wall face and a dot bedded to establish the position of the nib rule.

Lines can be snapped for the wall rules and 50 mm x 12 mm rules nailed and trapped with plaster dabs. The nib rule should be bedded to a tightly stretched chalk line held between the nib rule dots. The bedding mix can be fairly weak because it will later have to be shelled off and discarded. To stiffen

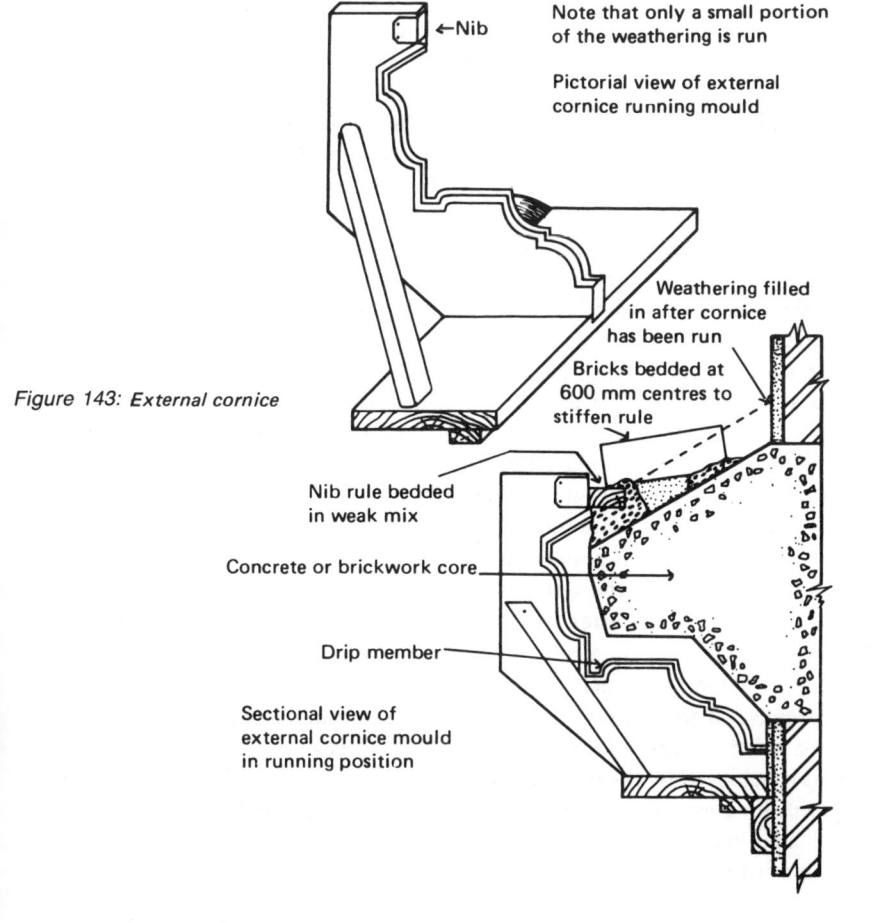

Figure 143: *External cornice*

and strengthen the rule, against any excessive pressure expected when running the mould, bricks are bedded at right angles to the rule at approximately 600 mm centres. The bricks should just touch the nib rule but not project beyond its face.

It is essential with this type of mould to core out within 6 mm of the finished profile, because of the special difficulties in running with sand and cement mixes. These mixes rely on suction, rather than on the setting action of plaster mixes, and big thicknesses can only be built up slowly.

Plaster muffles are unsuitable and would wear away too quickly. A sheet iron profile having a projection 6 mm greater than the original is nailed on the running mould until coring out of all the lengths is complete.

The coring out is applied in thin coats, passing over the mould regularly to ensure that thick places are not allowed to gather. Advantage must be taken of the maximum amount of suction from the background. Applications of thick coats of stiff stuff must be avoided as they will disrupt the material already applied. It is best to apply thin coats of fairly soft material and if no suction is available driers can be applied. This consists of a dry sand and cement mix, of the same proportion as the wet material being used. The dry mix can be thrown on to the wet surface or applied carefuly with a trowel or gauger. It is important that the driers should be allowed time to absorb moisture from the wet material previously applied, before further coats are added.

When the coring out is complete it should be well scratched to provide key for the finishing coat and allowed to harden until the following day.

It is advisable to complete all the coring out, including breaks and returns, before commencing the finishing coat.

The finishing coat should be two parts well graded washed sand to one part Portland cement, passed through a 3 mm sieve to keep out small stones or coarse grit. This mix is usually divided into small batches which are gauged with water as required to take advantage of the initial set, and to leave available sufficient driers for use later. The finishing coat should be applied fairly soft in consistency, in thin, tight coats, passing the running mould over regularly to cut off surplus material. Any wet places lacking suction can be dried off and filled out as previously described.

When the outline of all the cornice members has been filled out a final slurry of sand and cement is applied or thrown over the surface, with the toe of the trowel. Immediately the wet slurry has been applied it is followed by throwing on a coat of driers after which the mould is passed over. The object is to achieve a fine sandstone type of finish. Any torn arrisses or slight irregularities can be filled in, and touched up, with small plain or curved floats.

Because of the necessity of forming all external and internal mitres as the work proceeds, the rules must be fixed to project beyond *each* external angle. This is achieved by the use of a halving joint where the rules intersect. Notches are cut for clearance of the rebated stick nailed to the underside of the slipper. It is also necessary to plane off a portion of the underside of the slipper to enable it to pass over the projecting rule (Figure 144).

Figure 144: Notched rules at external angles

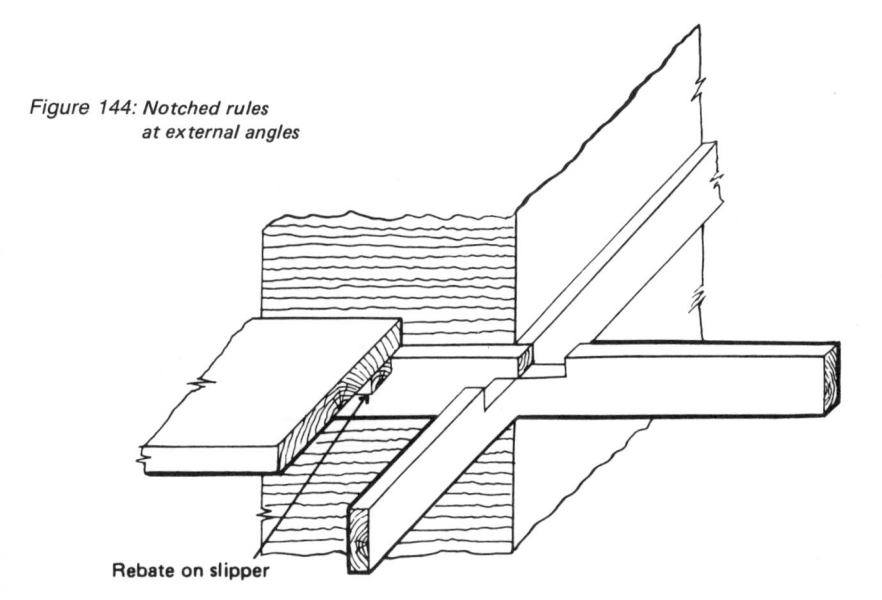

Rebate on slipper

Internal mitres are cored out and finished as the work proceeds by use of boxwood joint rules, and touched up if required by small floats.

Breaks and returns must also be formed and filled out *in situ*. The shape of the break or return is marked on the run length of cornice with a pencil or nail. A plumb bob is suspended from the top member of the cornice and measurements taken for the projection of each member in turn. Next a plumb line is marked on the face of the cornice and the projections transferred. By connecting up the points marked the outline of the break or return can be made.

This shape can now be cut with a small tool and the rest of the break or return cut back or filled in square to the run of the cornice. A timber square with a broad flat edge can be held to each member in turn and a joint rule traversed from its face. This method ensures that the break or return is kept square, and again the final touching up can be carried out with the aid of small floats (Figure 145).

ARCHES

Arches may be single or double sided, plain or moulded.

Plain arches with a square arris may be formed by the use of a template cut to the shape of the arch outline. Hardboard arch templates can be nailed to both sides of a double sided arch. Care must be taken to ensure that the templates are fixed linable with the springing line and centres on each of the two faces of the arch. The arch soffit can now be cored out and skimmed after first ruling in from the edges of each template. When the soffit has been finished the templates can be removed, and both faces of the arch skimmed using the arrises already formed to leave a good arch outline.

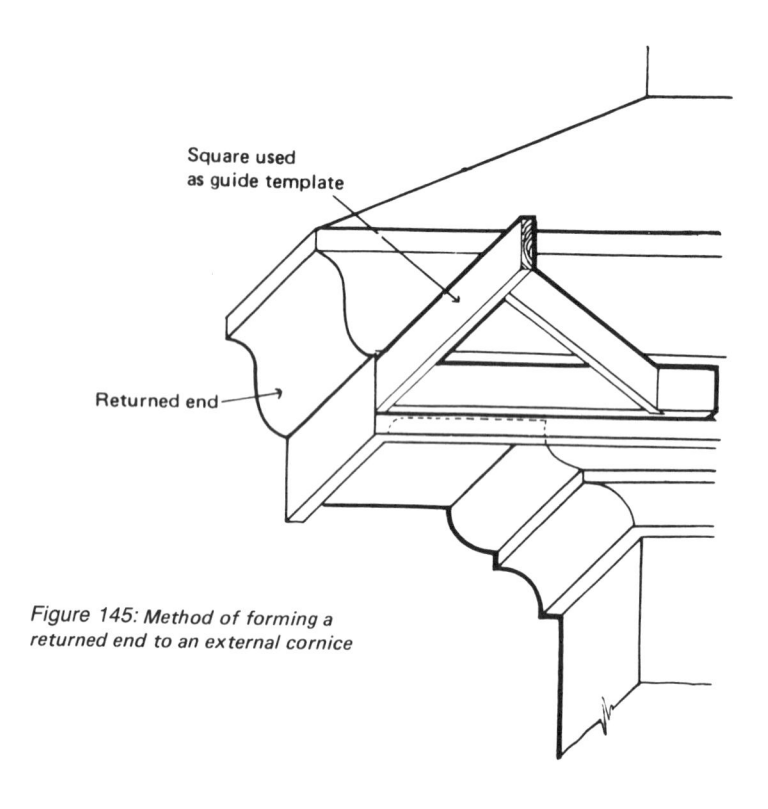

Figure 145: Method of forming a returned end to an external cornice

Expanded metal arch forms

These are preformed arches in expanded metal lath with a nosing bead guide to the required arch outline. The types available are semi-circular, bull's eye, elliptical and spandrel; the latter is a Tudor arch. The arch forms are supplied in two, four, six or eight parts, depending upon type. They are available in sizes to suit a number of opening widths up to a maximum of 3048 mm, except bull's eye window arch forms which have a radius of 229 mm only.

Semi-circular arch forms are nailed to each side of the opening, using masonry nails. Galvanised wire ties are used to unite the two sections around the soffit. If the wall thickness is greater than 230 mm then a strip of metal lath is used to bridge the gap. Bull's eye arch forms are fixed similarly but in four parts.

Arch forms for wide openings need to be in several pieces and the pieces are aligned by first inserting a 40 mm length of plastic dowel in the nosing of the fixed part. The nosing is then lightly pinched to secure the dowel until the next section is slotted over the dowel to be tightened in a similar manner. The soffits are then fastened as described previously.

The nosing of the arch forms should project 12 mm from the wall face to allow adequate coverage for subsequent plastering later. The recommended

ARCHES

rendering and floating coats are metal lathing plaster. The rendering should be applied about 6 mm thick above the surface of the lath and then scratched deeply for key. The floating coat should be filled out to the arch line nosing and then, when set, scraped back lightly to give skimming clearance, also keying for the finish plaster application.

Another type of ganvanised preformed arch frames, known as Truline, includes Spanish, Gothic, Arabian and Indian arch styles in addition to semi-circular and bull's eye types.

Moulded arches

Moulded arches can be run with the aid of a gig stick (single or double pivot), peg mould or trammel. Examples of each will be given.

In single sided archways, such as window openings, it is often possible to position the centre or centres to be used in the existing frame.

Open or double sided archways are much more difficult and a suitable framework must be fixed temporarily inside the opening. As it is also usual to

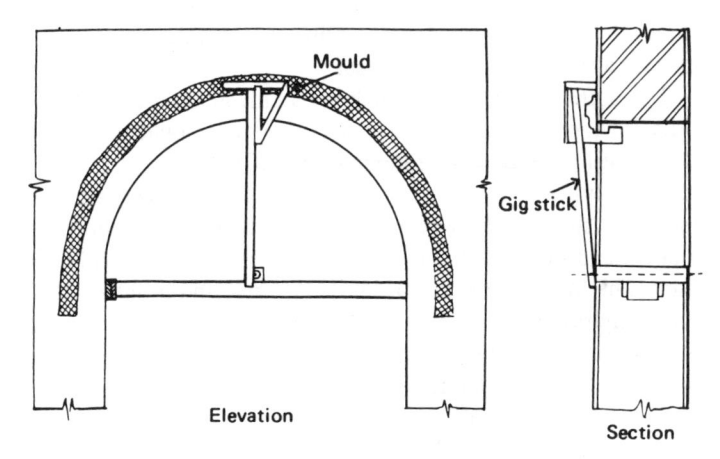

Figure 146: Semi-circular arches

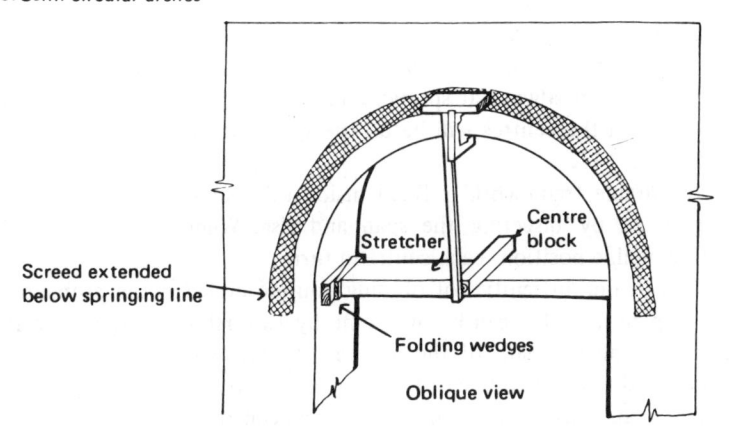

run the two arch outlines separately, then great care must be taken in ensuring that the two outlines are level, parallel and square. It is usual to fill in the arch soffits after the archivolts (arch mouldings) have been run.

In the example given in Figure 146 of a double sided semi-circular arch it will be noted that a stretcher of 110 mm x 75 mm timber is wedged about 25 mm below the springing line in a position midway between the two wall faces. The centre block is nailed securely at right angles to the stretcher at its mid point. The centre point is found by levelling through from the springing line and measuring the mid point of the span. A headless nail can now be driven into this point for use as the centre. The position of the centre for the second face of the arch can be found by levelling across the centre block and by squaring from the stretcher which must also be parallel with the two wall faces.

Vertical screeds should first be formed in line with the span and when straight and set they can be used as guides to rule in the arch screeds. The gig stick and mould should be set up as in the sketch taking care to adjust the mould to the correct rake and radius.

Mouldings of large sections should be cored out first. The archivolt moulding should be run below the springing line at each side of the arch and trimmed off exactly level when finished and set.

The soffit is filled in when both archivolt mouldings have been run. It is usual to include a skimming member in the mould profile nib to provide a clean finishing edge when filling in the soffit.

Other types of arch suitable for running with a gig stick include the segmental, horse shoe and bull's eye. The first two are merely shorter or longer arcs of a circle and methods similar to that described for the semi-circular arch can be used. The bull's eye arch is a complete circle and problems may arise regarding the withdrawal of the mould during the running process. Archivolt profiles with undercut members may require the mould to be withdrawn for cleaning, and at the finish, at one place only; and this place to be touched up later. Occasionally the mould can be withdrawn without damage to the run archivolt moulding by removing the gig stick at its pivot point first and then sliding the running mould clear.

Arch mouldings run from two centres, as with Gothic or Moorish arches, can be run in a similar manner with certain adaptations. The two centres of a Gothic arch are always at springing line level. In the case of a drop or obtuse Gothic arch the centres will be on the springing line *inside* the span, as shown in Figure 147.

A suitable framework is fixed inside the archway opening and the centres positioned by bisecting the span and rise. Where the bisector crosses the springing line positions each centre in turn.

One of the difficulties of running pointed arches is the problem of forming the top mitre. This can be overcome by traversing the mould on its gig stick and marking the path of the nib near the crown of the arch where the nib overlaps the opposite side of the arch. Marks are also made outside the finished screed at the springing line and the vertical centre intersection.

ARCHES

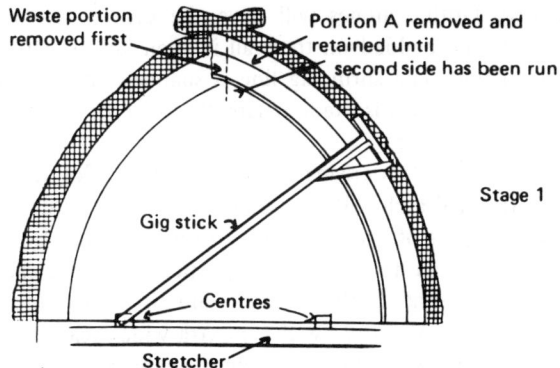

Figure 147: Obtuse Gothic arch

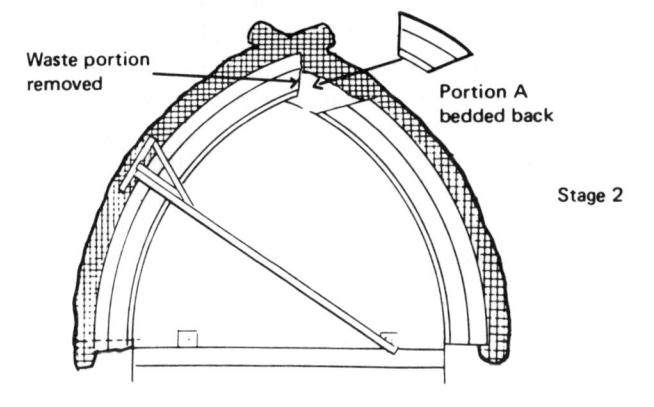

Between the nib clearance mark and the vertical centre line the surface to receive the archivolt moulding is lightly coated with putty lime. This side of the arch can now be run and finished. The surplus below the springing line and beyond the centre can be cut away. Next a saw cut can be made along the nib clearance line and the portion above carefully removed and retained for bedding back later.

The second half of the Gothic arch can now be run throughout its full length. Surplus material above and beyond the vertical centre line can be removed and the mitre insert previously removed bedded back to its original position. Any touching up required should be managed by aid of a small tool, because if done correctly the joints should only be the width of the original saw cuts.

Equilateral Gothic arches, where the radius is equal to the span, will probably require a small portion removed from just above the springing line in a manner similar to that described for the mitre inset. This is necessary to provide clearance for the second centre point after the first half of the arch has been run. Both upper and lower insets can be bedded back to their original positions after the second side of the arch has been run.

MOULDED WORK *IN SITU*

Lancet Gothic arches will have the centre points outside the arch span. These can be bedded on the wall in their appropriate position on the line of the springing. The springing line is snapped on the wall surface and then the span and rise are bisected. Where the bisector crosses the snapped chalk line positions the centre.

Another problem arising from this type of Gothic arch is that the first half of the finished archivolt moulding will often obstruct the gig stick for running the second half of the arch. This can be overcome by raising the centre blocks sufficiently high to allow the gig stick enough clearance to pass over the existing moulding. Suitable blocks are shown in Figure 148. It is important to bed the blocks firmly with an extra wide area of fixing material, to ensure that the blocks are not loosened during running.

Gig sticks with a double pivot point can be used to run compound curves. An example of this is shown in Figure 149. The pivot points are formed by cutting a V shaped notch instead of a hole as used in the previous examples.

The gig stick should be broad enough to allow the fish tail pivots to engage the pivot nails. This will require a cutaway clearance at the end of the gig stick to expose the profile or alternatively to use the gig stick as an extended stock or horse.

Figure 148: Lancet Gothic arch

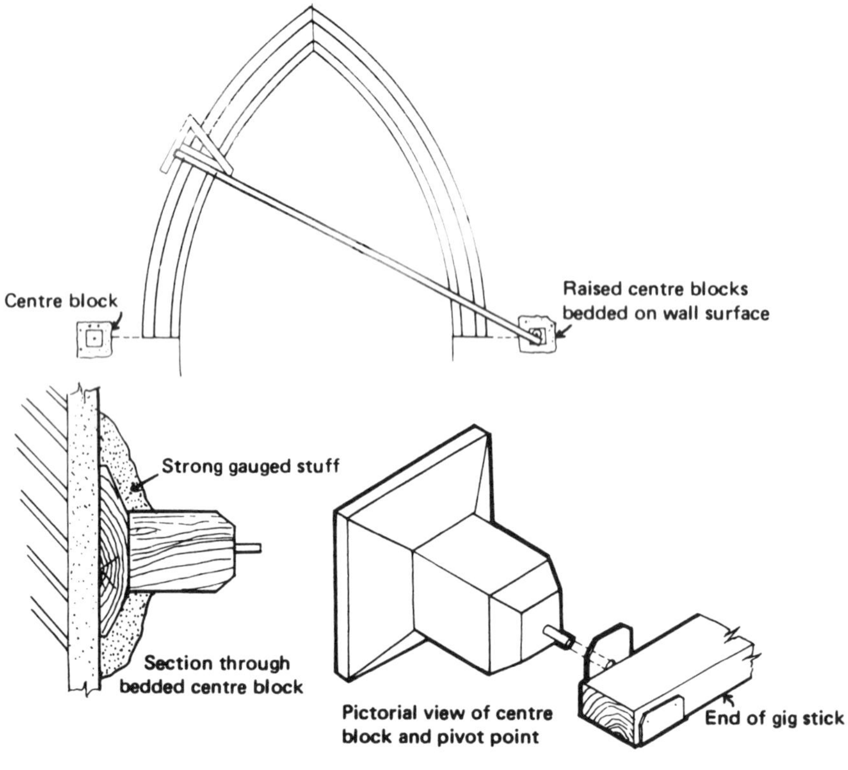

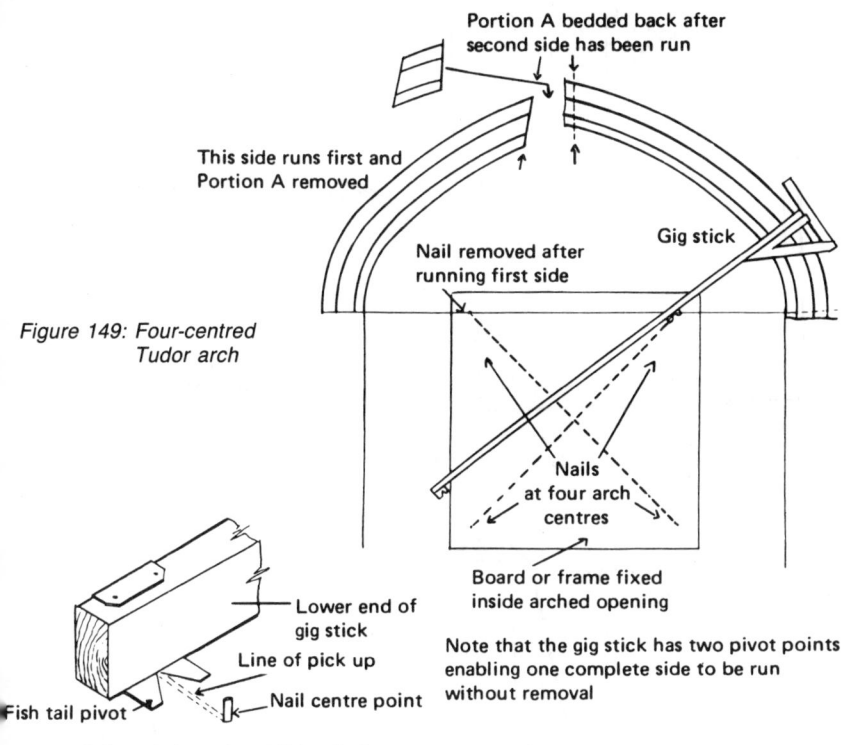

ARCHES

Figure 149: Four-centred Tudor arch

Labels on figure:
- Portion A bedded back after second side has been run
- This side runs first and Portion A removed
- Gig stick
- Nail removed after running first side
- Nails at four arch centres
- Board or frame fixed inside arched opening
- Note that the gig stick has two pivot points enabling one complete side to be run without removal
- Lower end of gig stick
- Line of pick up
- Nail centre point
- Fish tail pivot
- Enlarged view of end fish tail pivot essential for running a complete compound curve

In running the Tudor arch a plain board is fastened inside the archway opening and linable with the floated surface of the wall. On this board is marked out the position of the four centres. Round nails are driven into these points and left projecting about 12 mm.

Screeds are laid and the position of the top mitre inset, allowing clearance for running the second side of the arch, is marked. The area of the mitre inset is now lightly coated with putty lime to aid removal after running.

When running the first side of the arch the material is applied from just below the springing line to just beyond the centre. The mould is pivoted from the top centre first and as it swings through the arc the lower fish-tail pivot will engage and revolve on the lower centre nail. To allow the lower part of the gig stick to rotate freely, it will be necessary to remove the opposite upper nail until the first side has been completed.

On completion of the first side of the arch the waste portions beyond the centre and below the springing line are cut away and the top mitre inset removed carefully for refixing later. The mould will then need to be removed from the gig stick, the profile removed and re-nailed on the opposite side of the horse or stock. Also the fish-tail pivots will have to be reversed, and the mould nailed back to the gig stick with the profile in line with the pivot points.

171

MOULDED WORK *IN SITU*

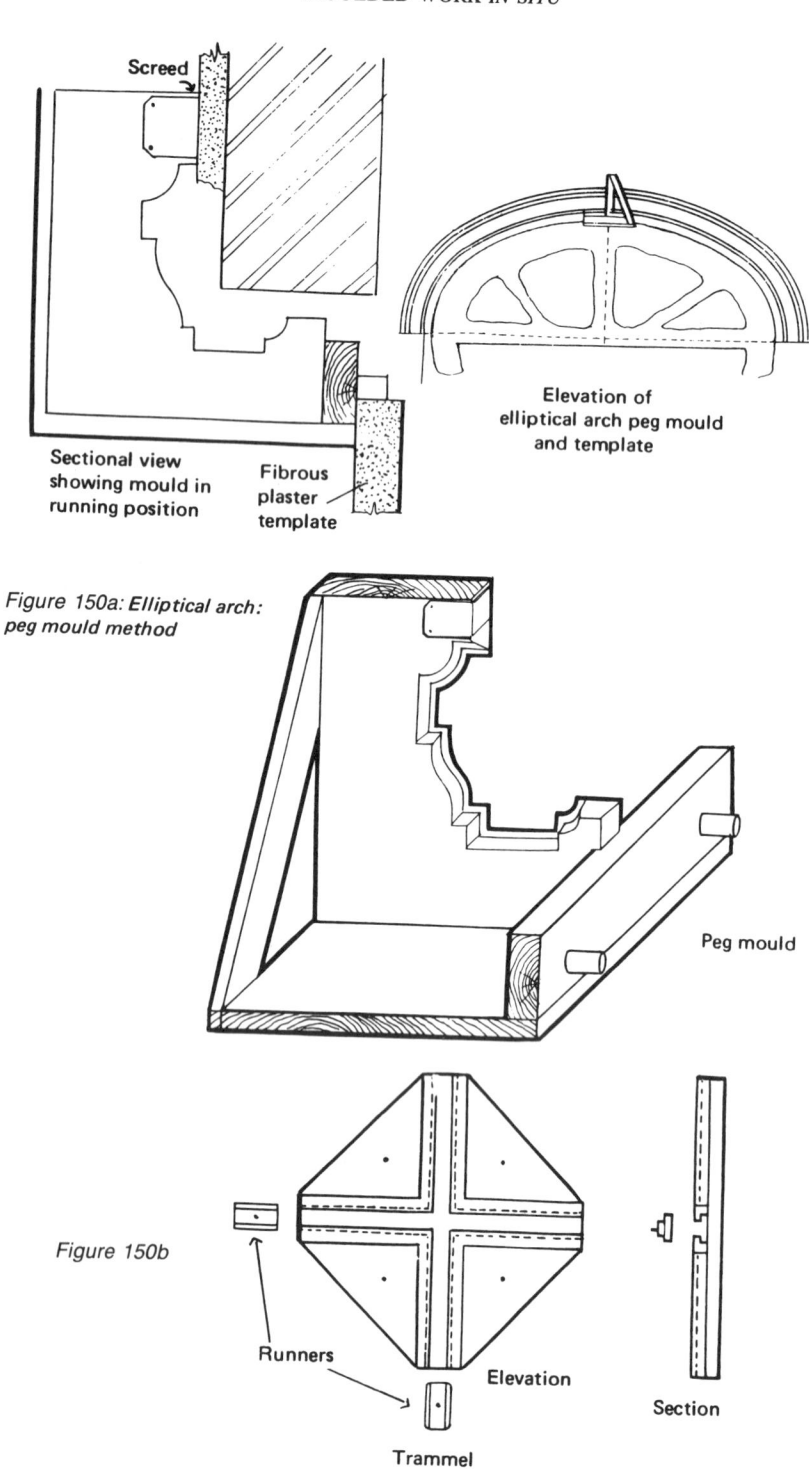

Screed

Elevation of elliptical arch peg mould and template

Sectional view showing mould in running position

Fibrous plaster template

Figure 150a: Elliptical arch: peg mould method

Peg mould

Figure 150b

Runners

Elevation

Section

Trammel

ARCHES

The second half can now be run after refixing the upper centre nail, and removing the opposite centre nail for gig stick clearance. When the moulding to the second half of the arch has been run the surplus material can again be cut away and the mitre inset bedded back to its original position.

Approximate elliptical arches are merely another type of compound curve and can be run with the aid of a gig stick. True elliptical arches can be run by the aid of a peg mould or a trammel.

The peg mould is a running mould having two wooden or steel pegs protruding from the ends of the slipper (Figure 150a). Many adaptations of this type of mould can be used according to the circumstances and requirements of each particular situation. Its use is based on the principle that a chord of a circle when bisected always gives a diameter. This means that when the two pegs of a peg mould bear against any circular curve the profile will always be radial to that particular curve. Similarly if the curve is elliptical or compound the profile of the peg mould will always be at a normal to the curve.

To run an elliptical arch with a peg mould, a template slightly smaller than the archway opening is required. This may be cut from timber or be cast in fibrous plaster against a curved lath on the bench. The template should be extended below the normal springing line to allow the extra bearing distance required for the peg below the line.

It will be seen from the elevation in Figure 150a that the template must be fixed below and inside the archway opening, leaving an equal margin around which the pegs will traverse. The peg mould will therefore bear forwards at the nib and slipper against the screed and template side respectively. At the same time the pegs bear down evenly on the curved surface of the template.

Running with a peg mould is usually easier than with the use of a gig stick and much better hand control will be experienced.

The other method of running an elliptical arch is by use of a trammel. This is a piece of apparatus in the form of a grooved cross on a board. An elevation and section of a trammel are shown in Figure 150b. The grooved recesses are usually formed by nailing fillets of timber to form the rebated channels as shown.

The size of the trammel required is determined by the difference between that of half the major axis and half the minor axis of the ellipse. In the case of a semi-elliptical arch, the minimum length of the trammel channels required is the difference in size between the rise and half the span plus 25 mm for the extra shoe traverse. For example, if the rise is 450 mm and the span 1.2 m, then the trammel channels need to be 600 mm minus 450 mm plus 25 mm equalling a minimum length of 175 mm (600 mm — 450 mm + 25 mm = 175 mm). So a trammel could be made on a board 375 mm square with the channels arranged as shown.

Two runners are made to a good sliding fit, and a trammel rod of suitable section cut to the length of half the span plus the length of the mould profile.

The trammel rod is fastened to the two runners with screws after first marking out the correct position on the rod. These positions are half the length of

MOULDED WORK *IN SITU*

Method of positioning runners to trammel

Figure 151: Elliptical arch: trammel method

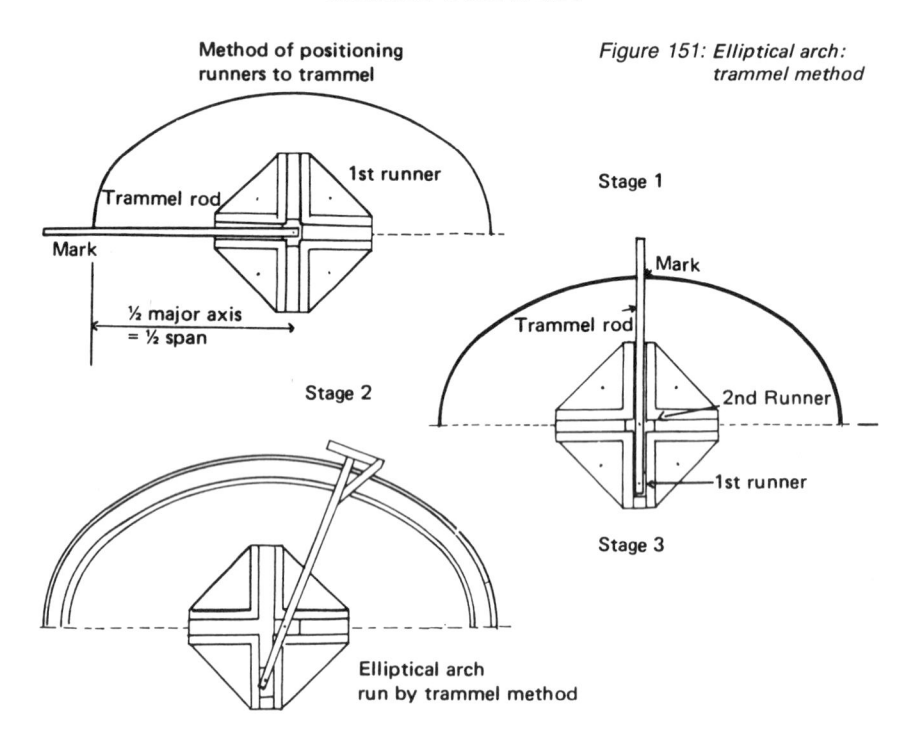

Trammel rod
Mark
1st runner
½ major axis = ½ span
Stage 1
Mark
Trammel rod
2nd Runner
1st runner
Stage 3
Stage 2
Elliptical arch run by trammel method

the span and the length of the rise from the inner edge of the archivolt moulding.

The trammel is nailed to a board or frame fixed centrally in the archway opening, linable with the wall face. The horizontal centre line of the trammel must be level with the springing line, with the vertical channels of the trammel in the centre of the span.

The running mould is held in the correct position at the springing line and the trammel rod nailed to the mould at one end. At its other end the trammel rod is screwed to a runner which is inside the vertical channel of the trammel at its centre point. The running mould is next traversed to the crown of the arch to the correct rise and the trammel rod screwed to the second runner. This runner must be in the horizontal channel of the trammel, and positioned at its centre point.

The mould can now be traversed around the outline of the arch, the horizontal and vertical runners sliding along their respective channels guiding the trammel rod to form a true elliptical curve for the path of the running mould (Figure 151).

ECCENTRIC RULE

When running a continuous compound curve it is not usually possible to use a peg mould, because the position of the pegs will cause the mould profile to move from the correct normal due to the differences in curvature. Running by

ECCENTRIC RULE

the eccentric rule method shown in Figures 152 and 153 requires two rules, one to follow the desired curve and the other plotted by setting out an eccentric curve drawn from a series of points positioned as follows.

A running mould is used and braced as shown in Figure 152. The inside moulding curve is set out and a series of normals projected from the respective centres of the different curves. The running mould is held with the profile on each normal in turn with the inner edge of the moulding section touching the inside moulding curve. A mark is placed at the heel of the slipper and this is repeated until the eccentric curve is plotted, Figure 153.

Figure 152: Sketch of running mould for eccentric rule

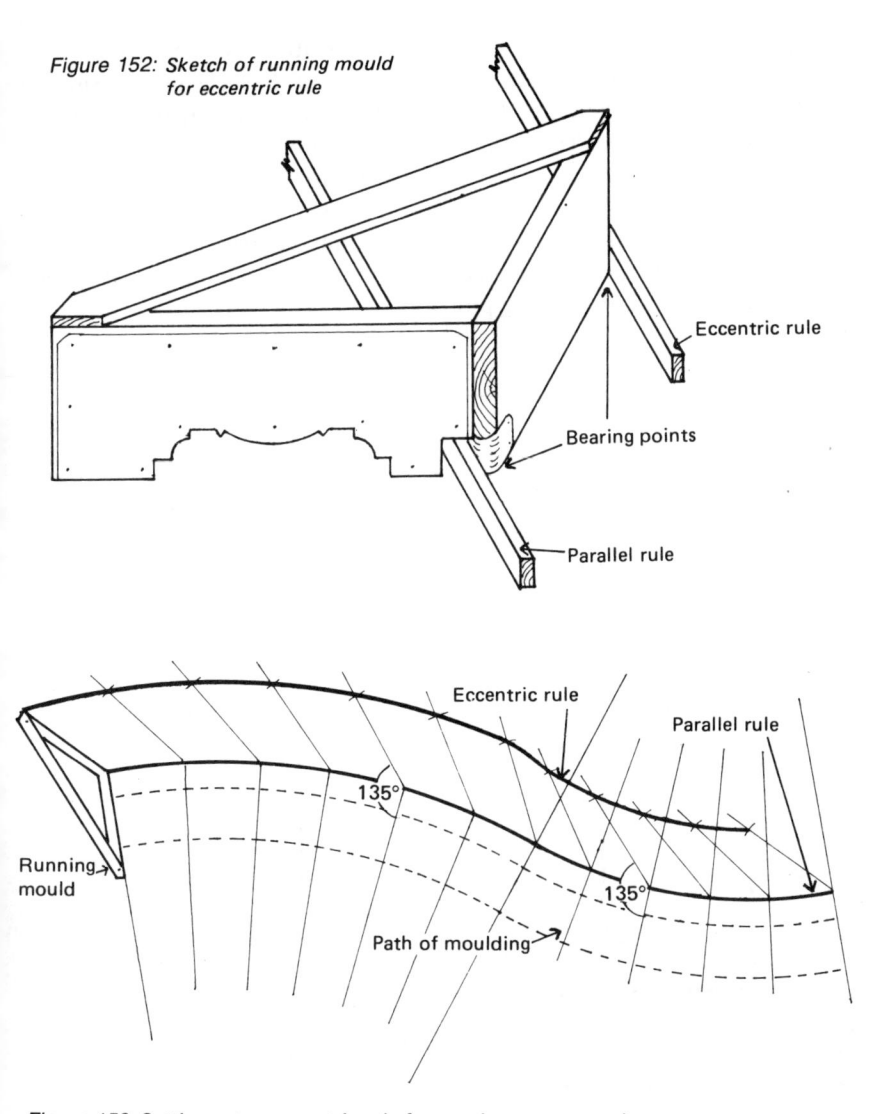

Figure 153: Setting out an eccentric rule for running a compound curve

175

MOULDED WORK *IN SITU*

Rules are nailed to the line of the two curves. The eccentric rule may need to be of square sectioned malleable metal to bend to the awkwardly shaped portion of the curve, or the shape marked on flat timber sections and cut to shape before nailing.

During the running process the mould is held forward against the inner rule and outwards against the eccentric rule. Bearing points on the mould and rules can be greased to reduce friction and prevent chattering.

NICHES *IN SITU*

A profile is cut to the desired vertical section of the niche and horsed up to a frame as shown in Figure 154. Strong, tightly fitting bolts are fixed to the top and bottom of the template at the centre line of the niche profile. Suitable brackets with socket holes, or hasps, are fixed to the wall at the crown and centre line of the niche. The sockets or hasps should be a good fit to the bolts.

The template pivoted from the sockets is turned in much the same way as opening a door through 180°. A muffle is first formed over the profile and the niche is cored out with the plaster or cement mix desired. The core is keyed with a wire scratcher, the mould muffle removed and the finishing coat applied and cut to shape with the turning profile.

Figure 154: Niches in situ

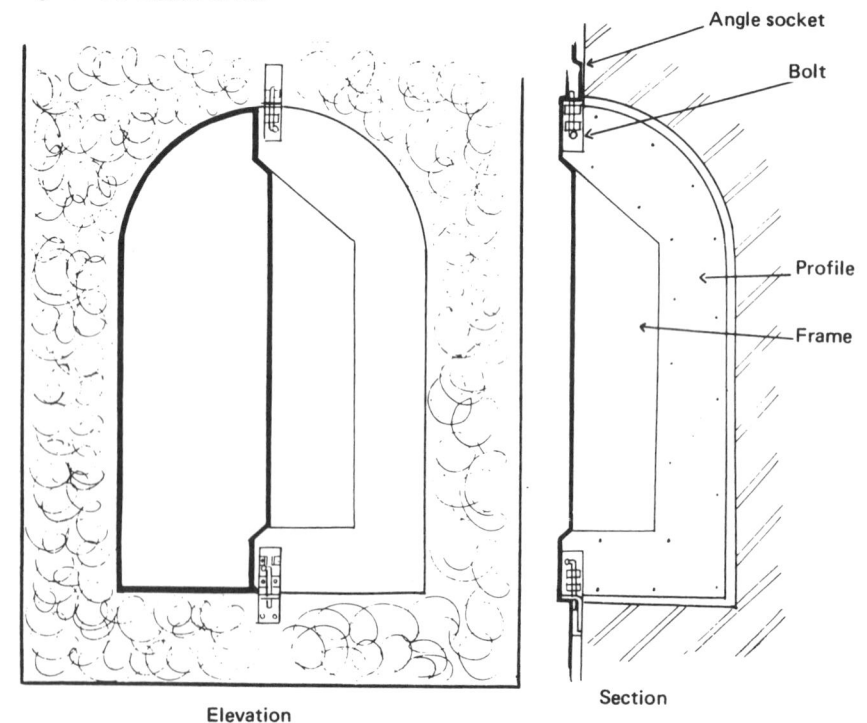

NICHES IN SITU

It is important that the floating on the wall face is plumb and linable. As the niche recess is being turned the outline arris is formed by cutting away any material gathering on beyond the wall face. This surplus material should be cut back with the blade of the trowel *away* from the arris. When the inside surface of the niche has been finished right up to the arris outline, then the wall face can be skimmed using the arris already formed as a skimming member.

Another method of forming niches with a semi-circular hood *in situ* is as shown in Figure 155, the sequence being as follows.

Strongly gauged putty lime and plaster screeds are formed on the wall surface around the niche. A centre line screed 50 mm wide is formed inside the shaft using a rebated template or gauge from the wall screed to ensure that the inside screed is parallel to the screed on the wall surface surrounding the niche. The pivot point (headless nail) is fixed at the centre point of the hood springing line. A turning mould having a profile of half the shaft section is used to spin the hood from the pivot point, see Stage 1 Figure 155a.

To fix the vertical running rules the mould is held horizontally from the pivot point and a mark made on the wall screed against the outside edge of the mould slipper. The rule is nailed plumb to this mark, and repeated for the rule on the opposite side. Next the pivot point is removed and the same mould can be used to run each half of the shaft in turn, the nib bearing on the centre screed. It is advisable to put a nail or block on the top of the screed to prevent the mould running past the springing line of the hood.

The shelf can be formed by running a length of suitable moulding returning the ends and planting level at the base. Finally the shelf is filled in, ruled off and finished to the top surface of the moulding.

Not all niches can be carried out by the previous two methods, a niche with an elliptical shaft and hood requires a different approach. The method used is known as the horizontal running mould system. The elliptical outline is set out on the profile and horsed up as shown in Figure 156. You will notice that the method uses a double slipper mould. Lime putty and plaster screeds are applied to the face of the wall where the mould is run, including an allowance for the running rules. The position of the niche is set out and the run of the left hand slipper marked plumb. Fix the left hand rule and offer up the running mould, marking the position for the right hand rule and from the springing line of the hood the extreme point of both slippers. Fix the right hand rule and fix blocks on the extreme slipper runs as shown in Figure 156a. A muffle is fixed to the mould, the core is run out in cement or plaster gauged coarse stuff. The muffle is removed from the mould and the finish run in lime putty and plaster. The method of running both the core and finish are the same. The mould is run up vertically until it strikes the end blocks, at this point the handle on the stock is pulled up and turns on the hinges. The hood is then turned out, see Figure 156b. On the final run the profile can be turned just back from the edge to form the hood arris. If carried out single handed a prop should be used to hold the mould in a temporary position against

MOULDED WORK *IN SITU*

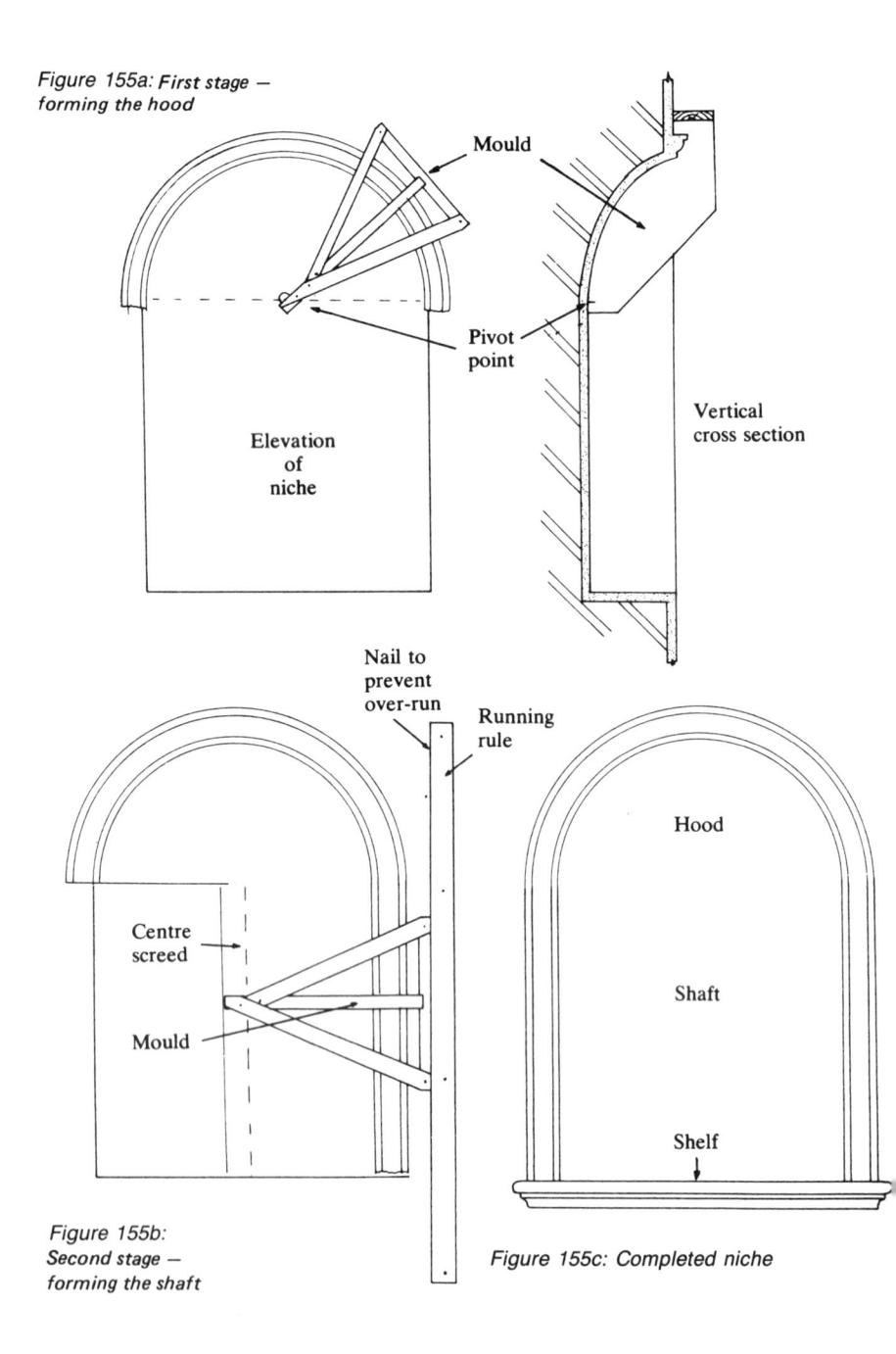

Figure 155a: First stage — forming the hood

Mould

Pivot point

Elevation of niche

Vertical cross section

Nail to prevent over-run

Running rule

Centre screed

Mould

Hood

Shaft

Shelf

Figure 155b: Second stage — forming the shaft

Figure 155c: Completed niche

178

(a) *Running the shaft*

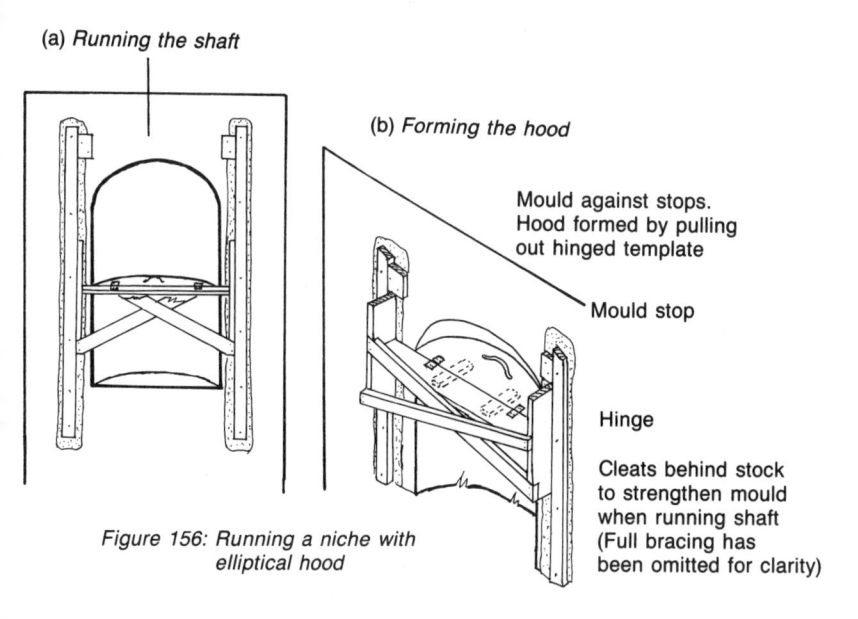

(b) *Forming the hood*

Mould against stops.
Hood formed by pulling
out hinged template

Mould stop

Hinge

Cleats behind stock
to strengthen mould
when running shaft
(Full bracing has
been omitted for clarity)

Figure 156: Running a niche with elliptical hood

the blocks on the hood line. If two operatives are working, one holds the mould while the other turns out the hood.

CLASSICAL ORDERS OF ARCHITECTURE

'Order' in architecture consists of a column plus entablature. The original Classical orders were three Greek types – Doric, Ionic and Corinthian. Later the Romans adapted these and added two other types Tuscan and Composite, (making five Roman types in all).

Each type is proportioned throughout by comparison to the lower diameter of its column shaft. This diameter is further divided into two which makes half a diameter equal to one module. The module is sub-divided into thirty parts or minutes. All projections and depths of the various orders, including mouldings, are proportioned from the lower diameter of the column and its sub-divisions or parts.

A column consists of base, shaft and capital. The column shafts are normally parallel for one-third the height and then entasised and diminished to five-sixths of the lower diameter.

The Tuscan column shaft is six diameters in height.
The Doric column shaft is seven diameters high.
The Ionic column shaft is eight diameters high.
The Corinthian and Composite shafts are nine diameters high.

A drawing of the principal characteristics of the Tuscan order with the main proportions is shown Figure 157. Included also is a pedestal which is an optional feature.

MOULDED WORK *IN SITU*

Entablature	2	Cornice	3/4
		Frieze	3/4
		Architrave	1/2
Column	7	Capital	1/2
		Shaft	6
		Base	1/2
Pedestal	2¼	Capital	1/4
		Die	5/4
		Base	3/4

TUSCAN ORDER

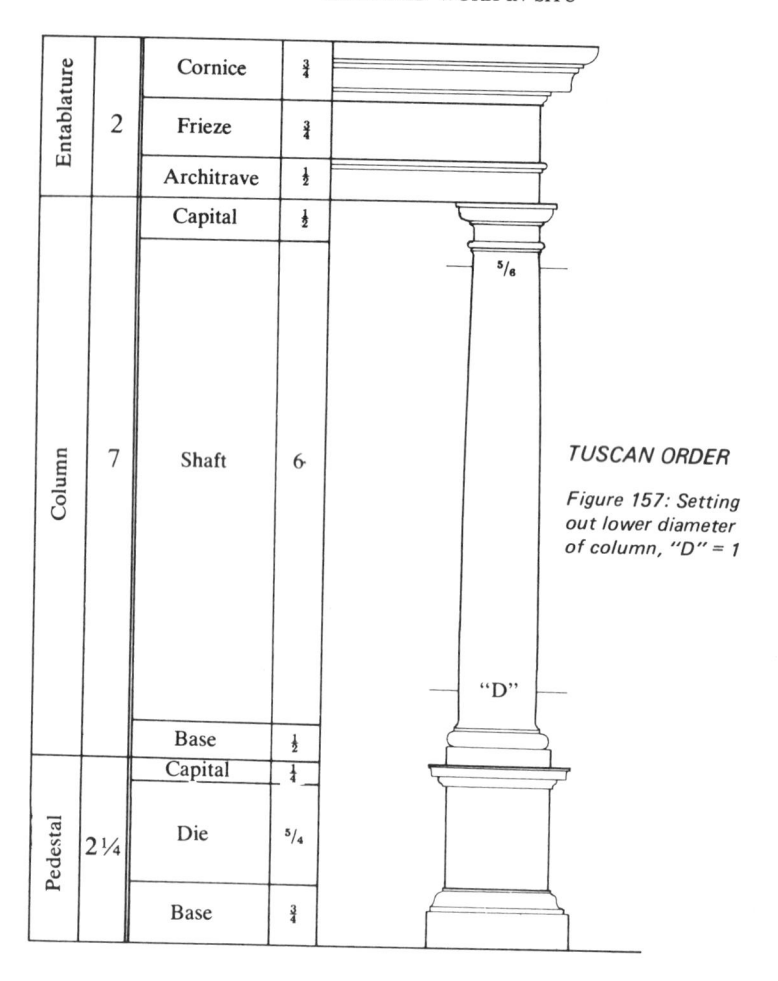

Figure 157: Setting out lower diameter of column, "D" = 1

For further details of the orders, including moulding members and proportions, a book on the classical orders of architecture should be consulted.

Two methods of forming the entasis to column shafts are shown. Figures 158 and 159.

ENTASIS TO COLUMNS – CUTTING ARC METHOD

The column centre line, also the upper and lower diameters are drawn. An arc with radius equal to half the lower diameter is drawn from the left hand edge of the upper diameter until it cuts the line at 'A' Figure 158.

A line is drawn from the left hand edge of the upper diameter through point 'A' until it meets the base line produced (of the entasised portion). This will establish the position of the polar point from which radiating lines can be drawn through the centre line of the column shaft.

Points on the entasis can now be established by marking along the

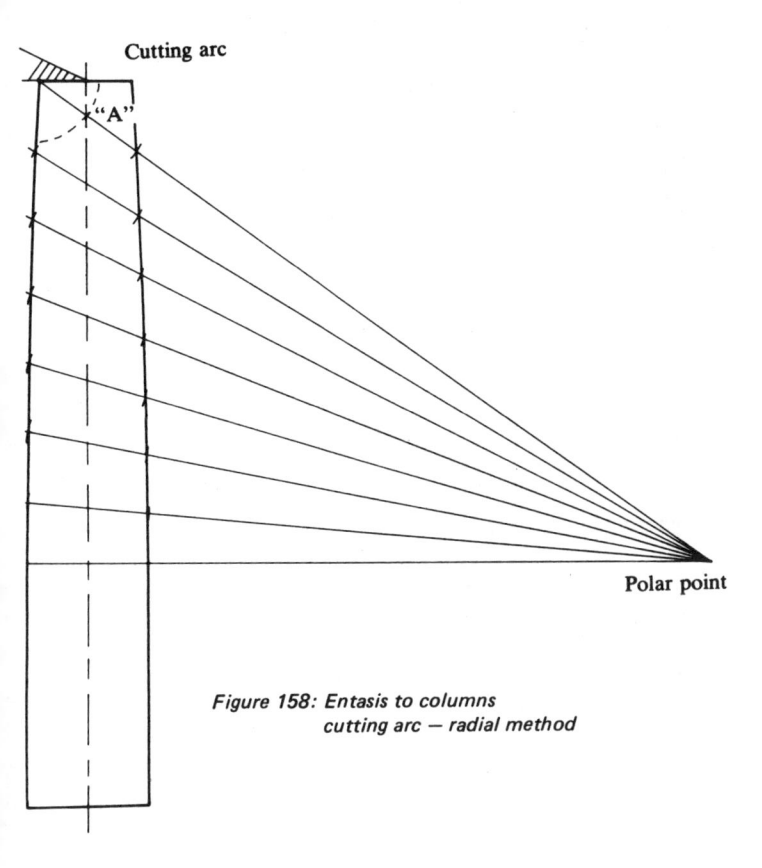

Figure 158: Entasis to columns cutting arc — radial method

radiating lines, in each direction from the column centre line, a distance equal to half the lower diameter.

ENTASIS TO COLUMNS – SEMI-CIRCULAR METHOD

Entasised columns are usually parallel for one-third the height. The upper diameter in the Tuscan order is five-sixths the width of the lower diameter.

A centre line is drawn and divided into six equal parts. The lower two parts will normally be parallel. On the base line of the entasised portion a semi-circle is drawn with radius equal to half the lower diameter. A distance of five-sixths of the radius is measured along the base line from the centre and a vertical line drawn to the top of the column shaft. The portion of the semi-circle to the left of the vertical line is divided into four equal parts.

Vertical lines are erected from these points to intersect with the relevant vertical divisions of the column shaft. These new intersections form the outline of the entasis.

The opposite side of the shaft outline can be found by transferring measurements across from the vertical centre line Figure 159.

MOULDED WORK *IN SITU*

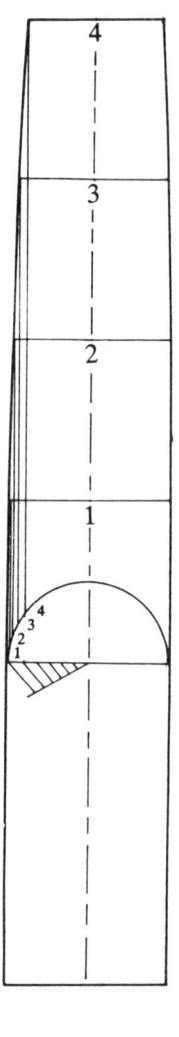

Figure 159: Entasis to columns semi-circular method

COLUMNS *IN SITU*

Columns may be square or circular, parallel or entasised, plain or fluted.

In most cases a column will occur under a beam and if it is to be a classical column it will need to be done to the correct proportions of its particular order.

Assuming that the column to be plastered is to be a plain Tuscan column without a pedestal the setting out is as follows.

Firstly the floor to beam soffit height is measured and divided into seven equal parts. This is the width of the lower diameter. Drawings are made of the cap and base to correct size and proportions for each moulding member. These proportions can be obtained from a book on classical orders of architecture.

COLUMNS *IN SITU*

The drawings are transferred to sheet metal, cut to shape, filed and horsed up as circular running moulds. The base mould is turned on a bench from a raised centre pin and the radius will be that of half the lower diameter. It should be run on a core to allow for a finished shell of approximately 25 mm of neat plaster. The capital is run in the same manner, though usually it is run inverted, also it will have a radius of half the upper diameter. (The Tuscan column diminishes in width at the capital to five sixths of its lower diameter.) The square portions of the cap and base can be cast separately and wadded to the circular portions before fixing *in situ*.

An entasised rule equal to the length of the shaft plus 50 mm or more for traversing should be marked out and cut to the reverse shape of the column shaft outline.

Two neat plaster circular collars are run on the bench, one to the exact size of the lower diameter and one to the size of the upper diameter. The sectional size of the collars should be about 25 mm square (Figure 160).

The column core may be a steel stanchion, concrete or brickwork. If it is a steel stanchion it may need to be furred out, metal lathed and then rendered.

The collars are bedded at cap and base levels but these must be fixed in line with previously plumbed dots.

Firstly the width of the upper shaft diameter is deducted from the width of the beam soffit and this difference divided into two. This latter measurement is the distance from the beam cheek to each of the two upper side dots. The upper collar can now be sawn into two halves and bedded into position tight to the two positioned dots.

Plumb gauges are made with a difference in length to the shoulder equal to the difference of half the diminish of the column. (If the diminish was 50 mm, then one plumb gauge could be 50 mm to the shoulder and the other would be 75 mm.)

The lower diameter dots are now plumbed and bedded from at least three points after which the lower collar is sawn in two and bedded to the dots.

These collars are now used as screeds and the column can be floated, ruling off the screeds with the aid of the entasised rule. It is important to keep the rule vertical when ruling off and to use as little traverse as possible. Wooden segmental guides are sometimes used to keep the rule vertical but it is easier without this restriction and just as effective if the rule is handled carefully. The entasised rule should be muffled for skimming thickness later by nailing a thin lath over the portion of the rule between the two bearing edges (Figure 161a).

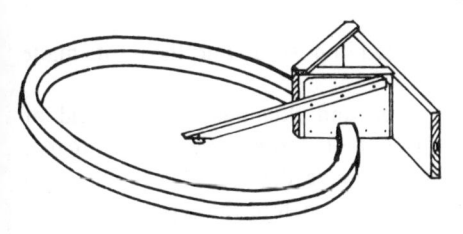

Figure 160: Forming the collar screeds

MOULDED WORK *IN SITU*

Figure 161a

Labels: Beam; Collar screed; Long gauge; Shaft core; Cap and base fitted last; Plumb line; Short gauge; Entasised rule

1. Plumbing the dots
2. Ruling the shaft

Figure 161b

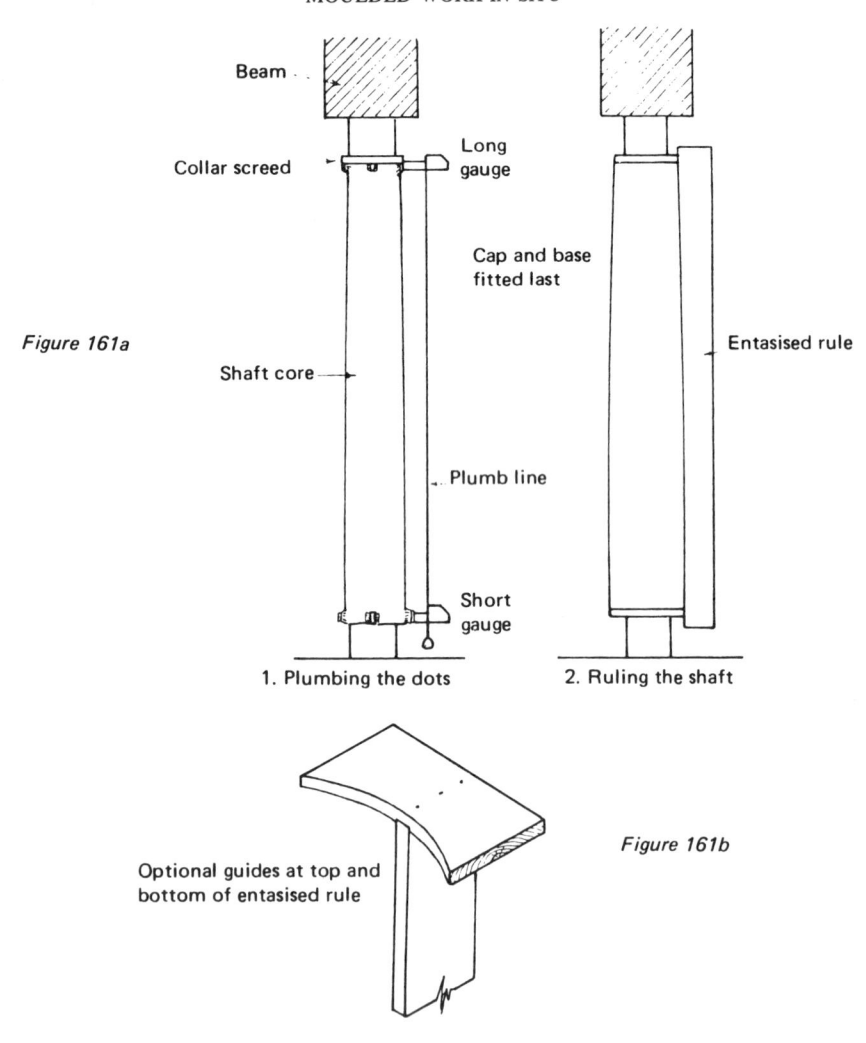

Optional guides at top and bottom of entasised rule

Some plasterers prefer to bed on the cap and base at this stage and to rule in the skimming from these moulded pieces.

More often the column shaft is skimmed first by removing the muffle on the rule and ruling in the skimming coat from the two collar screeds. The skimming is laid down by float, tightened in with the trowel and trowelled up to a good finish.

Cap and base are then sawn in two, bedded in position and the joints made good.

Circular column shafts which are floated and skimmed with the aid of segmental guides at the top and bottom of the entasised rule need a broader screed (Figure 161b). These can be formed above the collar at the cap and below the base collar, by ruling in off the collars before the segmental guides are nailed on. The false screeds will have to be removed after the shaft has been skimmed to provide fixing space for the cap and base.

ENTASISED FLUTED COLUMNS *IN SITU*

The flutes vary according to the type of architecture. Some meet at an arris, others are separated by a fillet. The number of flutes to a column is determined by the order of architecture to which it belongs, and may be either twenty, twenty-four or twenty-six.

The sections and setting out for the three types of flutes are shown in Figure 162. Fillets for the semi-circular flutes are usually one-third of the diameter.

Various methods of forming entasised fluted columns are used; this is determined in part by the size of the column shaft and the materials used.

Full-size sections or half the base and cap are made allowing bedding thickness for flutes. A sufficient length of fluting is run down on the bench for cap and base models. This will require two different sections of flute, with sufficient length to cut and plant about 100 mm long, around the perimeters of the cap and base half models. Returned ends are formed in the flutes and the models are moulded. Sufficient half casts of cap and base can then be obtained from the moulds using the material specified.

These casts are fixed carefully to dots previously plumbed making sure also that the flutes are also linable top and bottom. An entasised feather-edged rule with a 6 mm muffle is used to float the shaft ruling in from the upper and lower flutes. The muffle is removed and the finishing coat is applied and ruled off in a similar manner. Curved floats, cut to the sectional shape of the flutes at the upper, middle and lower portions of the entasised shaft, are used to rub up and fine down each individual flute. With sand and cement finishes the flutes and arrisses would be completed with the floats, but soft pliable busks or drags are used if a polished surface is desired.

Tall fluted columns may require intermediate fluted collars as bearings for ruling off and the entasised rule divided into two or more easily handled lengths.

Fluted collars can be made by cutting out the required half section in sheet metal and horsing to a running mould. The fluted half section can be run over a semi-circular core leaving a thickness of about 25 mm. Fluted half sections

Figure 162: Setting out for three types of flutes

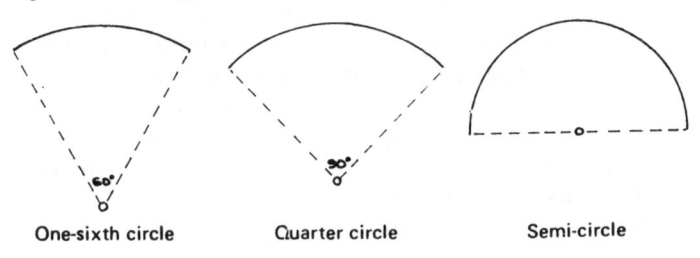

One-sixth circle   Quarter circle   Semi-circle

approximately 50 mm long are sawn off and bedded on the shaft core at the correct height, entasis and line. When the ruling in has been completed they are cut away and filled in with the finishing mix before completion.

In certain cases it may be possible to use fluted collars throughout and to fix the cap and base, also returned ends, afterwards.

REPAIR AND MAINTENANCE OF MOULDED AND ENRICHED SURFACES

Damaged *in-situ* moulded work, such as cornices can often be made good by use of a joint rule. If the damaged section is less than 300 mm then a joint rule of 400 mm or longer could be used to traverse over the missing or damaged portion. The length or area to be repaired should first be trimmed back to a clean sound piecing, and well damped to kill excessive suction. With normal shallow sections it may only be necessary to carry out the repair by applying a strongly gauged mix of lime putty and plaster, ruling off each member in turn to bring out the damaged portion in line with the original. The joint rule should bear on the original cornice on each side of the repair gap, working away from the arris of each member in turn. Care must be taken to hold the joint rule parallel to the run of the cornice, otherwise curved members will be cut or formed out of shape. Usually two or three mixed applications will be required to fill out the damaged cornice section to bring out to the correct shape, alignment and finish. The first gauging is made the strongest and is slightly stiffer than usual mixes. Succeeding mixes are best made slightly weaker and softer.

Larger section cornice repairs will have to be cored out first. A suitable coring out mix is sand/lime mortar gauged strongly with plaster. Cornice repairs to bracketed cornices may require replacement brackets and lathwork covering before coring out.

If the damaged cornice is too long to be traversed or formed with a joint rule, then a running mould will have to be made to match the original cornice. The shape of the original cornice profile can be matched by using a variable profile former. This is an instrument composed of a series of steel tines, similar to the teeth on a comb, held between two metal bands which, in turn, are fastened by wing-nut screws. The wing-nuts are slackened and the loosened steel tines are pressed forward against the profile of the existing cornice to each member in turn. When contact has been established over the whole contour of the cornice, the wing-nuts are tightened and the shape of the profile formed at the extremity of the tines is transferred to a piece of sheet metal by drawing round the outline with a pencil or scriber.

The more usual method of obtaining the required profile is to cut the end of the existing cornice to a square stopped end with a chisel. A piece of sheet metal can then be placed flat against the cut end and the shape of the cornice profile marked out in pencil on to the metal.

The sheet metal profile for the required mould can now be cut out and filed.

REPAIR AND MAINTENANCE OF MOULDED AND ENRICHED SURFACES

It should be tried and matched against the original cornice before horsing up as for a normal cornice mould.

Screeds are next formed at wall and ceiling bearing points. These are bands of strongly gauged lime putty and plaster applied about 50 mm wide at the points mentioned. The screeds are ruled in off the original bearing edges at each end of the cornice gap, so that the mould, when traversed, will start and finish from, and to, correctly matched profiles.

When the screeds are finished and set, the mould is placed over the original cornice at one side of the gap and the position of the lower bearing edge of the slipper (or horse) marked on the wall surface. The position of the nib slipper is also marked on the ceiling screed. This procedure is repeated at the other end of the gap and chalk lines snapped, usually on the ceiling only. A running rule is nailed to the wall mark at each end of the gap, holding the mould on to the matched original cornice to ensure correct alignment. The running mould is then traversed along the rule, lifting or lowering the rule to match the mould nib against the chalk line and nailing the rule securely when correct. Additional plaster dabs against the nailed running rule will avoid any risk of movement when running the mould.

The cornice mould can now be run off in the normal way, any making good at the intersection of the old and new being carried out by use of joint rule and small tool.

If the damaged cornice contained an enrichment such as a bead and reel, or egg and dart, then this can be repaired or replaced by a clay squeeze method. A short length of the existing enrichment, say egg and dart, is cleaned and dusted with French chalk. A small batch of clay is 'wedged up' to a non-tacky pliable consistency and this is pressed carefully, piece by piece, into the enrichment. The minimum clay coverage of the enrichment should be about 12 mm. The exposed surface of the squeezed clay should be straightened and the plaster margin of the cornice adjacent to the clay brushed with a film of oil, grease or lime putty. A layer of plaster is next applied to the back of the clay squeeze to a thickness of approximately 12 mm — 15 mm. Any overlap on to the plaster cornice surround will not gain adhesion because of the separating film previously applied. Before the plaster has finally set, the back surface should be straightened and, after a further period of hardening, the case can be removed. The clay squeeze should be removed carefully, if possible still retained in the plaster case. When the clay has to be removed separately it should be replaced in its correct position inside the plaster case. A certain amount of touching up may be required at this stage, after which the inside of the clay squeeze is filled with plaster, ruling in to a correct right angled edge from the plaster boundaries of the case.

The cast, when set, is removed from the case, and the ends cut and trimmed to ensure correct 'picking up' sequence when bedding later, i.e., one end should be a full dart and the other a full egg, each correctly shaped to match the next piece. The trimmed cast is then bedded into a right angled or L-shaped plaster trough as shown in Figure 163. Any necessary further touching up can be done at this stage, after which the trough and cast are soaked in water until sat-

Figure 163: L-shaped plaster trough with egg and dart bedded

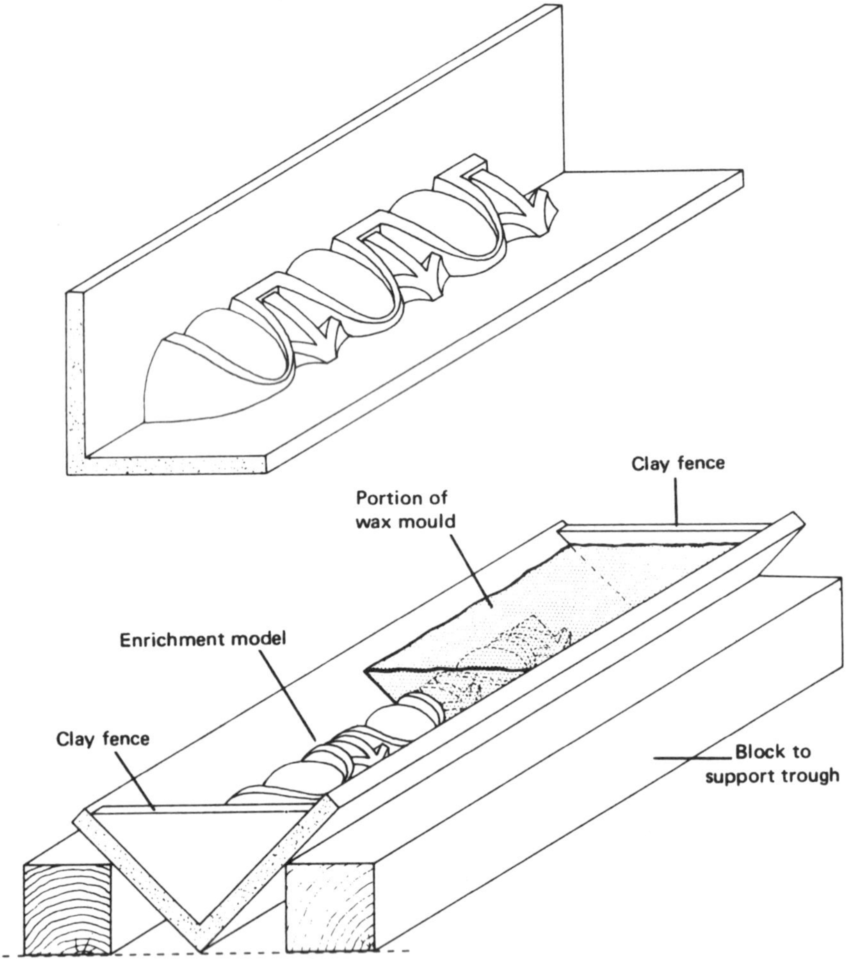

urated. Surplus water is then drained off, clay fences fixed at each open end of the trough and a wax mould made. See page 198 on wax moulding.

Sufficient plaster casts can be taken from the wax mould to be bedded in position after the cornice has been run. The cornice mould used must be adapted to provide a right angled bed to receive the egg and dart casts when the cornice has been run.

The egg and dart casts are bedded individually into the right angled 'bed' of the cornice, care being taken to ensure that the enrichment 'picks up' correctly at each intersection.

If there are a large number of casts to fix, it is best to mark the position of each cast in pencil on the length of the cornice before fixing. Any shortness or overlapping possibilities can be corrected in different ways. If the shortage to

REPAIR AND MAINTENANCE OF MOULDED AND ENRICHED SURFACES

be made good is 12 mm and there are twelve casts, then the joint in each cast can be opened up to 1 mm. After filling in and touching up, the extra width of the joint will probably be unnoticeable. Similarly, if the last cast is 12 mm too long then 1 mm can be rasped off each cast before bedding.

Another method of fixing enrichments to ensure correct picking up sequences is to use 'shrinkers and stretchers'. These are shorter or longer casts than the normal lengths. A normal egg and dart cast can be sawn through the centre of each egg in turn and the separate pieces reassembled and bedded in on an L-shaped plaster trough so that the new length is slightly shorter than the original. After touching up it is wax moulded in the normal way. A similar method is used to form the model for the 'stretchers' except that the joints between the separate pieces are opened marginally before re-bedding and touching up. This model is also wax moulded. When fixing, it is only necessary to use the required lengths from three different sizes to gain or lose position for a perfect final intersection.

Restoration of unsound existing work may have to be carried out by carefully sketching and dimensioning the pieces concerned, and then removing the damaged or unsound portions. Replacement pieces will then have to be made up by moulding or modelling from the information. If the repair is to be carried out *in situ* it may be possible to make running moulds to run parts in position and bedding others to complete the work. When the restoration is of fibrous plaster and includes modelled or scrolled enrichments, then a clay model facsimile of the original will have to be made before moulding, casting and finally fixing back as a replacement for the original unsound work.

CHAPTER TEN
Benchwork

Fibrous plastering is a form of precast plasterwork. The plaster casts are reinforced with open mesh hessian scrim and strengthened with timber laths. This results in strong casts that are light in weight. Another advantage of fibrous plastering is the saving in cost when reproducing several similar types of moulded work which can be cast from the same mould. These casts can be prepared in a workshop, as the building progresses separately. When the appropriate stage of the building is reached the casts can be fixed by means of galvanized nails and/or screws to timber joists or furring, and the joints wadded with scrim and stopped (filled in) and made good with plaster. In fireproof construction the casts are fixed by means of galvanized wire passed under the lath reinforcement of the fibrous cast and over the metal or fireproof framework. Further assistance in providing strength and rigidity of fixing is the use of plaster wads which are pieces of hessian scrim saturated in gauged plaster mixes of creamy consistency.

BENCHES

These are usually constructed *in situ* from timber with the bench top surfaced in plaster. Sizes vary according to the type of work undertaken but benches of 4 m in length by 1.5 m wide or greater would obviously have to be assembled in the workshop. The tops are sheeted in usually with rough boards or planks and a level framework fastened securely around the perimeter. This should project about 25 mm above the sheeting boards. The perimeter framework should also be carefully levelled.

The plaster top can be filled in a variety of ways and can be cored out with sand and plaster, held down with galvanized nails projecting 12 mm from the sheeting surface, or scrim can be used to strengthen the core which can also be neat plaster. The core surface should be deeply keyed with a wire scratcher, and then coated with a strong neat plaster mix. This finishing coat can now be straightened by ruling in with a straight edge from the top edge of the perimeter frame. When the plaster has set the surface should be scraped to a good finish with a drag and when dry, coated with several applications of shellac polish. Rules are normally nailed to the two long edges only.

Benches are occasionally constructed on brick pillars and the top sheeted in with hy-rib. After the perimeter fence has been fixed the hy-rib may be covered with coke breeze concrete and topped in the normal way or concreted and finished with a sand and cement granolithic surface. The latter types are used when specially hard wearing surfaces are essential.

Plaster topped benches when worn can be repaired or re-surfaced as necessary. This may merely mean cleaning out loose or worn patches, damping and refilling with a strong neat plaster mix. When dry the surface should again be sealed with several coats of shellac. Slightly worn or hollow places can be resurfaced by scraping off any remaining shellac, scratching for key, brushing clean and damping the surface well. A neat plaster mix can then be poured over the hollow surface and ruled in with a straight edge. When set the surface can be scraped to a good finish and again sealed with several coats of shellac.

Shellac used in benchwork is a solution of shellac flakes dissolved in methylated spirits. The function of shellac is to seal the surface of plaster models and moulds so that the grease, applied later as a separating film will not soak into the plaster.

When coating a plaster surface with shellac it is better to start with one or two very thin coats before trying to build up to a good surface. Each coat must be allowed to dry thoroughly. The thinner the solution of shellac, the greater its proportion of methylated spirits, and therefore the quicker the drying due to evaporation.

The number of coats required will be determined by the requirements of the job, some moulds may need to have up to six coats. The finishing coats should be of good strength shellac.

Neat plaster mixes are used in almost all benchwork jobs. Lime mixes are rarely, if ever, used. Neat mixes can still be made weak or strong by varying the proportion of plaster added to the water. Stiff mixes are difficult to use and cause excessive swelling. Thick creamy mixes are strong and are used to build up the main sections of the moulding required. Thinner, or more watery, mixes should be used to obtain an easier and better finish.

Plaster for any kind of cast work, wadding, or running moulds must be gauged with sufficient strength. A good guide is that it should be slightly thicker than the consistency of the cream on the top of a bottle of pasteurised milk. If it runs off the hand when mixed it is too thin and will also run down inclined or vertical surfaces when moulding or casting. Gaugings which are too thin will also run off the scrim when wadding.

When running mouldings on the bench it is important to grease the surface well if the moulding length is to be removed later. To prevent the moulding sliding off during the running process, small nails are inserted in the bench surface or at one end of the moulding run. The nails should be placed about midway in the intended section, with the heads protruding just less than half the height of the section. A small cone of clay can be formed around each nail head to ensure ease of removal for the moulding length when finished and set.

Occasionally lengths of moulding are intended to be used when fastened to the bench. In this case the nail heads are left exposed without the clay cones, and the set plaster under the nail heads provides a good anchorage and prevents easy removal of the moulding length.

Figure 164: Diminishing moulds

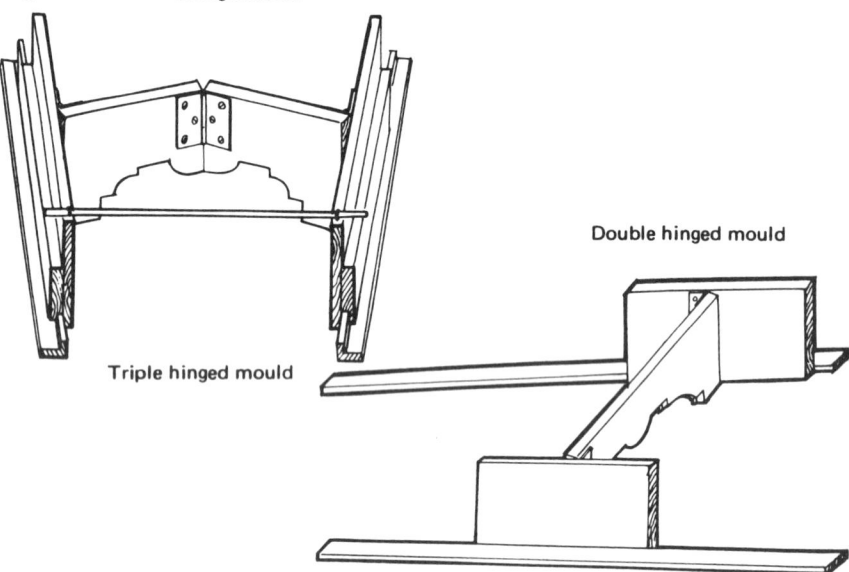

Double hinged mould

Triple hinged mould

When running mouldings of small section on the bench, it is best if the general shape can be filled out with the first gauging. Subsequent gaugings can be thinner and when fully filled out the moulding length can be run over finally with a very weak mix. When building up the shape of the moulding it is important to feed the highest members of the section and to prevent gathering on at the lowest points or on the bench surface.

Larger sections require much more care and skill in bringing the moulding to a good finish. This is because neat plaster mixes swell when setting, and the swelling causes the mould profile to tighten during the latter part of the running process. When this occurs the running mould jumps or 'shudders' slightly causing disfigurement to the moulding. This is known as 'chattering' and once this occurs it is difficult to remove. One method of removing the jump marks is to run the mould in the reverse direction. This method is known as 'backing off'.

The safest method to avoid 'chattering' of the moulding is to muffle the running mould with 3 mm — 6 mm extra thickness on the profile. The moulding length can now be cored out and scratched for key. When finishing the moulding sufficient plaster should be mixed to cover the core and fill out all the members with one thin gauging. In this way all the members will be formed together and little swelling will occur.

'Chattering' of the running mould can also be avoided, or minimised, by inclining the profile and horse so that it is leaning slightly forward. This only needs to be the thickness of a pencil line out of square.

Mouldings of large section are often cored out with pieces of old set material. The pieces are arranged dry in the moulding space on the bench and the running mould passed over to ensure clearance. A thin plaster mix is poured over the coring out pieces which then become part of a type of plaster concrete. The running mould is passed over frequently to clear away any surplus material. When set, the moulding length can be cored out to the muffle profile in the normal way.

Figure 164 shows hinged diminishing moulds for running mouldings which diminish in width.

Circular mouldings run on the bench can be carried out as shown in Figure 122. To set up a running mould for a curve of given radius, a nail with the head removed can first be driven into the bench surface and left protruding about 12 mm. A straight line can be drawn with a pencil from the nail head. The length of the desired radius can now be measured from the nail along the line and the position of the radius marked.

The running mould should now be placed with its profile along the line. If the radius marked is for the internal radius of the curved moulding, then the first or inside member of the moulding profile is placed on this mark. (When the external radius is marked, the outside or last member is placed on the mark.)

A suitable gig stick is placed so that it rests on the slipper and 25 mm beyond the nail centre. Also it should be in contact with the metal profile of the mould. The outline of a splay cut at the foot of the gig stick can be marked by placing a piece of 12 mm thick timber along each side of the gig stick foot in turn and marking with a pencil. This will give a line around the foot of the gig stick parallel with the bench surface. When this line has been sawn through the pivot point can be nailed to the splay formed.

A hole should be punched through the pivot point, tight to the side of the gig stick. This hole should now be enlarged until it will pass over the centre nail to a good tight fit. Any burrs at the back of the pivot point should be smoothed off with a file (Figure 123).

The pivot point can now be placed over the nail centre and the other end of the gig stick nailed to the running mould at the correct radius and rake. When nailed correctly the running mould profile should be radial to the centre and at the correct radius. The pivot point should also be in line with the metal profile of the mould and the pivot point, nib and slipper of the mould all be in contact with the bench surface.

The points made with regard to running straight lengths of moulding also apply to curved lengths. Larger sections, or others which have been gauged incorrectly may swell outwards away from the centre. The solution to this problem is again to muffle the mould, core out and to complete the actual finishing in as few gaugings as possible with mixes that are not too stiff.

Moulding lengths similar to those described can be used to make up models of various features. The outline of the feature required is marked out and the pieces of moulding cut to shape, planted and the joints mitred and finished. The finished model may then be moulded by any one of a number of methods, and the required number of casts obtained from the finished mould. This is the model, mould and cast process.

MOULDING TECHNIQUES AND MATERIALS

The simplest type of moulding method is that of a *straight reverse mould*. This can only be done with models having no undercut parts and is only really suitable for models having shallow sections and splayed members. 'Vertical' members are often slightly splayed to ensure easier removal of the cast; this is known as 'giving it draught'. The model is placed flat on the bench, moulded face uppermost and given at least three coats of shellac. When dry the model is greased and then a thin mix of plaster is brushed and splashed over its surface; when almost set a second coat is applied in the same manner and scrim pressed into the soft mix to reinforce the mould. A further application of the second gauging is then applied plus a second layer of scrim. The mould is then strengthened round the edges by applying a thickness of 12 mm or more of the setting mix. The back of the mould is roughly straightened and when set the mould can be removed and given at least three coats of shellac. The mould is now the reverse shape of the model. After greasing the mould, casts can be taken in a manner similar to that described for the moulding process.

The reason for two separate gaugings described is to make sure that the scrim is not pressed through to the surface of the mould or cast. This is prevented by allowing the first coat to become hard enough to prevent the scrim passing through, and yet soft enough to ensure that the second coat will stick to the first without having to scratch it for key. This method of gauging is known as 'firstings and seconds'. Usually the second coat is retarded to give the plasterer sufficient time to form the mould or fill in the cast.

One gauge mixing is used when the normal two gauge system of 'firstings' and 'seconds' is not practical, such as deep narrow work. The mix is retarded but is gauged thicker than usual and a coat is brushed and splashed on in the normal way. Dry plaster is folded between layers of scrim to form an open meshed canvas dusting bag, and dry plaster is sifted over the recently applied wet surface. When the surface has firmed up sufficiently a further coat of the wet mix is applied and the canvas can be applied and the cast completed as in the two gauge system.

Piece moulding is a type of moulding process used when a straight reverse mould is not possible. The mould is in several separate pieces which are held together by a fibrous plaster case.

In the example given in Figure 165 the various stages of piece moulding a modillion are shown. Firstly the modillion model is mounted on a suitable moulding block, which in this case can be a flat plaster slab with a further slab at right angles. The modillion model and complete moulding ground should be given a minimum of three coats of shellac and allowed to dry thoroughly.

Each exposed side of the modillion is moulded in turn as follows. Firstly a band of clay about 18 mm thick and 37 mm wide is placed just behind the external angles on the top and side. The clay band should be positioned so that it projects at an angle of approximately 45 degrees.

The exposed side of the modillion and moulding base should be well greased and the rest of the model and moulding ground protected with paper. A thin

MOULDING TECHNIQUES AND MATERIALS

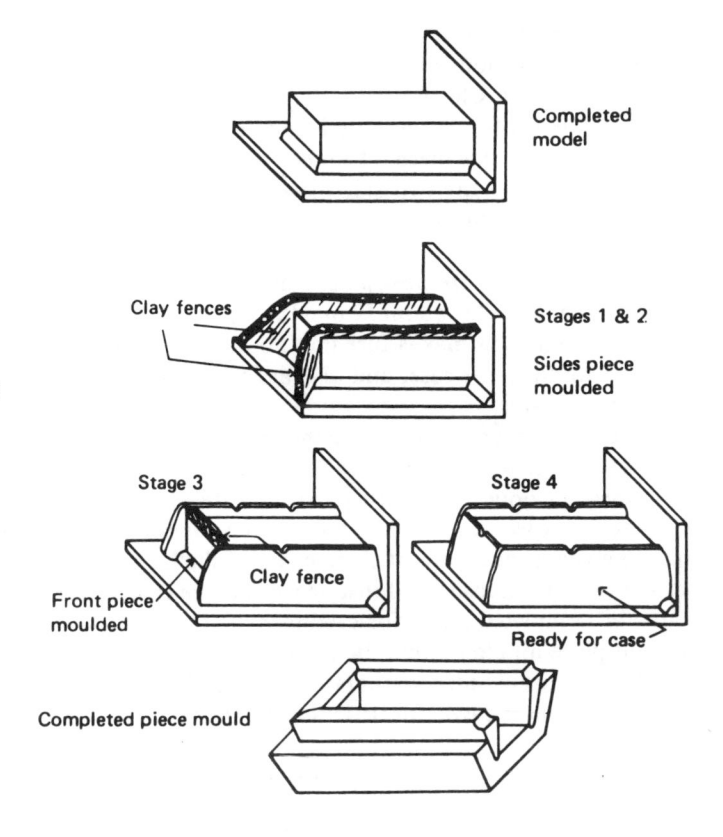

Figure 165

mix of neat plaster should now be brushed and splashed to the side of the modillion recently greased and as the mix sets it should be applied up to the full height of the clay band to a thickness of about 35 mm – 40 mm. The outer edge should be slightly chamfered inwards for ease of withdrawal later.

The clay band can be removed first and used again for the opposite side.

The side piece meanwhile should be allowed to set and harden after which it can be removed for trimming. The overlapping edges, formed beyond the external angles when on the model, can now be trimmed down accurately with a rasp or Surform type of plane. Top and side projections should be worked to a 45° chamfered ridge and the line of the ridge itself blunted to approximately 3 mm wide. One or two semi-circular joggle holes should be cut or formed with a half-round rasp on the upper edge of the ridge. The purpose of the joggle is to position the piece when the case has been formed so that unwanted movement of individual pieces will not occur.

When the moulded piece has been trimmed, three coats of shellac are applied and allowed to dry.

The opposite side piece is done in the same way, except that the joggle holes are cut differently to make easier the problem of identifying the position of each separate piece in the case, when this has been made.

After both side pieces have been completed and greased they are returned to their original positions on the modillion model and the front piece formed to the two side pieces. Clay is used to trap the two side pieces so that no movement is possible whilst the last side is moulded.

When set, the last side piece is trimmed at the top and back edge only. The edges in contact with the two side pieces should be left untouched. Joggle holes can again be formed in the top edge and the piece shellacked and greased as described.

All three side pieces are now placed in their correct positions around the model and a band of clay about 12 mm thick placed around the pieces on the moulding ground. This makes sure that no movement of the pieces will take place when forming the case, and also assists in easier removal of the case from the pieces later.

The case is now made by covering the exposed top of the modillion model and the side pieces with 'firstings and seconds' and strengthening with scrim (as described previously in making a straight reverse mould).

When set the case can be lifted off and the edges trimmed after which it should be given a minimum of three coats of shellac and then greased. The pieces can now be greased and inserted inside the case and the piece mould will now be completed ready for casting.

Casts in piece moulds should be removed when set by taking off the case first and then the side pieces separately. Many plasterers prefer to form the side pieces without removal for trimming during the piece moulding process.

WASTE MOULDS

These are used when only one cast is required, the mould being destroyed, and therefore wasted, in the process.

Usually waste moulding is used when a plaster cast is wanted of a clay model. The model itself is often an enrichment, or a special architectural feature or design in low relief made by the modeller.

Clay models in low relief will have one side only exposed, the rear, flat side being in contact with the modelling ground or base.

The clay if damp will require no pretreatment before waste moulding, but dry clay may want coating with shellac and oiling.

When waste moulding, the first coat of plaster applied to the clay model should be splashed on about 3 mm thick. It is essential that this first coat should be coloured and this may be done by mixing lime blue or other colouring pigment first into the gauging water. As soon as the first coat has set the exposed surface can be lightly greased, except for a few small areas about 25 mm square which will make sure that some adhesion between the first and second coats will take place. The second coat can now be applied about 25 mm thick and the top surface roughly straightened with a flat board. An alternative method is to well stipple the 'firstings' to provide key. When this coat has set a clay wash is brushed over the surface, then the 'seconds' applied as described.

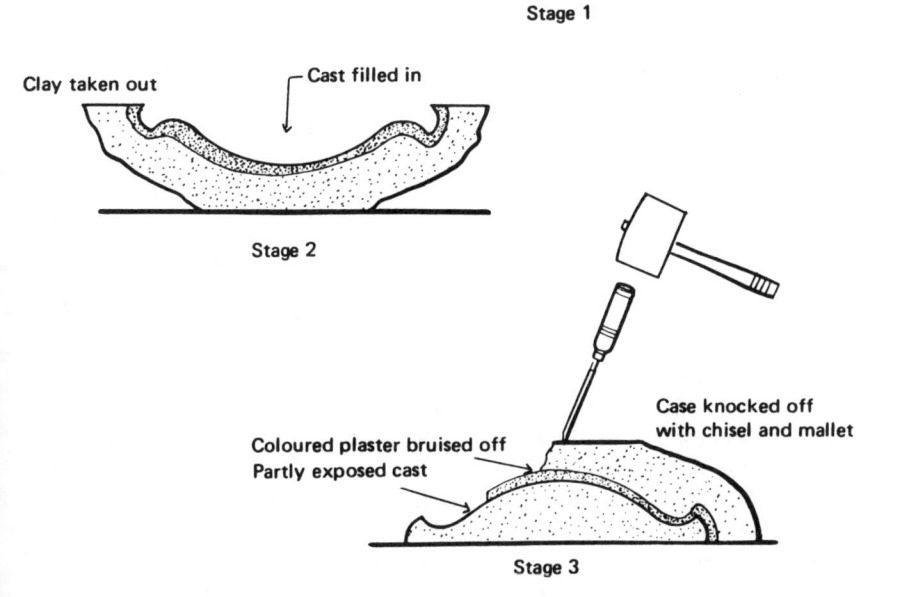

WASTE MOULDS

After this coat has set the mould and clay model can be removed from the modelling base and the clay peeled out from inside the mould. The coloured reverse interior of the waste mould should be cleaned out thoroughly, given three coats of shellac and then greased.

The waste mould is now complete and the cast, often made solid not fibrous, can now be filled in.

When the cast has set and hardened, the waste mould complete with cast can be turned over and the waste mould broken off to expose the cast. This can best be done with a plaster chisel and mallet. Small pieces should be broken off systematically working from the outside edge. Because the coloured first coat and second white plaster coats have been separated by greasing, the thicker second coat should break off cleanly leaving the coloured coat exposed. In places where this does not occur, the coloured plaster will act as a warning barrier to indicate that the cast is only 3 mm away.

The inside surface of the coloured reverse has previously been shellacked and greased, so easy removal of this coat without damage to the cast should be expected. It is often possible to shell off the coloured coat easily after first bruising its surface by light tapping with a blunt chisel (Figure 166).

Waste moulds of clay models 'in the round', such as a statuette or bust, need to be made in at least two pieces. This is to provide access for cleaning out the clay which would otherwise be impossible from the small area of the base.

Figure 166: Waste moulds

The method used to waste mould the clay model of a bust is first to divide the model into two halves by placing a fence of clay or thick metal foil around the imaginary centre line. The front half, including the face, could be waste moulded in the manner described previously after first protecting the rear side by covering with paper. When the front side has been done the fence can be removed and the piecing cleaned up. Joggle holes should now be cut at two or three points of the piecing, which can at this stage be shellacked and greased.

The rear side can now be waste moulded up to the piecing. The two halves of the completed waste mould should be separated by tapping a joint rule along the piecing. When separated, the clay is cleaned out, both sides shellacked and greased on the inside and piecings. The two halves can now be rejoined, positioned correctly by the joggles, and the piecing wadded on the outside with scrim.

If the cast is to be made hollow a thin plaster gauging is poured in the mould, which is then turned to feed the plaster to all parts of its inner surface and then emptied out again. This process is repeated until a reasonably thick layer of plaster has been built up inside. Scrim is pressed into the sides of the cast when the second gauging is introduced. For casts from waste moulds a minimum of 25 mm thickness may be necessary and solid casts advisable in certain cases.

Large, flat waste moulds, such as coat of arms or similar, can be moulded hollow and have scrim reinforcement. When the mould has been completed, the clay is removed and the waste mould shellacked and greased. A strong fibrous, cast is placed in the mould and, when set, the waste mould can be removed from the cast by separating carefully at one corner and then pulling off the waste mould using the scrim reinforcement as an aid for gripping.

WAX MOULDS

These are used when clean sharp details are desired from enrichments such as egg and dart, etc. The enrichments required must not be undercut.

Wax moulds are made from beeswax and resin, mixed in the proportion of 1 to 1 or 1 to 2, beeswax to resin. Beeswax is fairly soft and pliable, so resin is added to give more strength and rigidity.

The mixture is melted in a pan on a hot plate. The enrichment to be moulded must consist of untreated plaster and should be placed in a suitable frame and soaked in water until saturated. Surplus water should be drained off and melted wax poured over the enrichment in a continuous stream.

If the enrichment has been painted then the paint *must be removed*, and the model given a clay wash before the wax is poured. Erratic pouring should be avoided because the cooling wax at the limits of the pouring would form seams in the finished mould.

The moulding wax should be at the right temperature when pouring. If it is too hot, steam would be formed on the surface of the wet model causing air holes in the face of the mould. If the wax is too cool when poured seaming will occur on the mould surface, due to the rapid cooling.

The correct temperature can be found by dipping a finger first in water and then quickly into the melted wax. If the heat can just be borne the wax is at the right temperature for pouring.

Wax moulds should be removed from the plaster model as soon as they are cool, and the inside face of the mould cleaned out with clay slip and water. This acts as a mild abrasive and cleans and polishes the mould face.

GELATINE MOULDS

This is a flexible moulding material made from a superior type of glue. Gelatine can be bought in cake form. It should be hard and brittle. To prepare the gelatine for moulding it should first be soaked in water until pliable, after which the surplus water is drained off (to be used if required, for making size water). The soaked gelatine (or jelly as it is often called) is melted in a double container. The inner container is filled with the gelatine and the outer container with water. This water is brought to the boil by placing over a stove and the gelatine melted by the surrounding heat from the water.

Gelatine is an organic material and if melted and allowed to cool naturally, will go mouldy and start to smell badly if kept for long periods. This can be cured by mixing into new gelatine melts a small quantity of carbolic acid to kill the living organisms. An addition of 0.5 litre of glycerine to 9 litres of jelly improves its flexibility when cool.

Because of its nature gelatine is easily softened back by heat and it will also readily absorb moisture in its natural state. Surfaces of gelatine moulds therefore should be pickled or seasoned to harden them. Firstly the mould is cleaned of the oil accidentally transferred from the model, by brushing over the mould face with methylated or white spirit. This can now be dried off by use of a dry cloth or blotted off with tissue or newspaper pressed in with a tool brush. When dry the mould should be dusted out with French chalk and brushed clean with a dry tool brush. A saturated solution of alum is then splashed over the clean mould surface, allowed to crystallise and then cleaned off. Finally the mould is oiled and is then ready for casting.

One of the problems of casting from a gelatine mould is that setting plaster generates heat and this tends to soften the surface of the gelatine mould. So the thicker the plaster cast and the longer it is in contact with the gelatine, the more damage it can cause. In extreme cases, particularly in hot weather, the cast may be removed with a coating of gelatine attached and of course the mould ruined.

The solution to this problem is to keep the gelatine mould well seasoned with alum and to extract the casts from the mould as soon as possible.

PVC MOULDS

This is also a flexible moulding compound and is the modern artificially manufactured counterpart to gelatine. PVC is the abbreviation for polyvinyl chloride and can briefly be described as a flexible plastic, or a type of artificial rubber.

PVC moulds can be retained for long periods without the necessity of pickling or seasoning in any way. The finished moulds are very strong and tough, and will resist rough usage. Because they do not absorb moisture they can be used for cement casting when the casts are in contact with the mould for long periods.

The temperature of the melting point varies with the different grades and manufacturers, but most types begin to melt around 130°C and to burn at 180°C. This critical temperature range can only be overcome satisfactorily by having a special heating apparatus (usually electrically heated) controlled by a thermostat to prevent temperatures in excess of the burning point.

The majority of electrically heated melting containers take up to three hours or more to melt the flexible moulding compound, but special quick melting apparatus is available in the more expensive types. The problem of moulding with the high temperature of the molten PVC is less important with models composed of high density materials such as metal or damp clay. Little or no suction can take place from these materials and therefore little or no air can be drawn from them by the hot PVC to leave air holes as a disfigurement on the finished mould surface.

Plaster models should not be sealed with shellac because the high temperature of the PVC will melt and destroy it. Certain proprietary sealants are made and recommended by different manufacturers and these can be used. A method favoured by many is to soak the clean plaster model in water until saturated, then dab off the excess water. This is similar to the method described for wax moulding.

Polyurethane based sealants give good results. Two coats are normally given with an interval of twenty-four hours between coats and then allowed a further twenty-four hours for maximum sealant effect.

Faults which can occur in PVC moulding on wet plaster models may be due to drying out too soon, possibly caused by differences in thickness in the model. Excess water left on the model surface will be turned to steam which will condense, leaving small air holes when dry. Air can be drawn through from the porous moulding ground. Also if the molten PVC is poured too slowly, particularly in case moulds which have had time to dry out, the temperature drops rapidly at the thin edges and seaming will occur.

PVC moulding compounds are manufactured in coarse, medium and fine grades. The coarser grade gives greater stiffness for a mould with large projections. Fine grades are more suitable for finer detail in *models* of low relief.

COLD CURE RUBBER MOULDS

Silicone rubbers are suitable for use in moulding processes at normal room temperature without the use of special heating equipment or apparatus. The rubbers cure by catalytic action and the curing time varies with type and amount of catalyst used.

For use in flexible moulds with little or no undercut sections a free flowing, easy pouring type can be used. A stronger type is used for moulds with deep undercut or for skin moulds. This type has high strength and tough handling properties.

Recommended catalysts to be used for each type are supplied with the rubber to be used. The amount of catalyst used is normally one part catalyst to ten parts rubber by weight.

The catalyst must be mixed thoroughly. The catalyst and rubber are in contrasting colours and mixing should continue until an even colour is obtained. Excessive mixing should be avoided as this may accelerate curing and also introduce trapped air.

During the moulding process any plaster models used should be sealed with shellac. Clay, metal or similar non-porous models do not require to be sealed. Silicone rubber will stick to Plasticine, which must be coated with shellac and treated with a parting agent. Rubber will also adhere to rubber, so two piece moulds will require the first piece surface to be treated with a release agent. A suitable release agent can be made from 2% petroleum jelly in white spirit. This latter mixture can be made by standing a container of white spirit and petroleum jelly in a large container filled with hot water. Stir well after the jelly dissolves.

During the moulding process the model can be given a brush coat before pouring on the remaining rubber. Brushes can be washed out in Silicone thinning fluid, which may also be used to thin Silicone rubber if required.

The catalysed rubber keeps flowable for approximately one and a half hours but curing times vary considerably, from six to twenty-four hours depending upon thickness and recommended catalyst used.

The mould may be used immediately after curing but full rubber strength may not be achieved for up to seventy-two hours in certain cases.

FLOOD, OPEN OR FENCE MOULDS

On low-relief models on the bench or on a plain face moulding ground, models in 'L' shaped plaster troughs as illustrated in Figure 163, flood moulding is the quickest and simplest method of moulding. Any flexible moulding compounds can be used. A fence or frame in timber, clay or metal is positioned around the model to project at least 25 mm above the highest point and should be on a level surface. All joints and junctions should be sealed with plaster or clay before, or after, any prior preparation of the model is carried out. The mould can now be poured and should commence in one spot, the pour should follow the flow until the model is covered with at least a minimum of 13 mm. After the compound is solid and cooled a case can be cast if required over the back of the compound. The case is struck off flat enabling the mould to lie flat when removed from the model, this system is often referred to as back casing. The case can be extended down the sides of the mould if the frame is first removed, the case should not extend down to the bench line as it can make removal of the cast difficult.

SKIN MOULDS

This is a compromise between the fence and case mould methods. The prepared model, which should be fairly shallow in section, is surrounded with a fence about 50 mm in height. If the outline of the model is curved a clay fence is best. The fence should be positioned about 6 mm away from the model outline and leaning slightly inwards towards the model.

Melted moulding compound is poured carefully over the surface, only sufficiently thick to ensure complete coverage. Any bare high points can be given a further application as the mould cools. A fibrous plaster case can now be formed over the moulding compound skin whilst it is still in position on the model. The back of the case should be straightened.

When set the case is removed and inverted. Now the moulding compound skin can be peeled off the model and placed in its original position in the case. This will hold and support it to the correct shape.

CASE MOULDS

These are used when the model is unsuitable for moulding with an open fence mould. Case moulds are therefore used to make sure that the moulding compound thickness is the same throughout the mould.

When forming a case mould for an enriched plaster truss the model is firstly fastened to a moulding ground and shellacked or coated with a suitable sealant. The surface of the model should now be covered with paper and then with a layer of clay 12 mm to 15 mm thick. The bottom edge can be worked to a slight groove around the perimeter.

A fibrous plaster case should now be formed over the clay, strengthened with laths if necessary and the top edge straightened with a flat board. Its position is now marked on the case and the moulding ground. When set the case is removed and a 25 mm circular hole cut in the lower part. Ventilation holes to enable air to escape are drilled at intervals around the case particularly in the highest part. The inside of the case is now sealed and prepared. See the section on PVC moulds. The clay and paper are removed from the model which can now be prepared, after which the case is replaced over the model to the position of that previously marked. The case is fastened down with wads and the joint with the moulding ground sealed with clay. A funnel is now fixed over the pouring hole with a plaster wad. A suitable funnel is about 250 mm high, 100 mm wide at the circular top narrowing to a 25 mm diameter circle at the bottom.

When pouring the moulding compound, small knobs of clay should be available to seal the vent holes as the moulding compound flows through.

The flow of moulding compound in the case should not be impeded by compressing the air inside. If the ventilation holes are correctly positioned the air will be pushed out completely without offering any resistance to the incoming moulding compound.

When cool the funnel is removed by cutting through the moulding compound at its connection with the case. The case is released and lifted from the moulding compound, assisted by pressure through the pouring hole. The moulding compound can now be peeled off the model, and the mould then placed in the plaster case. It will be noted that the grooved edge formed in the clay when preparing the case now forms a lip in the moulding compound at the top perimeter of the mould enabling it to obtain a firm grip to the case at this point.

CORNICE CASE MOULDS

Clay case

This type of moulding process is suitable for cornice sections with two or more enrichments. An example is given in Figure 167. The moulded part of the cornice is run down on the bench, allowance being made in the mould profile for bed moulds for enrichments. The enrichments are cast separately from wax moulds, or similar, and bedded in position on the cornice model. All joints are made good and the model surface coated with shellac or other suitable sealant. Paper is placed over the model and a 15 mm layer of clay pressed over this. A ridge, or lip, is formed along the long edges of the clay. See Figure 167. A fibrous plaster case is now formed over the clay, strengthened with laths and straightened on its upper surface.

When set the case is removed and pouring holes are cut at 900 mm intervals in its length. Suitably spaced ventilation holes are also bored at intervals along its length. The inside of the case should also be sealed and prepared. See the section on PVC moulds.

The clay and paper are now removed from the model which can now be prepared, and the case replaced over it. The openings at each end of the cornice should be covered with a board and sealed. Joints along the long edges between the cornice model and the case are now sealed with clay. Finally the case is wadded, or wedged, down to prevent separation during the pouring process.

The pouring funnels are fixed next, after which the moulding compound is poured commencing at one end of the cornice. As the moulding compound passes the second funnel the pouring can commence from there and be repeated along its length. Vent holes should be made good as the moulding compound oozes through each one in turn. When the moulding compound is at least one-third of the way up each of the pouring funnels, pouring should cease.

When cool the case can be removed and inverted to receive the flexible mould after this has been peeled from the model.

The moulding compound is held in position in the case by the lip formed

BENCHWORK

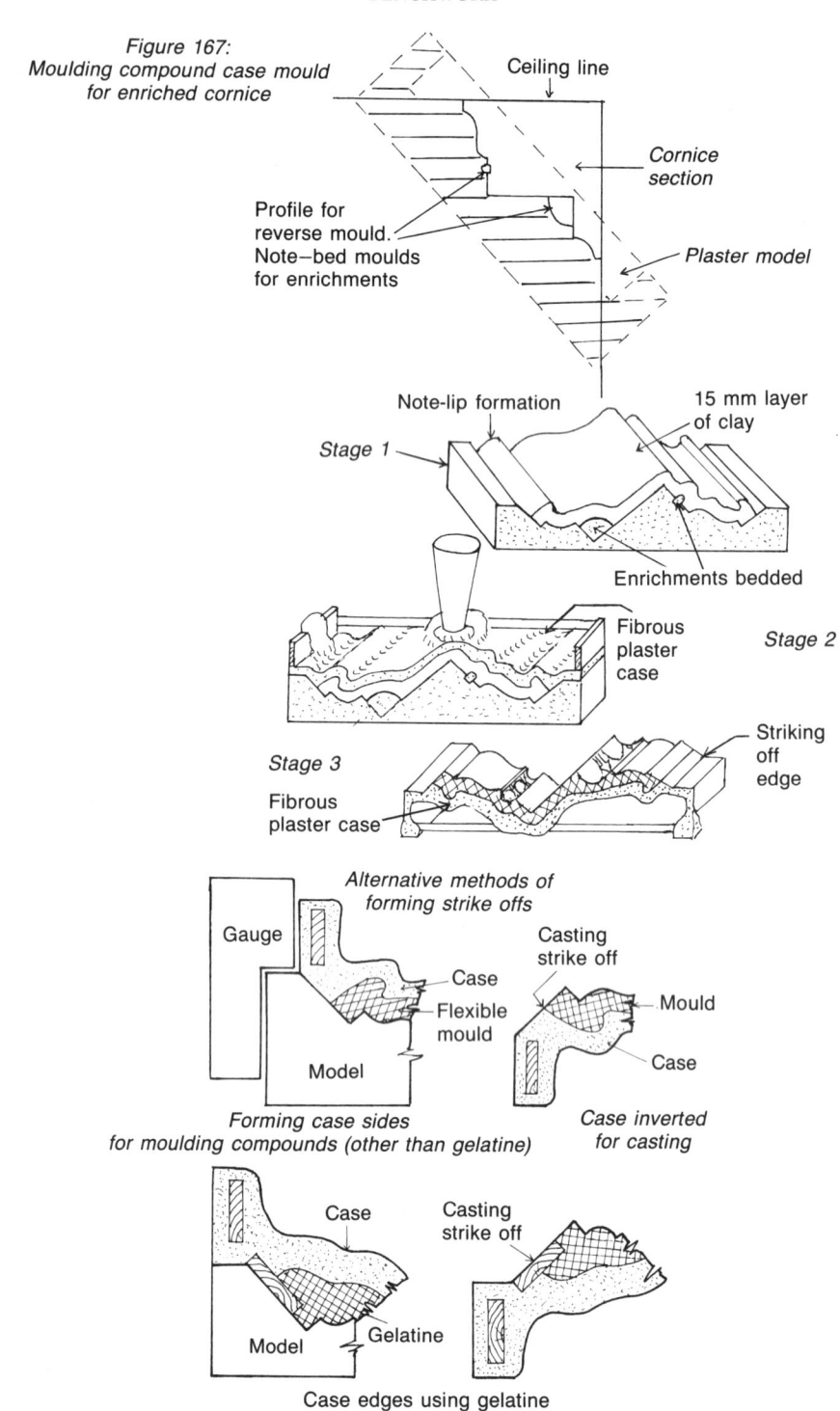

Figure 167:
Moulding compound case mould
for enriched cornice

Ceiling line

Cornice section

Profile for reverse mould. Note—bed moulds for enrichments

Plaster model

Stage 1

Note-lip formation

15 mm layer of clay

Enrichments bedded

Stage 2

Fibrous plaster case

Stage 3

Fibrous plaster case

Striking off edge

Alternative methods of forming strike offs

Gauge

Case

Flexible mould

Model

Casting strike off

Mould

Case

Forming case sides for moulding compounds (other than gelatine)

Case inverted for casting

Case

Model

Gelatine

Casting strike off

Case edges using gelatine

INSERTION MOULDS

along the edges. Ribs can also be formed across the back if excessive movement is anticipated on large moulds, unless shrinking or stretching is intended.

INSERTION MOULDS

These are a combination of plaster and moulding compound reverse moulds. Plaster is used for the moulded portion of the cornice and moulding compound for the enrichment. Three methods are used. In the first method a model of the actual cornice required is run down on the bench and the enrichments bedded as described for the enriched moulding compound case moulds. With this method however, only the enrichments are covered with clay, which must be splayed at the back to ensure easy withdrawal later. A fibrous plaster case is now made over the cornice model and clay core. When this has set it is removed and pouring holes bored opposite the enrichment. The clay core is removed from the model and after coating with shellac, or other sealant, and oiling the case is fixed back on the model. Funnels are fixed to the pouring holes, the ends and sides sealed and moulding compound can now be poured.

When the moulding compound has cooled the case is removed and inverted face uppermost on the bench and the moulding compound strip inserted in the channel formed by the clay core.

The second method is an alternative to the clay core and is simpler to carry out. The model is shellacked, other than the ornament. The ornament is treated according to the moulding compound being used. Next, clay or timber fences are fixed up on the model either side of the ornament. Moulding compound is poured as a flood mould to form the insertions. The fences are removed and when the compound has hardened the reverse mould is cast directly over the model and insertions, see Figure 168. The insertions are encapsulated in the reverse mould.

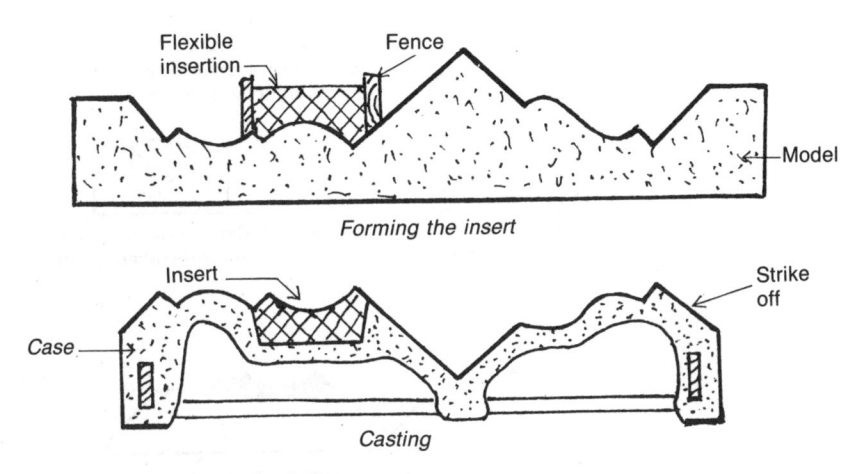

Figure 168: Alternative method of forming flexible insertion

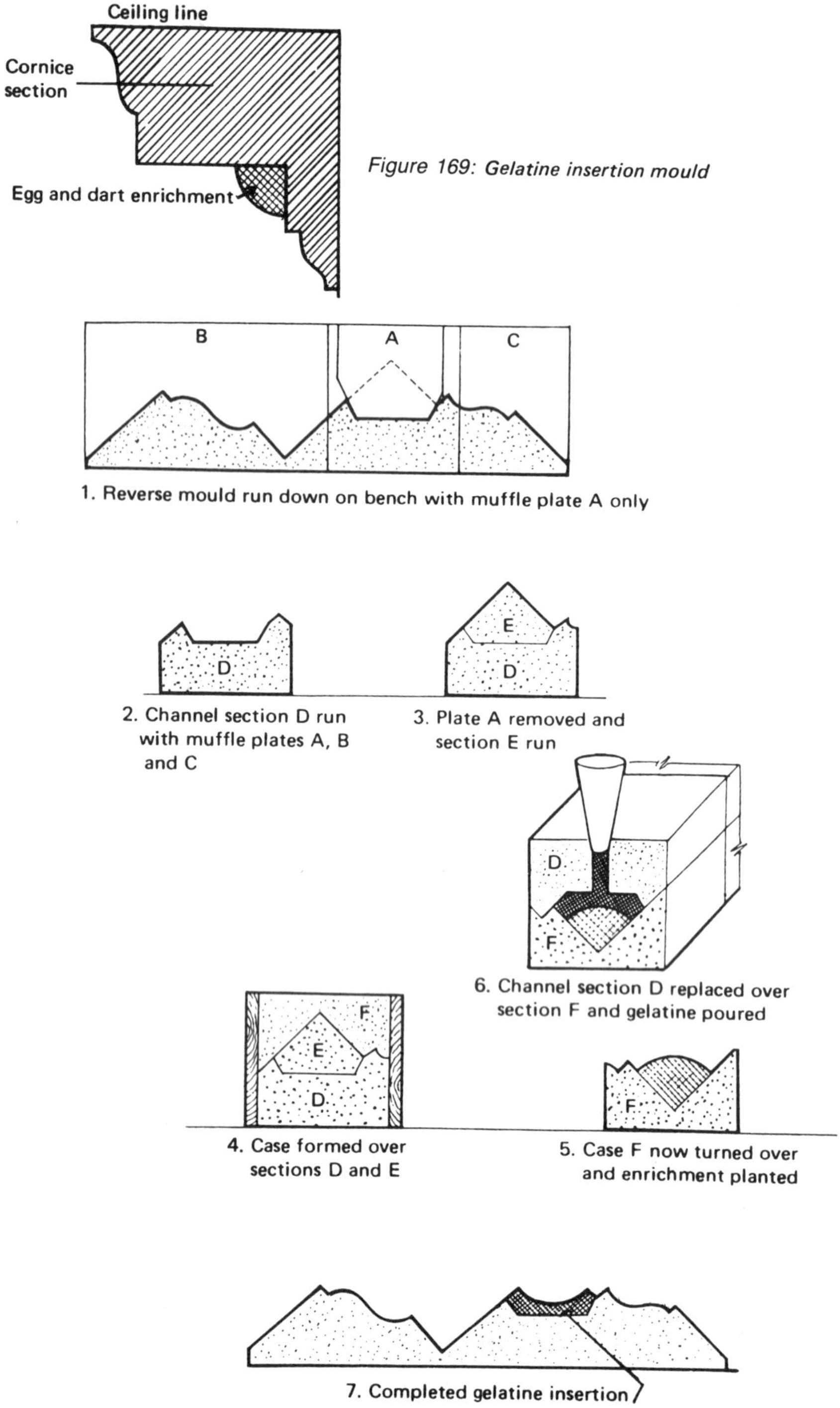

Ceiling line
Cornice section
Egg and dart enrichment

Figure 169: Gelatine insertion mould

1. Reverse mould run down on bench with muffle plate A only

2. Channel section D run with muffle plates A, B and C

3. Plate A removed and section E run

4. Case formed over sections D and E

5. Case F now turned over and enrichment planted

6. Channel section D replaced over section F and gelatine poured

7. Completed gelatine insertion

206

The third method was specially designed for use with gelatine and is shown in Figure 169. Firstly the normal reverse profile is cut and horsed omitting the enriched portion. This is covered with a muffle plate 'A', and the required length run down on the bench. This section can now be shellacked and greased.

Two further muffle plates 'B' and 'C' are now nailed to the running mould and the channel section 'D' is run on a separate part of the bench. This section should now be shellacked and greased.

The muffle plate 'A' is next removed and the section 'E' run and formed over the channel section 'D'. This section is also coated with shellac and greased.

Pieces of timber are fastened to each side of channel section 'D' and the top section 'F' filled in. When set, section 'F' is lifted off and turned over on to the bench. The enrichment can now be bedded into the right angled portion of its upper surface. When the separate pieces of enrichment have been bedded and the joints made good, the completed section should be shellacked and greased.

The section 'E' can now be removed and discarded. Channel section 'D' is drilled for pouring and ventilation holes and then fixed upside down over the section 'F'. Funnels are fastened, joints and ends sealed and the gelatine poured.

When cool the gelatine insertion can be extracted and pickled, after which it is placed in the channel formed in the main reverse mould. The completed mould is now ready for casting, and the sections 'D' and 'F' retained for further use until the gelatine insertion needs replacing.

RUN CORES

An alternative to clay cores on case moulds is a run core. This is formed by running in plaster over a protected run model, Figure 170.

A separate running mould may be used to run the core, or the original running mould used to run the model may be adapted. In the latter case the running mould can be made to fit the profile of the cornice model outline

Figure 170: Cornice mould with run core

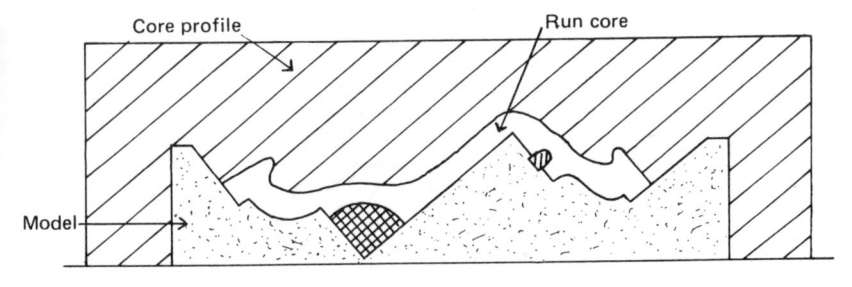

and adapted later. After the model has been run the enrichment is bedded to correct alignment and spacing. The model surface is then sealed with shellac. Damped paper is laid over the model surface, overlapping the joints the way the core is to be run to prevent plaster being forced under the paper on to the model. An alternative to paper is to use thin gauge polythene sheet fastened at the sides with a lath nailed into the edges.

The running mould is now adapted for running the core by removing the model profile, marking out, cutting away the timber and fixing on the core profile, Figure 170. The new profile should allow a clearance of approximately 15 mm for the core thickness.

Plaster used to run the core should be gauged to a weak strength for ease of removal later.

After the core has been run it is sealed with shellac, greased and a strong reinforced plaster case formed over it. When set the case and core are removed. The core is normally locked into the case because of the undercut grooves at the end and can be removed by splitting down the middle with mallet and chisel. The procedure is then as described previously.

A simplified alternative to using an adjusted or separate running mould to run the core is to use a hardboard or plywood template for profiles which are not too complex. The template is drawn along the model length, keeping the gauge end tight to the model edge. It is a cruder method but a great time saver if done carefully, Figure 171.

RUN CASES

Instead of forming the case over a core a case mould can be run separately. This has the advantage of cutting out the process of forming a core. A disadvantage is that it is difficult to reinforce adequately to make long lengths strong enough to handle. Large section timber used in a solid run case can cause problems with subsequent swelling though suitable metal reinforcement can be used. A further important consideration is that any irregularities in the straightness of bench surface or running rule will affect the perfection of fitting the run case to the run model.

Figure 172 shows the profile suitable for running a case mould to fit the cornice model drawn in Figure 170.

Figure 171: Cornice model with core formed by gauge

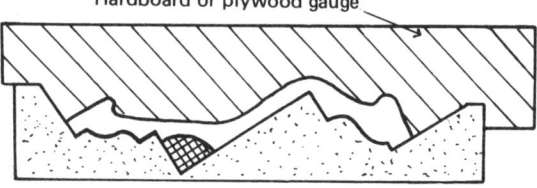

Figure 172: Run case section and profile

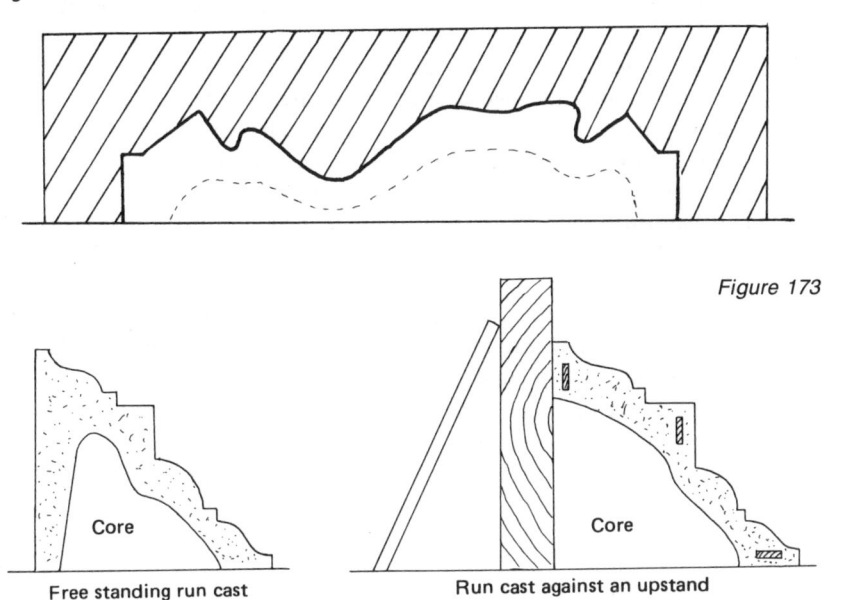

Figure 173

Core

Free standing run cast

Core

Run cast against an upstand

RUN CASTS

These are run lengths of moulding suitably reinforced for subsequent fixing on site. Their use is normally confined to situations when this method is quicker and saves the time and expense of making a reverse mould and casting. It is also a suitable method to use when the back of the moulding is required to be moulded or smooth as well. Two examples are shown in Figure 173.

Run casts are normally run over a shellacked and greased core, allowing sufficient thickness of the run cast for scrim and timber reinforcement necessary for fixing later. The positioning of the timber reinforcement is as usually required for a normal cast.

During the running process plaster used to place the scrim and laths will need to be retarded and clearance left for finishing off the running with unretarded plaster.

A disadvantage of run casts is the weight due to extra thickness. This can cause problems when fixing, particularly with large section mouldings.

REVERSE MOULDINGS

This type of moulding cuts out one of the processes required in the model, mould and cast methods. A reverse mould is made directly without the use of a model. A simple type of reverse mould is that used for making *plain face fibrous plaster slabs*.

A frame of the desired size is made by nailing a timber fence to the smooth bench surface. The depth of the fence can be approximately 30 mm. Scrim should be cut to cover twice the area of the slab and laths as shown in Figure 174.

The reverse mould is greased previous to the cast being filled in. A coat of 'firstings' is brushed and splashed over the inside area and fence sides of sufficient thickness only to ensure complete coverage. The 'strike offs' should be scraped clean at this stage. When the 'firstings' has stiffened slightly the 'seconds' can be applied over the area and into this is pressed the scrim, brushing in with a splash brush. The laths are next positioned after first brushing on a coat of 'seconds'. Next, another coat of 'seconds' is applied and the scrim is turned over, brushing in well.

A stiffer mix is used to form a fixing band about 35 mm wide around the perimeter, and over the centre lath reinforcements. These bands should be ruled off level from the top edges of the frame fence.

When set the plain face slab can be removed and the face scraped with a flexible busk to a good surface.

Slabs with a rebated margin are formed in a bench mould with pieces of suitable section nailed inside the fence perimeter. See Figure 175.

Figure 174: *Plain face fibrous plaster slabs*

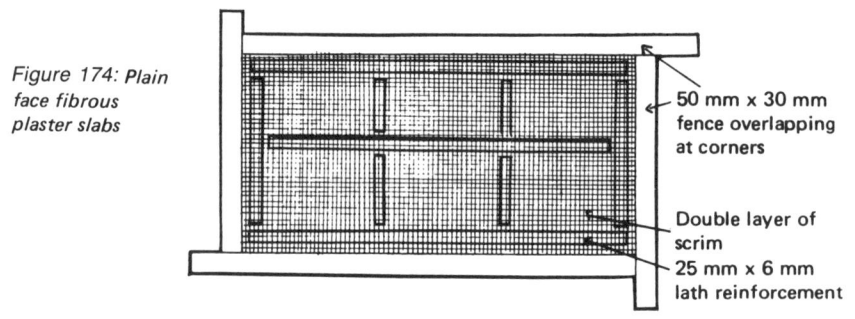

- 50 mm x 30 mm fence overlapping at corners
- Double layer of scrim
- 25 mm x 6 mm lath reinforcement

Figure 175: *Forming plain & rebated joints to sides of casts*

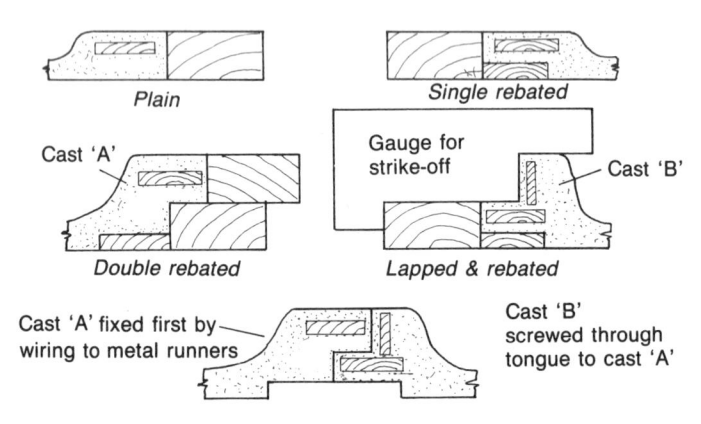

Plain

Single rebated

Cast 'A'

Gauge for strike-off

Cast 'B'

Double rebated

Lapped & rebated

Cast 'A' fixed first by wiring to metal runners

Cast 'B' screwed through tongue to cast 'A'

Rebated casts when fixing to metal

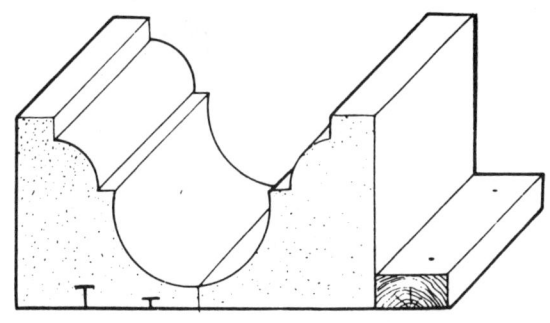

REVERSE MOULDINGS

Figure 176: Rib moulds

One half of mould section may be anchored to bench, if required, by projecting nails

L shaped timber sections screwed to mould and nailed to bench

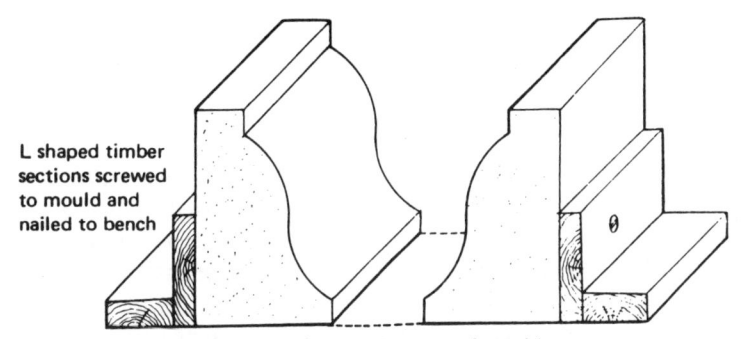

Bench surface used as part of mould

Figure 177: Cornice reverse mould setting out

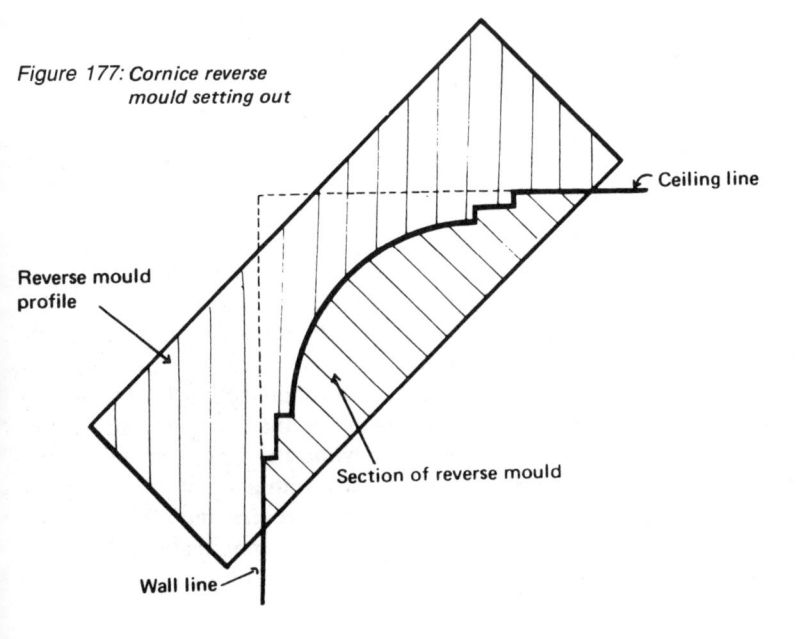

Ceiling line

Reverse mould profile

Section of reverse mould

Wall line

Figure 178: Cornice reverse moulds

Striking off edges

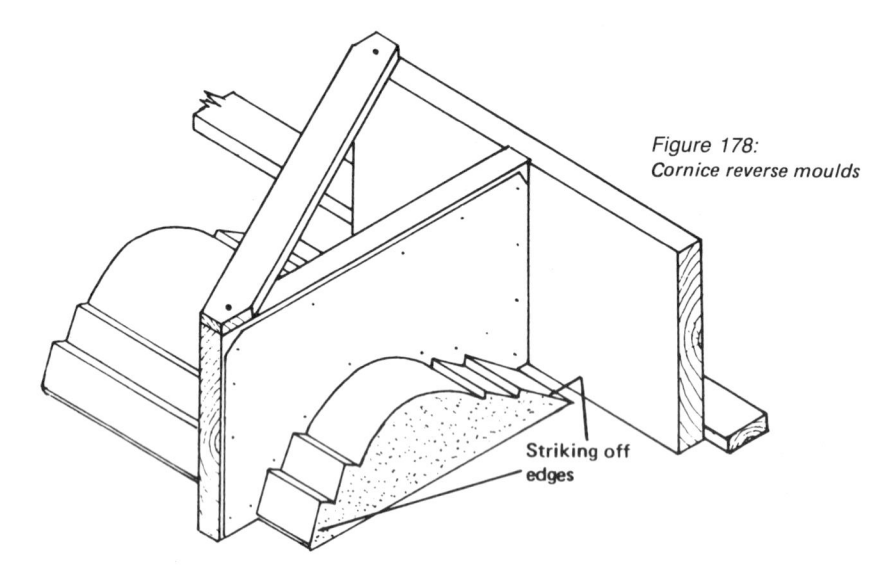

A reverse mould for a rib moulding is shown in Figure 176. A reverse profile of half the finished moulding section is horsed up and a length of the reverse section is run down on the bench. This can be cut into two halves and assembled as shown in Figure 176. If a number of casts are required, one side of the reverse moulding can be anchored to the bench with protruding nails.

Cornice reverse moulds without undercut sections are made as described in Figures 177, 178. The reverse profile is marked out after first extending the wall and ceiling lines, as these are used later for striking off lines when casting from the finished reverse mould.

Mitred and returned stopped ends can be used in reverse mouldings, so that the lengths can be cast to any given length with the appropriate external or internal mitre, or return, at each end.

A solid cast, about 25 mm thick, is made and taken from the reverse mould. This is now cut to the reverse mitres required both for left and right hand edges. Each piece need be only 50 mm — 75 mm wide and often left and right hand mitres are cut on opposite sides of one piece. The pieces are shellacked and greased and placed on the reverse moulding to the correct mitred angle and spaced as required. Square stops are also used for piecings.

Reverse returned stopped ends can be formed by scribing, cutting and bedding a short length of the reverse cornice moulding on to the main length of greased reverse mould. The reverse returned end should be bedded square to the main reverse mould, each member being in line and the ceiling member of the returned end at the correct projection beyond the wall member. As the bedding coat sets it should be mitred correctly with a small joint rule and when set the returned stopped end can be slid off the reverse mould. After shellacking and greasing it can be returned to the reverse mould and arranged at any given distance along its length. Left and right hand returns may be necessary.

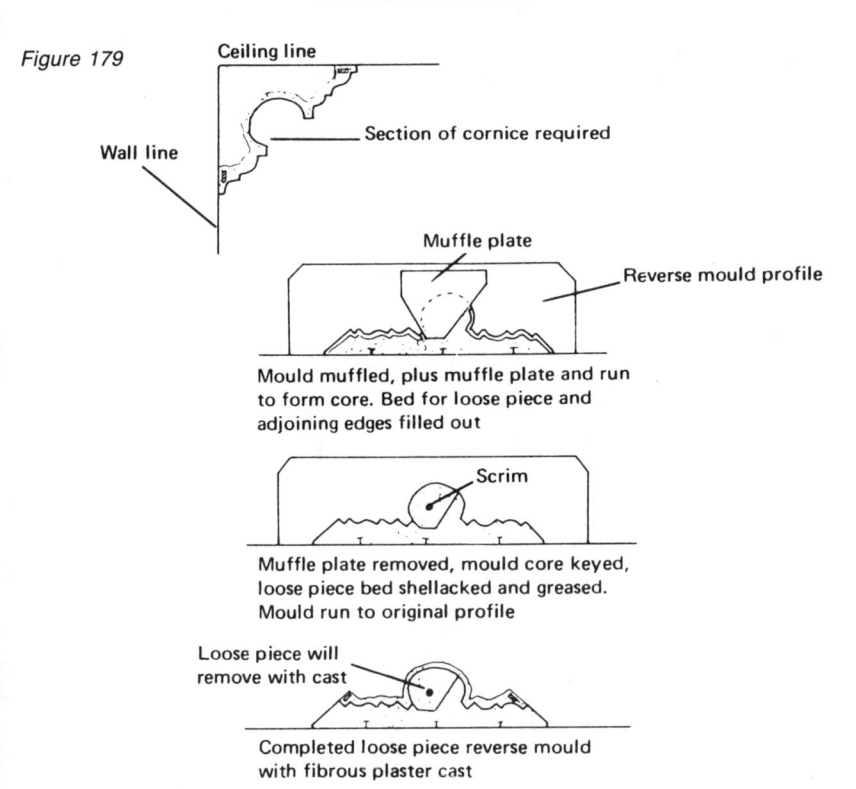

REVERSE MOULDINGS

Figure 179

Ceiling line

Wall line

Section of cornice required

Muffle plate

Reverse mould profile

Mould muffled, plus muffle plate and run to form core. Bed for loose piece and adjoining edges filled out

Scrim

Muffle plate removed, mould core keyed, loose piece bed shellacked and greased. Mould run to original profile

Loose piece will remove with cast

Completed loose piece reverse mould with fibrous plaster cast

Loose piece moulds are used in certain instances when a cornice contains an undercut section. The normal reverse profile is cut and horsed up as described for ordinary cornice reverse moulds. A muffle plate is cut and fixed to blank out the undercut section, as shown in **Figure** 179.

The rest of the profile is now muffled with plaster to within 3 mm of the metal muffle plate. Sufficient length of the cornice required is now run on the bench bringing the bed mould for the loose piece and the two 3 mm members at each side to a good finish. This portion should be shellacked and greased and the remainder of the reverse cornice core scratched well for key. The metal and plaster muffles on the running mould should now be removed and the loose piece is formed, first filling out the thicker part with a roll of scrim dipped in the plaster mix. When the loose piece is cored out, a wet mix is applied over the whole reverse mould and this is now brought to a good finish. The loose piece usually swells slightly and can be removed easily leaving a clean arris along each edge.

After shellacking and greasing, the cast can be filled in. When set the cast should be removed with the loose piece until clear of the undercut section, after which the loose piece is allowed to slide gently back into the bed mould recess.

Circular cornice reverse moulds (see Figure 180).

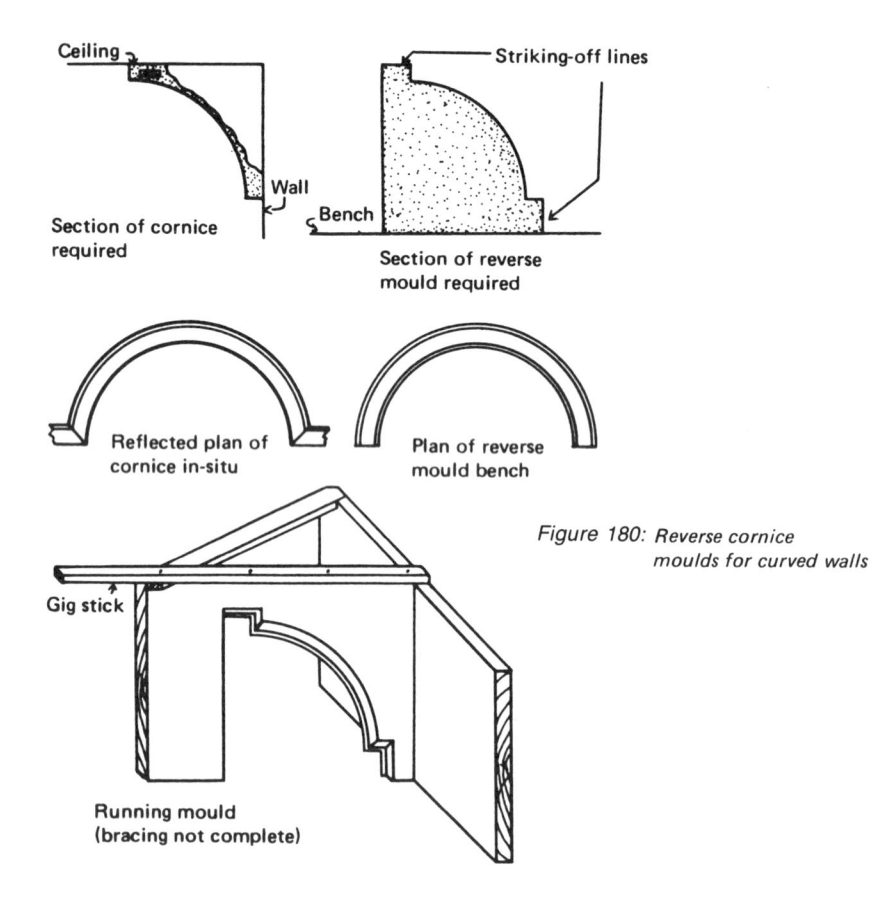

Figure 180: Reverse cornice moulds for curved walls

BEAM CASINGS

Figure 181 shows a method of forming a fibrous plaster casing to a beam. In this instance the reverse mould is in three separate pieces of run moulding. The soffit reverse is often anchored down to the bench. Both side pieces can be formed with one running mould.

The prepared moulding lengths should be assembled as shown, bracing the side pieces to the bench to prevent movement during casting.

When casting, gauges may be used in forming a fixing member for the top edge of the beam cheeks. A simple gauge can be made by nailing two short pieces of 50 mm x 12 mm timber at right angles to a 150 mm length of similar section. The two short pieces should be nailed so that the distance between the two right angled pieces is the width of the side reverse moulding at the top plus the width of the intended fixing thickness of the cast. When forming the fixing edge the gauge is drawn along both top edges and held tight to the outside edge of the beam cheek reverse.

Short pieces of lath are placed inside the case diagonally when casting for

BEAM CASINGS

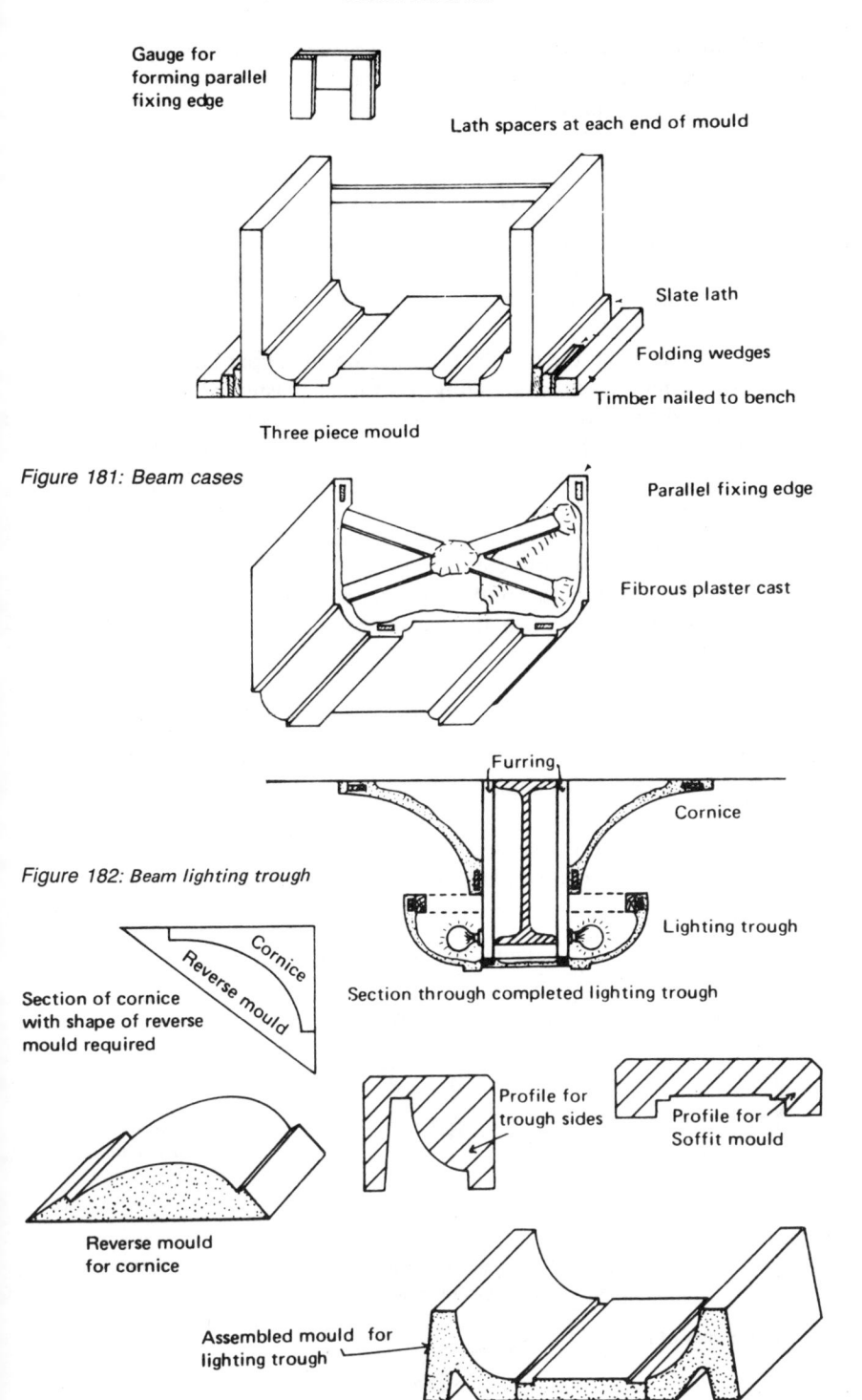

Gauge for forming parallel fixing edge

Lath spacers at each end of mould

Slate lath

Folding wedges

Timber nailed to bench

Three piece mould

Figure 181: Beam cases

Parallel fixing edge

Fibrous plaster cast

Furring

Cornice

Figure 182: Beam lighting trough

Lighting trough

Section through completed lighting trough

Cornice

Reverse mould

Section of cornice with shape of reverse mould required

Profile for trough sides

Profile for Soffit mould

Reverse mould for cornice

Assembled mould for lighting trough

215

use as temporary braces to hold the casing square during storage and transport to the site. When ready for fixing the braces are removed and the casing nailed to the timber furring with galvanized nails.

Fibrous plaster lighting troughs are made in a similar manner, Figure 182.

HAND LATHES

These are a simple type of lathe consisting of a spindle with a square section, tapered in length. It is pivoted at two mid points on a frame and revolved by hand by means of a cranked handle. A sketch of such a turning box is shown in Figure 183.

It will be noted that the profile, mounted on the upper lid, does not move during the formation of the moulding. To 'turn' a plaster baluster, the spindle is first greased then wads of scrim soaked in a thin plaster mix are wound round the spindle. Further applications of plaster are built up as the spindle is revolved. To assist in cleaning during the turning process the profile lid is hinged. When completed the plaster baluster can be removed by tapping the narrowest end of the spindle on a timber block.

Due to the fact that the spindle is tapered, release is fairly easy. The spindle is made square in section to avoid the plaster baluster slipping and refusing to turn. This would occur when the plaster begins to swell and the baluster model tightens against the profile.

TURNING MOULDS

Turning moulds of a different kind are used to form mouldings of deep section. In Figure 184 an example is given of a mould suitable for forming the model of a small dome.

The profile is cut to the desired profile and horsed up as shown. A centre block is fastened and strutted to the bench and this can be covered with paper

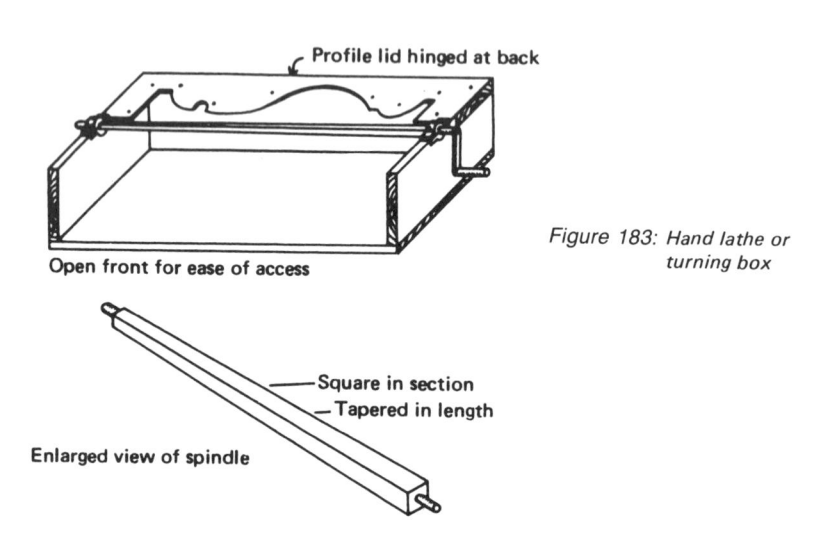

Figure 183: Hand lathe or turning box

TURNING MOULDS

Figure 184: Sectional view of turning mould for forming dome

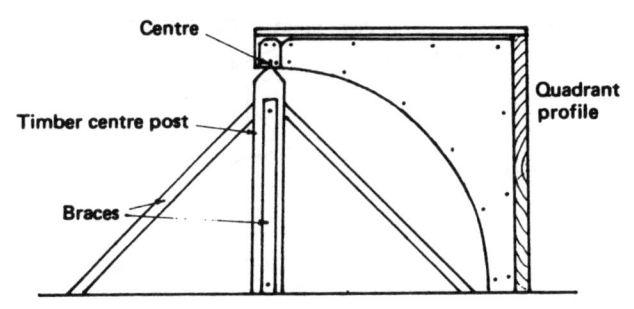

[Diagram labels: Centre; Timber centre post; Braces; Quadrant profile]

and cored out hollow. If the dome shell is required for lifting the mould muffle should be about 20 mm thick.

The *mould* in this instance is pivoted around the centre block. When the core has been filled out small joggle holes should be cut in its surface before shellacking and greasing. This is to prevent the dome shell from slipping round on the core when this is being turned.

Other adaptations of this type of mould are used for turning circular cap and base mouldings for columns, reverse moulds for niches, etc.

One method of forming *plain niches* in fibrous plaster is to run the shell of a dome, as previously described, to the required section and radius. The finished dome shell should now be cut accurately into two equal parts through its centre point. The two half shells are now fastened to the bench, with wads around projecting nails, so that they are the depth of the niche shaft apart. They must be at right angles to a centre with the curved part of the dome shell outward (Figure 185).

The inside, recently cut, edge of one of the dome shells should be smoothed straight with a rasp or plane, and shellacked and greased. This will be the reverse edge of a skimming member to form the niche shelf later.

Between the dome shells a core can now be formed either hollow or solid with bricks, old plaster pieces, etc. The core should be left 6 mm short and clear of the shell recently prepared for the shelf reverse.

A feather-edged rule should be used to form the barrel-shaped drum by traversing from the top edge of the half dome shells. The surface can be scraped to a good finish with flexible busk. At this stage the half dome shell with the shelf reverse edge is broken and removed, leaving a good edge. This edge

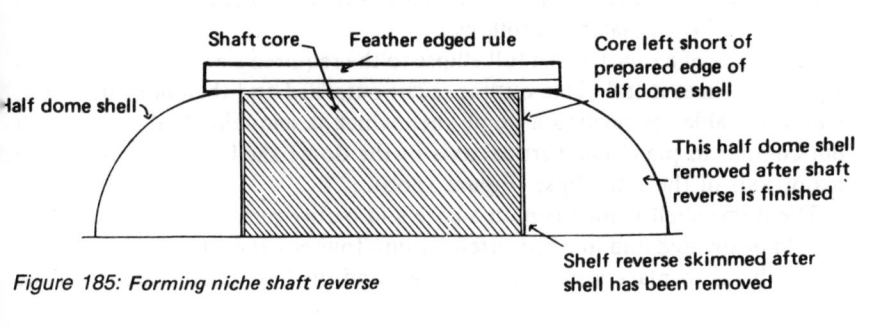

[Diagram labels: Shaft core; Feather edged rule; Core left short of prepared edge of half dome shell; Half dome shell; This half dome shell removed after shaft reverse is finished; Shelf reverse skimmed after shell has been removed]

Figure 185: Forming niche shaft reverse

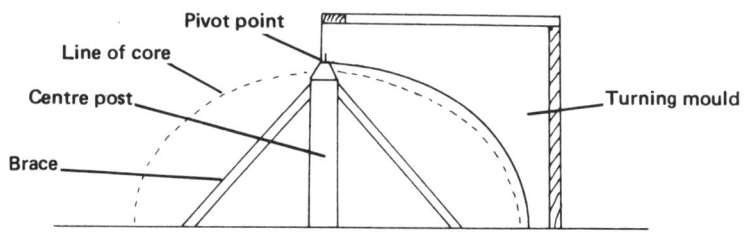

Figure 186a: Section of semi-elliptical dome for fibrous plastic niche with semi-circular hood

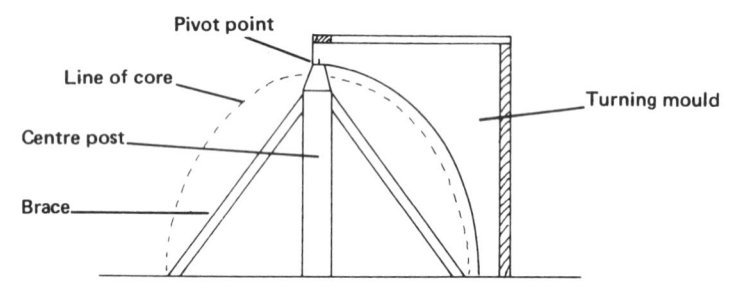

Figure 186b: Section of semi-elliptical dome for fibrous plastic niche with semi-elliptical hood

should now be used as a skimming member to fill in the vertical end exposed because this when finished is the niche shelf reverse.

The completed niche reverse can now be sealed with shellac, and then surrounded on the bench by a fence of about 30 mm thickness. A minimum space of 100 mm margin should be allowed around the niche reverse which, after greasing, is now ready for casting.

Kerfed laths should be used around the curved parts to strengthen the casts. These are laths bruised with a lath hammer until they assume the curvature required.

Fibrous plaster niches which have a shaft semi-elliptical in section and a semi-circular fronted hood, can be made in a manner similar to that described for the semi-circular sectioned niche. The only difference being that the turning mould will have to have an elliptical quadrant of the required size instead of a circular quadrant (Figure 186a).

Plain niches semi-elliptical in section and with a semi-elliptical hood can be formed in fibrous plaster as follows.

The shape of a quarter of a full ellipse to the required size is marked out on a piece of sheet metal. This outline is cut out, filed and then horsed up to a mould suitable for turning a dome shell which is semi-elliptical in elevation but circular in plan. The vertical axis of the dome shell is equal to half the major axis of the full ellipse (Figure 186b).

The dome shell is then run on a core and before removal from the core it should be divided and marked carefully into four equal parts. One method of doing this is to place a piece of string around the base circumference of the dome, mark its length, remove it and divide the full length into four equal parts. The string can then be replaced around the base perimeter and the four marks transferred from the string to the base of the dome shell. The turning mould can now be replaced on to the pivot point with the profile touching one of the base marks and a pencil line marked on to the dome shell using the profile as a guide line. This is repeated on each of the three other marks. After removal the shell can be sawn into four equal quadrants along the lines previously marked.

The positions of the shaft, shelf and springing line for the hood are marked out on the bench surface. The quadrant shells are paired together and fixed at shelf and springing line points by wadding to projecting nails on the bench to the inner faces of the shell pieces. The paired shells should be fixed outwards at the lines stated, with one of its long sawn edges in contact with the bench surface.

The succeeding procedure for completion of the model is as described for the semi-circular niche except that the joint between the two elliptical quadrants for the hood reverse will have to be touched up with a small tool and flexible busk.

Plain and also *fluted niches* may be formed by an alternative method preferred by many. In this method a semi-circular bench running mould is made first, using two identical quadrant profiles linked together. (The quadrants should be held together in a vice while filing.)

The semi-circular sectioned reverse shaft of the niche is run down first on to a core, which has been built up on the bench. The ends are next cut off square and plumb. One of the ends is skimmed and brought to a good finish, to form the reverse niche shelf.

Following this, the mould is cut exactly into two halves and a dome shell turned as explained previously. It is very important that the centre point be fixed at exactly the right height otherwise the resulting dome will be too high or low, too wide or narrow, when matching takes place.

The dome shell, when finished, should be sawn into two equal halves one of which can now be 'married', or joined, to the finished shaft reverse. The joint can be filled in and finished with a thin flexible drag or busk, after which normal mould preparation for casting should be carried out.

Fluted niches are also made using a quadrant fluted mould only. The dome shell is turned in the normal way, and the shaft reverse is run in two halves by running the mould along a bench with the nib bearing against an upstand fixed at right angles to the bench surface (Figure 187).

The plan of the shaft, shelf and springing line of the hood are marked on the bench surface. Half the dome shell is bedded and wadded to projecting nails on the springing line with the hood reverse outwards. The straight run lengths for the reverse sides of the shaft are cut to the correct size and bedded on to a previously prepared core, to match the contours of the fluted hood reverse. The shelf reverse is formed by cutting and surforming the ends of the shaft reverse pieces to a good square edge, and using these as a skimming member to skim the shelf reverse.

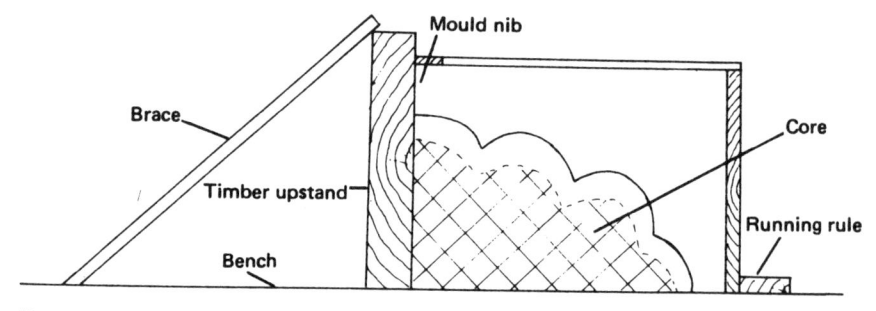

Figure 187: Section showing method of forming sides for fluted niche reverse mould shaft

FIBROUS PLASTER ARCHES

Fibrous plaster arches are formed on reverse plaster drums of the required shape and size.

A plain semi-circular double-sided arch can be made by running a semi-circular L shaped sectioned moulding on the bench as shown in Figure 188, stage 1. This is sawn to a true semi-circle, shellacked and greased, and a rule nailed to the springing line of the semi-circle. Two fibrous plaster casts with pierced sides are made as in Figure 188, stage 2. These casts are then fixed upright on the bench the correct distance apart, with the rebated sides inwards, and then braced together. Plasterboard is cut to fit between the two rebated parallel edges, scored across its width, bent and nailed to the rebate around the semi-circular perimeter (Figure 188, stage 3).

The fibrous plaster semi-circular drum now formed is laid flat on to the bench surface and a suitable timber fence nailed as a surround on the bench surface (Figure 188, stage 4). The drum and bench surface are shellacked and greased, and a fibrous plaster cast taken of the semi-circular side of the drum and the bench surface, striking off at fence and top edge of the curved drum rim. Reinforcement lath and braces are positioned as shown in Figure 189, stage 6.

The fibrous plaster cast when set is taken off the mould, the reverse drum removed from the bench and the cast replaced in the original position, but upside down. The bench surface around the replaced half cast in greased, and the second face of the cast is made on the bench surface, overlapping the scrim reinforcement on to the curved soffit of the first cast. Bracing is made to fasten the two casts together (Figure 188, stage 5).

The stiles are made by making a box mould to the required size, as shown in Figure 188, stage 7. Two casts are taken from the mould and these are wadded and braced to the arched section, preferably laid flat on the bench to ensure true alignment. Finally, the joints are stopped, and made good to produce the plain, semi-circular, double-sided arch as shown in Figure 188, stage 8. A method of forming a keyhole arch is shown in the sketches in Figure 189, stages 1 – 8.

FIBROUS PLASTER ARCHES

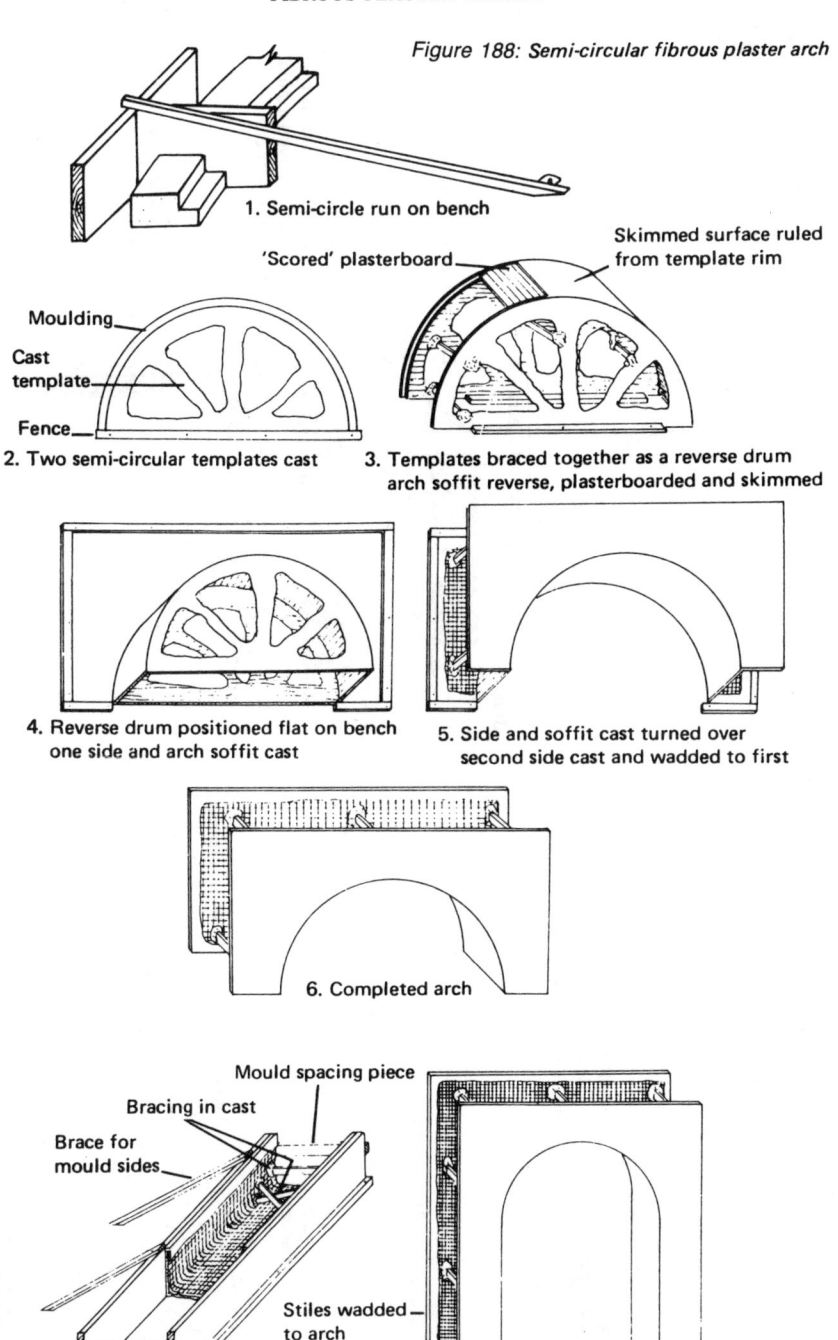

Figure 188: Semi-circular fibrous plaster arch

1. Semi-circle run on bench

Moulding
Cast template
Fence

'Scored' plasterboard

Skimmed surface ruled from template rim

2. Two semi-circular templates cast

3. Templates braced together as a reverse drum arch soffit reverse, plasterboarded and skimmed

4. Reverse drum positioned flat on bench one side and arch soffit cast

5. Side and soffit cast turned over second side cast and wadded to first

6. Completed arch

Mould spacing piece
Bracing in cast
Brace for mould sides
Stiles wadded to arch

7. Stiles cast in box mould

8. Arched doorway

BENCHWORK

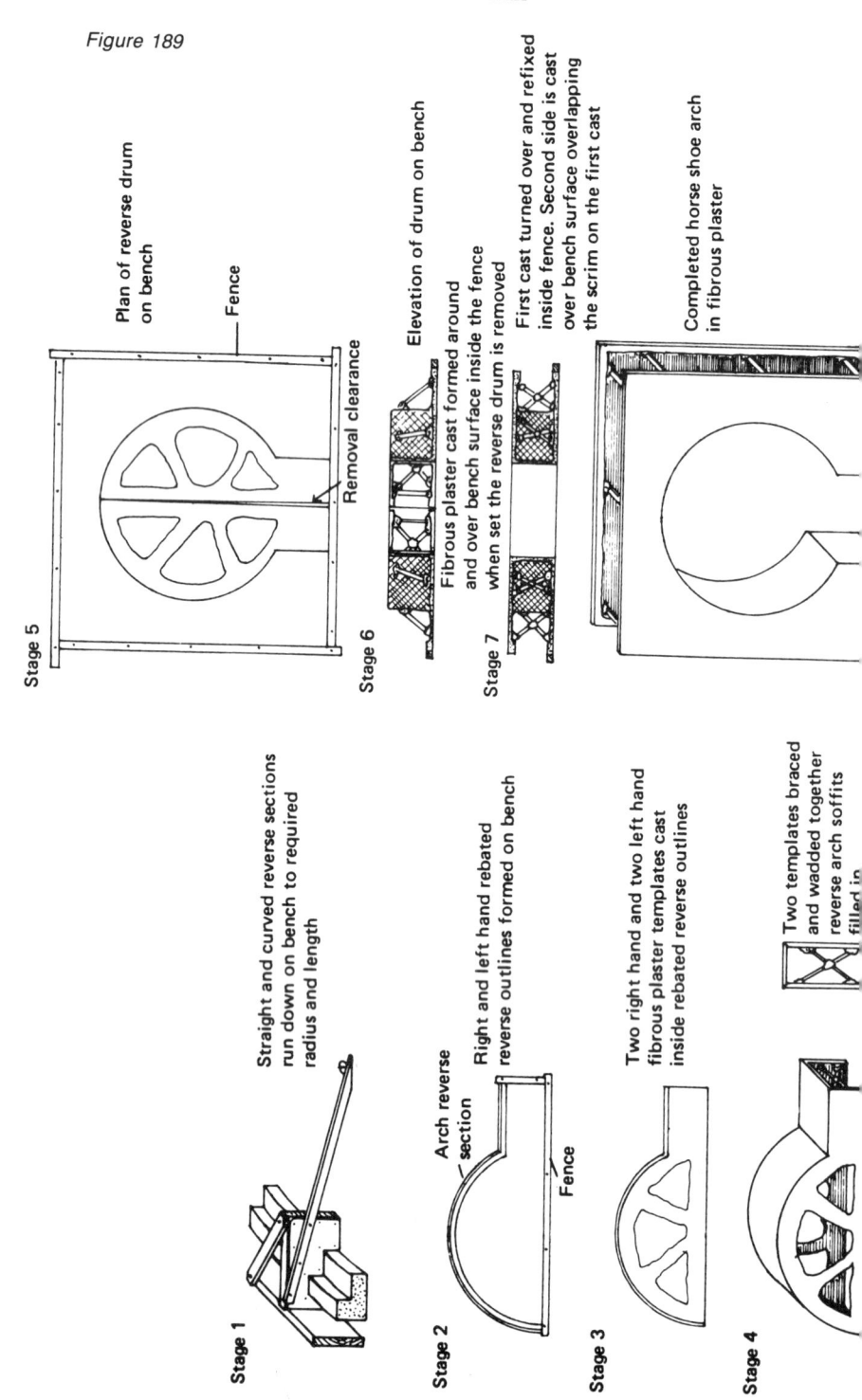

Figure 189

Stage 5 — Plan of reverse drum on bench; Fence; Removal clearance

Stage 6 — Elevation of drum on bench; Fibrous plaster cast formed around and over bench surface inside the fence when set the reverse drum is removed

Stage 7 — First cast turned over and refixed inside fence. Second side is cast over bench surface overlapping the scrim on the first cast

Completed horse shoe arch in fibrous plaster

Stage 1 — Straight and curved reverse sections run down on bench to required radius and length

Stage 2 — Arch reverse section; Fence; Right and left hand rebated reverse outlines formed on bench

Stage 3 — Two right hand and two left hand fibrous plaster templates cast inside rebated reverse outlines

Stage 4 — Two templates braced and wadded together reverse arch soffits filled in

222

BARREL CEILINGS – FIBROUS

Barrel ceiling sections of suitable size are cast from a drum which is a reverse mould for the curved ceiling (Figure 190).

Smaller semi-circular or segmental barrel ceiling drums can be formed by use of a gig stick to run the reverse outline. Larger curves may need to be run with a peg mould.

The reverse outline moulding, when run, needs to be cut to the segment shape required, shellacked, greased and a fence placed at the open side. Side and middle framework casts are taken from the segment frame mould. These are erected and braced together as shown in Figure 190. Spacing for the intermediate frames is approximately 600 mm. Laths are nailed into the rebates around the curved frame. The lathwork in turn is covered with canvas, cored out, then filled in and ruled off a section at a time to avoid excessive swelling. The curved surface, when set, is brought to a good smooth finish by use of a 300 mm joint rule and weakly gauged plaster.

The drum reverse mould now formed is shellacked and greased. If a rebated joint is to be formed at the sides then a smooth timber lath, bruised on the underside, is nailed around the side frames.

Casts are made with the normal lath reinforcement at right angles to the curve, bruised laths around the curve and, if required, overlapping timber battens (approximately 15 mm x 40 mm) wadded as shown in Figure 191 for strength in handling and fixing.

Figure 191a shows a method of forming a rebated joint in the casts and also adjacent casts with rebated joints.

Figure 191b shows a method of temporary bracing to barrel ceiling casts which can be useful in case of long storage or difficult transport. The temporary frames are sawn off on site, previous to fixing, and returned to the workshop for re-use with other casts.

LUNETTES IN BARREL CEILINGS – FIBROUS

Semi-circular lunettes can be formed by running a semi-circular reverse outline, and then making a frame mould as described for barrel ceilings. Two semi-circular framework casts are taken from this mould and they are erected and braced as shown in Figure 192 against a mainly completed barrel ceiling reverse mould.

Laths are nailed around the semi-circular rebate as shown and fastened at the barrel ceiling drum by nailing or wadding. Coring out is as described previously. The ruling off is carried out by a straight edge cantilevered from the paired semi-circular templates as shown.

BENCHWORK

Figure 190: Barrel ceilings — fibrous plaster

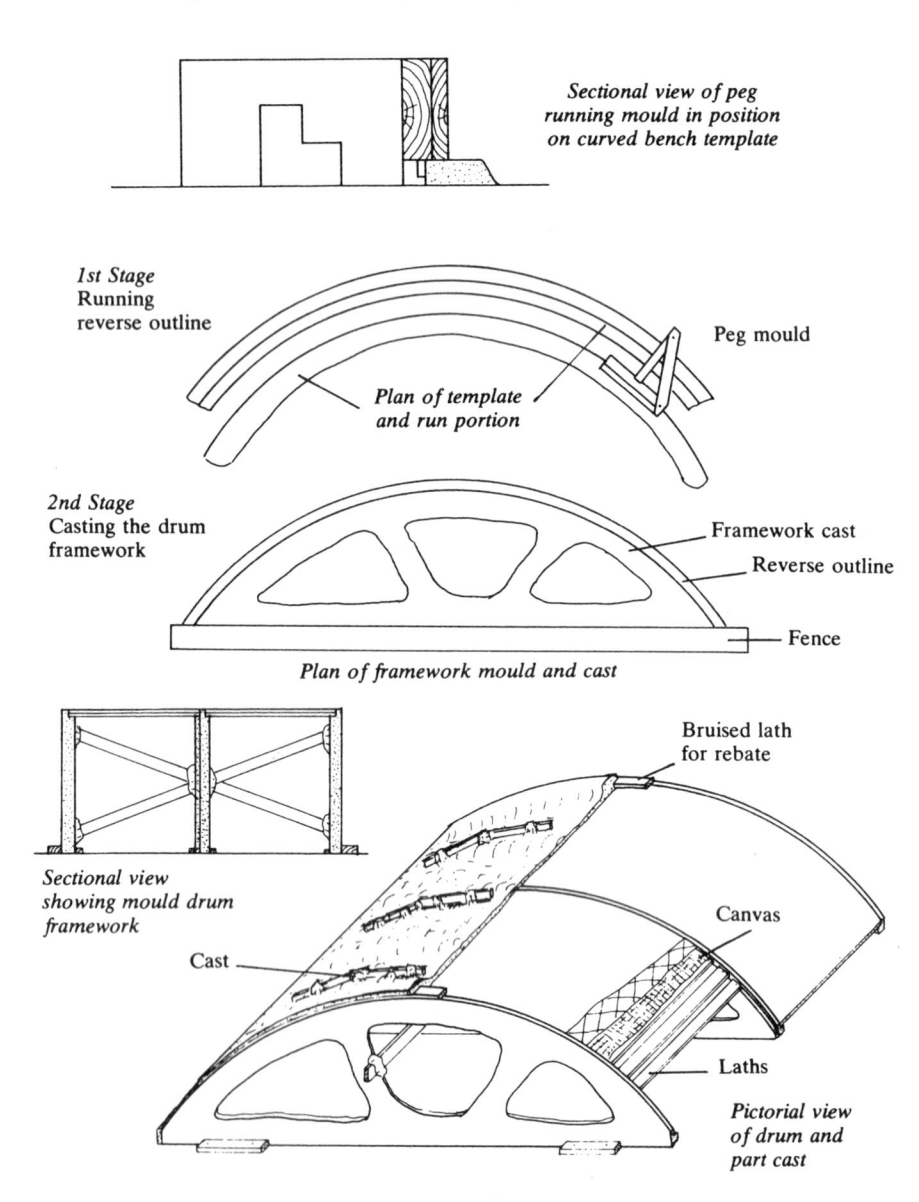

Sectional view of peg running mould in position on curved bench template

1st Stage
Running reverse outline

Peg mould

Plan of template and run portion

2nd Stage
Casting the drum framework

Framework cast
Reverse outline
Fence

Plan of framework mould and cast

Sectional view showing mould drum framework

Cast

Bruised lath for rebate

Canvas

Laths

Pictorial view of drum and part cast

224

BARREL CEILINGS

Figure 191a

Laths can be "on end" if fixing space allows

Section showing adjacent casts with rebated joints

Cast

Method of forming rebated ends in barrel ceiling casts

Bruised timber rule nailed to drum side

Drum side frame

Figure 191b

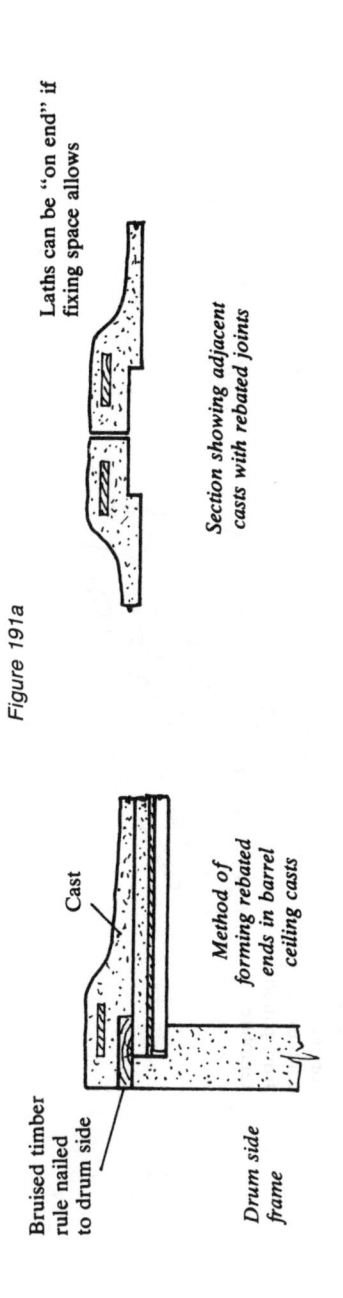

Temporary frame removed on site

Method of fixing temporary frame to casts for transport

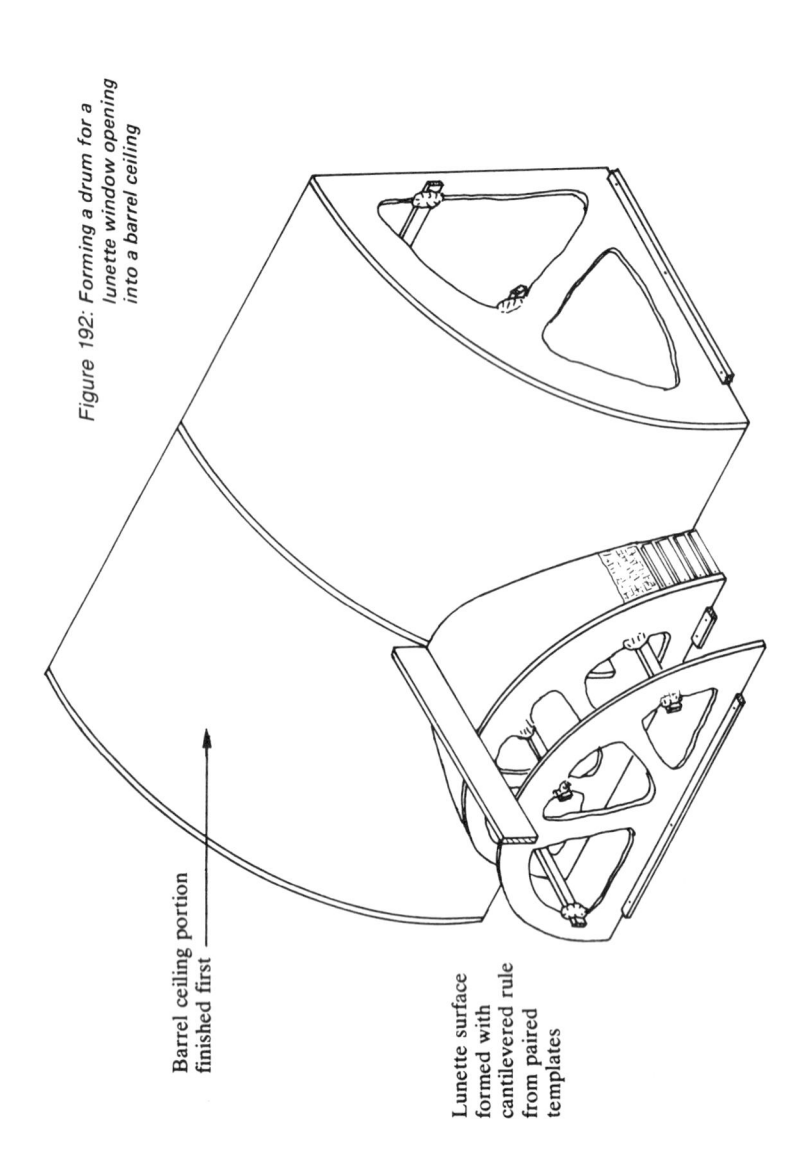

Figure 192: Forming a drum for a lunette window opening into a barrel ceiling

Barrel ceiling portion finished first

Lunette surface formed with cantilevered rule from paired templates

CIRCULAR DOMES – FIBROUS PLASTER

Plain semi-circular or segmental domes, which are circular on plan, can be done in a wide variety of ways. Small domes can be cast from a single hemispherical drum spun on a bench or floor. A centre post is erected, braced and a suitable quadrant or segmental profiled mould horsed up for spinning from a pivot point (Figure 193). When spun the pivot point is removed and the drum shellacked and greased. Fibrous plaster casts can be taken as required.

Medium sized domes may be sectionalised and casts made for each particular section, as shown in Figure 194. The dome cap reverse mould drum is run and the cast taken in one piece. Subsequent zone sections coming down from the dome cap are cast from separate drums, approximately 1.25 metres high and up to 2 metres long, depending upon the size of the dome. It is often convenient to divide the dome into six or twelve equal sectors, or gores, for convenience of setting out.

Top and bottom fences can be run in the mould. Side fences can be formed by nailing adjoining narrow laths to form a rebated edge, if so required.

Dome casts are best strengthened with 25 mm x 50 mm timber battens on edge to prevent distortion of the casts and also to improve handling and fixing.

Stopping of the rebated joints should include keying the recesses and scrimming with plaster soaked scrim. The finishing is done with a curved joint rule and flexible busk.

Whenever possible the bottom row of casts should include a base rim to provide a good finishing edge for the ceiling.

Larger domes can be divided into a number of convenient sized casts which can be taken from a single drum. Plain semi-circular or segmental domes which are circular on plan will have the same curvature throughout its surface. Therefore casts taken from a single mould will have the correct curvature for any part of the dome. It is essential to mark out carefully the required area and correct perimeter shape.

The upper portion can be spun and cast from a single drum as described for medium sized domes. Casts for the lower portions can be taken from the crown drum. The perimeter of the dome panel required can be set out on the crown drum from a centre line (Figure 195).

Fences are fixed around the panel perimeter and casts taken as required.

BENCHWORK

Figure 193: Fibrous plaster domes

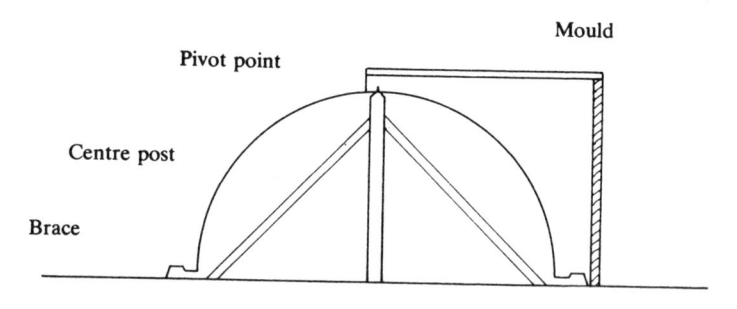

Mould
Pivot point
Centre post
Brace

Sectional view of mould in position for turning a small reverse dome drum

Mould for middle section
Mould for dome cap
Section centre post and braces
Mould for base section

Figure 194: Sketch showing method of sectioning medium sized plain domes

CIRCULAR DOMES

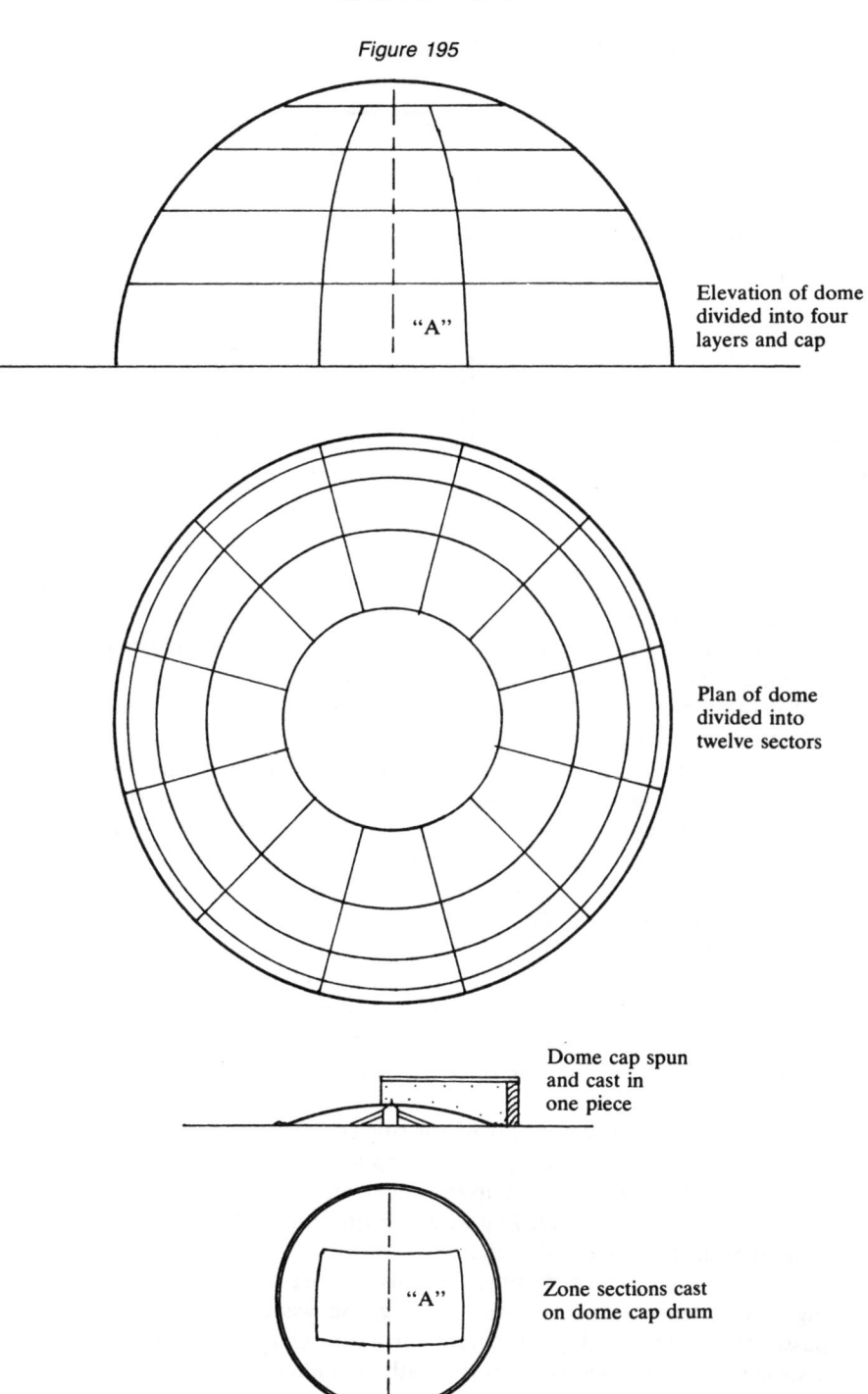

Figure 195

Elevation of dome divided into four layers and cap

Plan of dome divided into twelve sectors

Dome cap spun and cast in one piece

Zone sections cast on dome cap drum

229

Figure 196: Column reverse mould

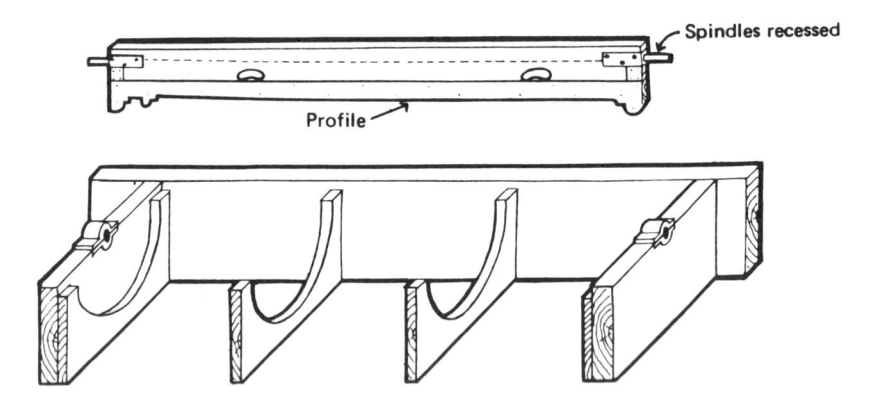

Spindles recessed

Profile

Mould box frame (side removed)

FIBROUS PLASTER COLUMNS

Plain columns, parallel or entasised, can be turned on a hand lathe. The resulting column model can then be half moulded, from which casts are taken to be wadded together or fixed in pairs on site.

It is often more convenient to form a reverse mould without the use of a column model. With this method an entasised rule is made as shown in Figure 196. The outline of half the column profile in sheet metal is nailed along one edge. The centre line of the proposed column is also marked on the rule and strong turning pieces screwed into the rule at each end of this line.

A mould box framework is made to the correct length and strong metal pivot points fixed at the top centres of each end. The entasised rule is pivoted between these points and a base framework as shown in Figure 196 is formed allowing approximately 18 mm clearance. This framework is covered with laths, scrim and plaster to within 6 mm of the desired outline, by muffling the entasised rule which is then traversed around the inside of the mould.

The top edges of the mould must be levelled all round from the mid section of the pivot point, so that the finished mould will be exactly half the completed column.

The muffle is next removed from the rule and the core deeply undercut for key. A finishing coat of neat plaster is applied and brought to a good finish with the pivoted rule, working up the upper edges to a good arris. The inside of the mould is finally cleaned up with a flexible busk, after which it is given a minimum of three coats of shellac.

If the casts are to form dummy columns they are paired up soon after casting and fastened together with screwing and wadding. The wads can be passed through the shell of the column on a stick and then brushed down with a splash brush fastened to a short length of timber.

Casts to be fixed around a stanchion need to have a rebate formed along the back of alternate long edges. The first half of the column is screwed to the firring, and wadded inside if required. The second half is screwed into the rebates of the first and the holes and joints 'stopped', finishing off with a flexible busk.

Entasised fluted columns in fibrous plaster

Several methods are used to form columns of this type, each method requiring great care at each stage of the process if satisfactory results are to be expected.

One method is to cut and horse up a reverse mould profile as shown in Figure 197. Allowance is made for cap and base moulds which can be turned separately as loose pieces. A bed is also formed in the column reverse mould to receive the reverse loose pieces for the flutes of the shaft. The notches in the shaft bed reverse are for tapes to pass under the loose pieces of flute reverses. The tapes are used to assist in lifting the cast and also to prevent breakage of any piece which may loosen itself as the cast leaves the mould.

It is usual practice to form the bed moulds for the cap, base and flutes reverses first. This base is then coated with shellac and greased. The muffle plates for the cap and base are removed, also the projection nibs and shaft profile. This allows the reverse mould profile, Figure 197, to turn easily when forming the cap and base reverses. When set these can be removed, sawn carefully into two or three radial parts, coated with shellac, greased and then replaced in their appropriate beds.

It may be desirable, in certain cases, to omit the moulded portion of the cap above the astragel and to turn the rest of the circular portion of the cap separately on the bench. This portion can be attached to the square abacus, piece moulded and the casts fixed to the rest of the column later. With this method any adjustment for height of the complete column, when fixing, can be made by varying slightly the length of the cap necking where the piecing will be made.

The reverse flutes can be formed as follows:

A profile is cut as in Figure 198 and this is horsed up to a diminishing running mould or a thumb mould as shown in the sketch (Figure 198).

A line is marked out on the bench for the long centre line of the flute reverse and diminishing rules fixed to the correct entasis in width for a single flute. The parallel portion can be run with a normal running mould and attached later.

A trial length of the entasised portion is run, keeping the curved ends of the thumb mould tight to each of the rules (Figure 198). Short sections of the moulding run in the trial length are now cut at various points along its length to check for accuracy of width. Another method of checking is to assemble a number of short pieces of similar width around the inside perimeter of the column reverse mould to prove that they fit accurately to exactly half a column. Any adjustments to thickness and correct entasis must be made at this stage.

Figure 197

- Muffle plate to form bed for cap loose piece
- Turning piece
- Centre line of column
- Notches to provide clearance for tapes in mould
- Entasised rule for fluted column mould
- Handhole

Figure 198

- Thumb mould
- Reverse flute moulding

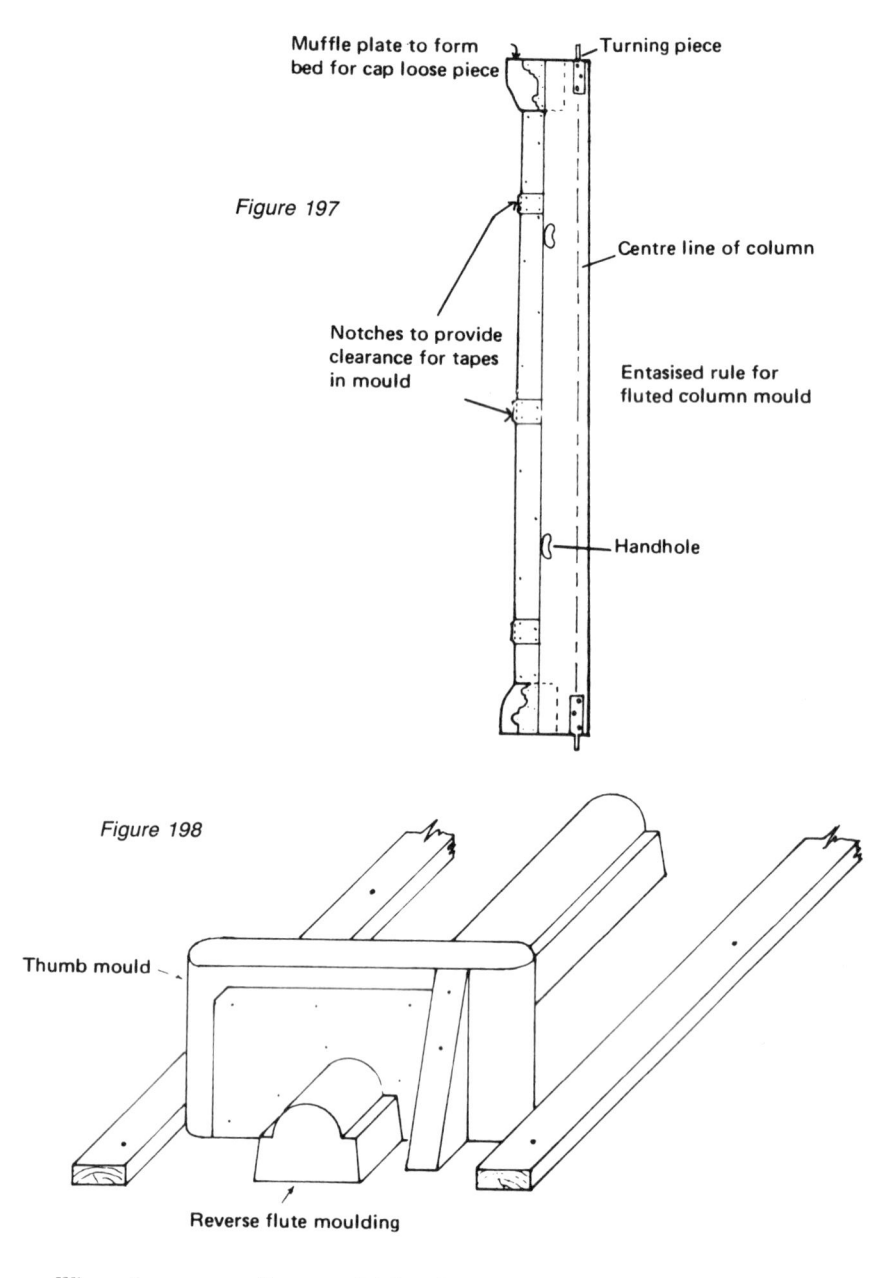

When the reverse flute model has been satisfactorily run and tested it is measured and cut to the correct length. Adjustment for entasis in the *depth* of the flute can be made by planing carefully to the marked diminished end, checking with a special entasis rule. (The latter rule will only diminish approximately one-sixth of the flute.) Returned ends can now be carved on the reverse

FIBROUS PLASTER COLUMNS

Figure 199

Mould
Reverse flute section

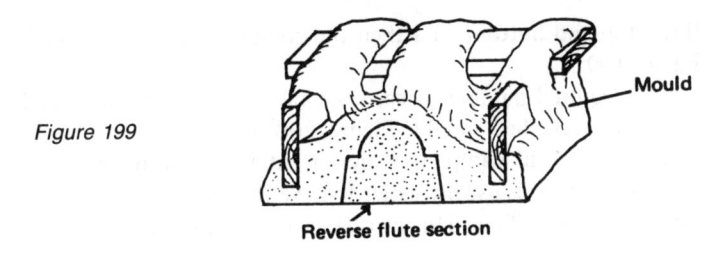

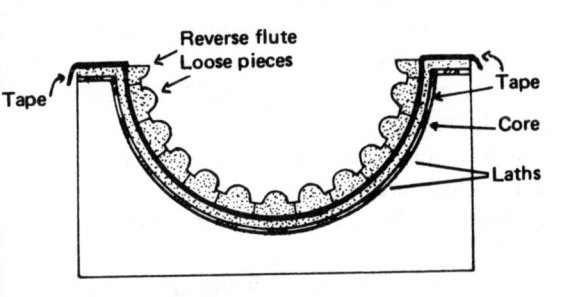

Reverse flute Loose pieces
Tape
Tape
Core
Laths

Figure 200: Section through a reverse mould for a fluted column shaft

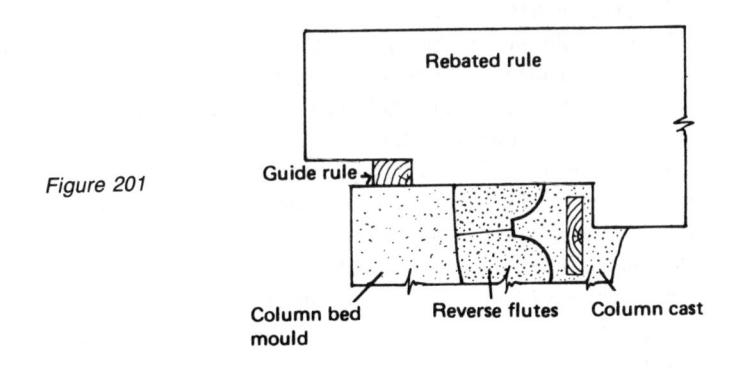

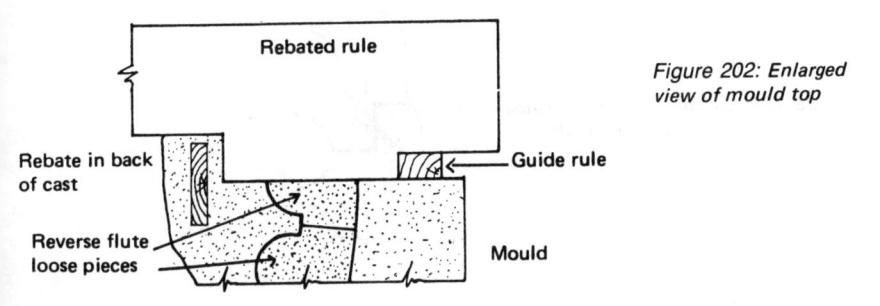

Figure 201

Rebated rule
Guide rule
Column bed mould
Reverse flutes
Column cast

Rebated rule
Rebate in back of cast
Guide rule
Reverse flute loose pieces
Mould

Figure 202: Enlarged view of mould top

233

flute, top and bottom, and then the completed model is moulded as shown in Figure 199.

The required number of casts are taken from this mould and placed in the column reverse bed mould, Figure 200, where they will bend to the correct entasis while still damp. It is important that exactly half the column is cast each time and it is usually easier to have the piecing in the middle of the flutes, rather than the fillets. This can be arranged as shown in Figure 200 by halving the top two reverse flutes.

The sketch, Figure 201, shows a method of forming the rebated edge to the column cast. (This rebate need not necessarily be full length of the column.) A corresponding rule in edge for the other half of the column cast to fit into the rebate of the first half can be formed as shown in Figure 202. The guide rule should be nailed parallel to the column edge.

The inside surface of the column mould and all loose pieces are well coated with shellac and greased. Tapes, which may be bands of scrim, are placed in the notches under the loose pieces and left projecting over the edges of the mould, which is now ready for casting.

Half column casts, when set, can be lifted out with the aid of the tapes, replacing the flute reverses in the bed mould after extraction from the cast. The two half column casts should be screwed, or fastened together with a tourniquet, as soon as possible after casting to prevent possible distortion on drying (Figure 203).

Dummy columns can be wadded together by use of a splash brush fastened to a lath after placing the two half column casts in a cradle. The scrim is also passed through the column shell on a lath, placed over each joint in turn and brushed well down with plaster.

A more reliable, though costly, method is carried out by forming a complete model of the fluted shaft and piece moulding this in two separate half moulds. The column shaft is turned on a spindle, the entasised portion being filled out to the outline of the fillets, but the straight portion is rebated 25 mm.

Half sections of the straight portion of the fluted shaft are run down with a normal running mould over a core on the bench. These fluted half shells are now bedded carefully in the rebated portion of the turned column.

Figure 203: Section through a rebated joint in column casts

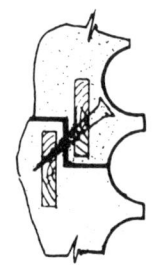

FIBROUS PLASTER COLUMNS

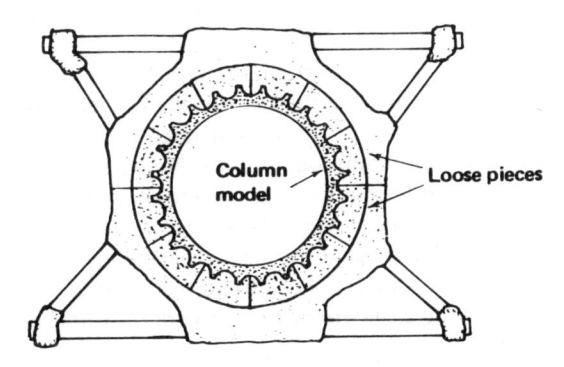

Figure 204: Fluted column model enclosed by two half moulds

The end sections of the flutes are now marked off at the diminished end of the turned column shaft and the lines of each fillet marked along the shaft all round the entasised portion. The individual flutes are next carved out carefully, checking each with an entasis rule for accuracy. This part of the job is often entrusted to a carver.

On completion the model is piece moulded. This is best achieved by dividing the model horizontally along the centre line of a flute on each side of the shaft. Two or more loose reverse pieces are formed on each side of the upper surface and then covered with a strong fibrous plaster case. The mould and shaft model are now turned over exposing the other half of the fluted shaft, which can now be moulded in a similar manner (Figure 204).

The completed half moulds, after coating with shellac and grease, are both used for casting separate halves of the fluted shaft. These half column casts are made with a rebate on the back edge, as described previously, and screwed together immediately after casting to prevent distortion which may occur in storage or transport.

Fibrous plaster casts and fixing

The most useful type of *scrim* used in fibrous plasterwork is 825 mm wide, 100 grammes to the metre and usually supplied in rolls of 250 – 300 m.

Timber laths and battens used in casts vary from 25 mm x 3 mm to 50 mm x 18 mm for normal sections, but 75 mm x 25 mm or even larger sizes may be necessary for bigger benchwork jobs.

Wadding is the fastening of the framework to the back of the cast or mould by the use of scrim saturated in gauged plaster. It is also the method used in the welding together of adjoining casts when fixing. Scrim, 200 mm or more wide, is dipped into gauged plaster of creamy consistency and pasted over the back of the casts, thereby bridging the joint. If access to the back of the cast is not easily obtained then a *hand-hole* is cut in the joint so that the wad can be pasted down into the back. If the section is not large enough to allow a hand-hole, then the wad is worked into the joint with a small tool.

Fixing casts to timber is carried out, in the normal way, by nailing through the timber reinforcement in the casts to the timber in the background. Nails used should be galvanized to prevent later corrosion and the lengths vary from panel pins to 150 mm round heads. Normal sizes 50 mm — 75 mm long. Screws are used for fixings that are unsuitable for nails, and these too are usually galvanized but occasionally brass screws are used.

In fixing to timber ceilings the casts are nailed into joists or bridging pieces. Fastenings to the wall can be into timber grounds and sometimes plugs. Around beams the casts are fastened to cradling and in other framework to timber firring.

Temporary supports or rules are often used to hold the casts during the provisional fixing and the nails left projecting. The casts are then adjusted for level, plumb, square or linable, and then the nails driven home and countersunk with a punch. This is followed by wadding, making good any damage, stopping or filling the holes, joints and mitres.

Fixing casts to metal runners in fireproof construction is carried out by use of galvanized wire and wadding. Casts for this type of work are made usually with lath reinforcement placed on edge with several clear spaces between the cast and lath. When fixing, the wire is passed under the lath reinforcement in the cast and over the metal runner or channel section. When adjusted to correct alignment the wire must be tightened by twisting with the aid of a tourniquet. A wad dipped in strongly gauged plaster is wrapped over the metal construction and around the wires, pasting the ends down to the back of the cast. The wad assists in preventing corrosion of the wire and also stops any untwisting of the wire which may occur if it is subject to extra strain.

PORTLAND CEMENT CASTS

Two alternative techniques may be used to obtain Portland cement casts from plaster piece moulds. They are the wet and semi-dry casting methods.

In the wet method a wet mixture of sand or granite chippings and Portland cement is placed over the surface of the mould to a thickness of approximately 25 mm. This can then be backed up with a fine concrete mix to within 25 mm of the top of the mould, then filled out and finished with a layer of the original mix. In larger section work reinforcement bars may be necessary.

The casts may be demoulded after twenty-four hours and then cured until hard enough for 'dressing'. This may consist of dry grinding with abrasive discs on a power drill or wet rubbing with a carborundum stone. Holes or slack places in the cast may be filled in with a strong sand and cement mix after the initial grinding or rubbing. A further period of hardening is allowed and then a final grinding or rubbing with finer abrasive discs or carborundum stone.

Wet method casting with Portland cement has many disadvantages when used with plaster moulds. The difficulty of vibrating the mould, or the newly placed cast, means that the mix must be wet enough to fill all parts of the mould easily. Air holes can form on the face of the cast as excess water dries out and uneven shrinkage often causes distortion to the subsequent shape of the cast. The plaster moulds also deteriorate quickly when subjected to repeated periods of wet casting, even when the contact surface has been sealed with several coats of shellac.

In the semi-dry casting method the Portland cement and sand are mixed together with water until the mix will stick together when pressed in the palm of the hand, without being able to squeeze out any excess moisture.

The plaster piece mould, previously sealed, should be dusted out with French chalk before casting. The semi-dry mix is applied in layers of approximately 50 mm thick over the inside surface of the mould and compressed tightly by tamping thoroughly with a wood block or mallet. Small casts are usually filled in solid with the semi-dry fine mix but a coarser mix may be used for the core if desired. Hollow casts can be made by filling in the middle of the cast with damp sand. The sand is tamped in the same way as the rest of the cast but is easily removed when the cast is set, leaving a cast with a hollow core.

It is often convenient to use the same mould many times in the same day with the semi-dry casting method. When the mould has been filled a piece of timber, called a pallet, is placed over the top surface of the cast and mould side pieces. The piece mould and cast should now be turned over on to the bench keeping the pallet tight to the back edge of the cast. Next, the mould case can be lifted off and the pieces carefully withdrawn from each side. Any irregularities in the surface of the cast can be touched up at this stage by use of small hand floats.

The semi-dry casts must be cured, and this can be achieved by placing slats of timber suspended above the upper surface of the casts and then covering completely with damp sacks. After twenty-four hours the casts should be immersed in water for at least seven days.

GLASS-REINFORCED PLASTICS (GRP)

Moulding and casting techniques used in hand lay-up methods for the production of fibreglass-reinforced plastics are similar to that used in fibrous plasterwork. Certain exceptions are necessary.

Plaster moulds used should be shellacked and coated with wax polish. A suitable proprietary release agent should be applied to the mould by brushing or spraying a thin film over the whole surface. Casting should not begin until the parting agent is dry.

An accelerated polyester resin gel coat should be mixed thoroughly with 2% — 3% catalyst (22 cc — 34 cc catalyst to 1 kg resin). This mix is applied evenly over the surface of the mould with a 75 mm or near-sized paint brush.

This coat is allowed to set to an almost tackfree condition at which stage a second gel coat is applied. When the second gel coat has set to a rubbery texture, a liberal coat of catalysed lay-up resin is applied and while this is still wet the chopped strand mat is placed to cover the mould area. A further application of lay-up resin is brushed over the chopped strand mat, using a stippling action with the brush to ensure thorough wetting of the glass fibre mat and total impregnation of the resin into the mat strands. Approximately 1 — 1.5 kg of lay-up resin will be required for each square metre of chopped strand mat. The setting time for the lay-up resin normally varies between one and two hours after incorporation of the catalyst 0.5 — 1.0% (7 cc — 11 cc catalyst to 1 kg resin).

If extra strength is required further layers of glass-fibre mat are applied repeating the procedure as described above.

The chopped strand mat should be rolled carefully with a suitably contoured roller to remove all entrapped air from the cast. Any trimming required at the edges of the cast should be carried out when the laminate has just set, using a sharp knife or similar cutting tool. The cast will set and harden more quickly at higher temperatures, a minimum curing temperature of 15°C is recommended.

When fully hardened the cast should be released from the mould and the parting agent washed from the surface of the cast with warm water.

All tools and brushes used for the application of resin should be cleaned immediately after use with acetone cellulose thinners, etc. Parting agent brushes can be cleaned in soapy water. It is good practice to wear plastic gloves or alternatively to use barrier cream on the hands when carrying out this type of work, and also to apply a hand cleanser cream afterwards.

GLASS-REINFORCED GYPSUM (GRG)

Glass-reinforced gypsum consists of high-strength glass fibres in a gypsum plaster mix. The resulting fibrous material has strength properties well in excess of unreinforced plasters.

The glass fibres should be uniformly distributed in a strong plaster mix for maximum final strength. Mixes containing a low water/plaster ratio will give greater strength on setting, but such mixes may be too thick for practical usage on many casting jobs.

For normal hand mixes chopped glass strands of 10 mm up to 45 mm are used in varying percentages from 3 — 6 per cent of glass to weight of plaster. Chopped strand mat can also be used as added reinforcement.

The introduction of random glass strands in a high water/plaster ratio mix (such as used in normal fibrous casting) will give greater dispersion than if used in thicker mixes. High water contents in a glass-reinforced gypsum mix will result in a weaker final strength. Excess water can be removed by suction or pressure before the setting action is complete and this system is used by some manufacturers of GRG by use of special vacuum and pressure equipment.

GLASS-REINFORCED GYPSUM

Spray fabrication techniques are also used for GRG and these are carried out by simultaneously spraying chopped strands of glass fibre rovings and plaster slurry on to the mould surface. Special spray and roving chopper units are required for this method.

Increases in strength vary from three or four times the strength of normal plaster for glass-reinforced gypsum with glass contents of up to 7 per cent using hand methods, and up to thirty times the strength of ordinary plaster for glass contents of 12 per cent using compaction methods in casting.

CHAPTER ELEVEN

Mechanical Plastering

A variety of different types of plastering machines are being operated in present-day plastering. Each machine has its own features and advantages for particular types of work. Machines are available for mixing, pumping, throwing, spraying, vibrating, trowelling, grinding and polishing.

Mixers include pan and roller types for grinding and mixing, drum type concrete or cement mixers, revolving paddle, etc powered by petrol or diesel. Small mixer agitators for stirring small batches can be operated from compressed-air lines.

Mechanical pumps for plastering mixes are often combined with mixing machines, and can be used to pump the mixed plaster to great heights on multi-storey building sites. Plasticisers have often to be used with cement mixes for pumping and subsequent spraying. The use of pumps on multi-storey blocks eliminates the necessity of a series of bankers and storage bays for material on each floor.

Throwing machines include the hand operated Tyrolean machine, already described. The Lubecker plaster thrower is mechanically operated and can be used for applying one or two coat work. The freshly applied coats should be straightened by ruling in, and gang organisation for this and other essentials for continuity are most important.

Spraying machines are usually composed of a pump with delivery pipe and a compressed-air pipe leading to a nozzle. As the material escapes through the nozzle it is atomised by compressed air from the air tube. The controls for air and material are situated on the spray gun and the supply can be stopped or started by the operative as desired. Different sizes of nozzles are available for undercoats or finished work. Spray plastering after application can be ruled in and finished by traditional methods or by adjusting the machine it can be used for application of textured surfaces. Spraying machines are manufactured for the spraying of plaster, lime/sand or cement/sand mixes or multi purpose. As previously stated it is recommended that plasticisers be used with cement mixes for this type of work. The smaller types of mechanical spraying machines are mostly powered by electricity.

Machines for vibrating and consolidating concrete work include poker vibrators for insertion in the freshly laid concrete and clamp-on vibrators for attachment to concrete formwork. In each case the concrete is vibrated to adjust itself into the densest possible mass and because of their use leaner

MECHANICAL PLASTERING

mixes with a lower water content may be utilised. Vibrators may also be attached to special straight edges for ruling in and consolidating large plain concrete areas.

Mechanical trowels are available for wall and floor trowel finishing. Some of these are multi bladed appliances, the blades circling around the central pivot point. For the consolidation and finish of concrete floors power floats can be used. These machines have a float plate of about 600 mm in diameter made of hardened steel. They can be powered by petrol or electricity, with the controls situated on the handlebars.

When a machine of this type is used for the surface finish a lower water/cement ratio is recommended for the concrete mix. The concrete should be ruled in to correct levels, and the power float used to compact and float the surface to a good condition as soon as the surface is stiff enough to prevent deep penetration of the operative's feet. All types of concrete and granolithic floors can be surface finished by this or similar methods.

Interchangeable fittings are also made for attachment to power floats to enable them to be used for texturing and also for the cleaning of base slabs before topping. Other attachments manufactured can be used for grinding floors to levels and for exposing aggregates to obtain terrazzo type finishes.

Most of the machines available in this country for pumping and spraying of plastering mixes are foreign, mainly German. Only one current machine of this type is wholly British, although other types of British machines are available.

Machine application of floating coats has been used extensively in the U.S.A. on suitable sites. Its development in this country will depend upon the arrangement of suitable site conditions and a ready acceptance of the advantages of mechanical plastering by operatives. These advantages are only worthwhile if a continuous series of large plain areas can be made available for the plastering crew, without interruption. Plastering gangs should also be trained in spraying, straightening and finishing so that the jobs are interchangeable among the members of the crew.

The size of gang required for mechanical spray application of floating mixes varies according to the type of machine used. On the largest machines a recommended number of up to eight men, comprising two men on the machine, one labourer, two men ruling in or darbying the applied plaster and up to three plasterers for rubbing up and finishing of small areas if necessary. An average size gang for a spray machine is three or four plasterers and one or two labourers.

On better-class work metal screeds can be fixed before spraying commences, and the sprayed plaster ruled in off these screeds which are removed and filled in before rubbing up.

The usual method is to spray a panel about 900 mm wide floor to ceiling, if this is not too high. Whilst this material is still soft it is straightened and given a provisional filling in with a darby. Meanwhile the machine nozzle is directed to the adjoining 900 mm wide panel, and on completion of this, a thin filling-in spray can then be directed again over the first panel to fill in any slack places. The plasterers darbying and finishing the floating coat must co-operate closely with the spray operative, to ensure that the spray application is uninterrupted before completion of the scheduled work.

Small areas, particularly awkward places such as reveals, narrow margins or cupboards may best be applied by hand as the spray plastering progresses.

Stopping of the spray machine for long periods can present problems, particularly with gypsum plaster mixes. Whatever the mix used, long stoppage periods must be avoided to prevent segregation of the different materials in the mix, and to prevent possible blockages due to drying materials in the nozzle. The nozzle can be kept moist by placing in water if the machine is only stopped for a short time, but long breaks may require a complete pumping out of the mix from the machine and delivery pipe.

Some machines are only capable when pumping mixes with lime, but others can cope with lightweight aggregates, sand/lime, cement/lime/sand, cement/sand plus plasticiser or cement with lightweight aggregates.

Any mix which is to be used for pumping and spraying must remain in a plastic state as it proceeds through the delivery pipe. Fatty mixes, although easy to pump and resistant to segregation, are not usually desirable because they increase the cost and reduce the quality of the finished work. In general a well graded sand is best, avoiding sands with an excess of any one particle size. Unsuitable sands which have been found difficult to pump can be corrected by adding up to 25 per cent of fine soft sand. Other methods of correction include the use of a plasticiser, care being taken to avoid an excess addition as the extra air entrainment may cause trouble. Pressure in the pipelines can result in air being forced out of the mix, resulting in a stiffening of the mix and finally segregation.

In extreme cases of segregation the materials will eventually be discharged through a delivery pipe as separate plugs of neat cement, sand and water.

It is recommended that a slurry of lime should be passed through the pump and delivery pipes before pumping sand/lime/cement or sand/cement mixes. If the pump is to be stopped for longer than 45 minutes with cement mixes a lime slurry can be pumped into the pipeline. This lime slurry can be discharged into a container, for use later, after which the machine will be ready to resume normal spraying.

When the system needs clearing out at the end of pumping operations the mixing hopper should be emptied and then a lime slurry pumped into the delivery pipe. The pump should be stopped, the pipe disconnected, and the pump washed out. A sponge or newspaper is next placed in the delivery pipe behind the slurry and pumped through the water. A sponge or newspaper can also be pumped through the pipelines to remove any sand left in the pipelines due to segregation from previous mixes.

Any advantages of mechanical plastering over hand application are dependent upon continuity of operations. Unless this continuity can be guaranteed with reasonable assurance, plus efficient organisation, workmanship and plastering machines, then its rapid introduction to this country will be confined to selected sites.

ONE-COAT PROJECTION PLASTERING

This is a system of spraying on and finishing the plastering in one application, instead of the normal two coat method of floating and finishing coats. The plaster used is a gypsum plaster with additives only to control the setting time, improve water retention and workability. No aggregate as such is used so no segregation can occur to cause blockage problems.

The projection plaster is supplied in 40 kg paper sacks or in bulk delivery to silos on site. Sacks, if used, are placed on the grill of the machine, slit with a knife and allowed to feed into the machine by gravity. The machine has the water pressure balanced either direct or with a small compressor in addition. The dry plaster only mixes at the base before entry to a small tube before delivery. This also helps to avoid risks of blockage.

Application of the projection plaster is best carried out by spraying in a series of zigzag horizontal ribbons about one metre wide in close contact and approximately 12 mm thick. When a sufficient area has been applied and whilst still soft, a long metal feather-edged rule is used to straighten the surface. Any slack places or hollows can be filled in by using the surplus removed from the feather edge, or further spraying from the machine may be necessary. As the plaster stiffens the feather-edge rule is used to straighten the surface free of bumps, hollows or blemishes.

The plaster is then allowed a further period of stiffening, during which time the internal angles should be planed out square by using a special angle plane having a series of angled blades set in a float frame.

Water is then lightly sprayed over the surface and a hand or power float rubbed over the surface. The hand float has a sponge rubber face and the rubbing action on the wet surface will work up a slight 'fat' on the surface. A special two-handled trowel is used to trowel the recently softened surface to a flat even texture. When this has hardened further it can then be given a final trowelling off, using a normal plastering trowel.

Sketches of tools used in projection plastering are shown in Figures 23 to 28.

The potential of spray plastering machines has always been appreciated in the British Isles but their use has only been limited. With both the special plasters and external renders designed only for spray application, also the use of smaller machines, there is now more chance of success on a larger scale. When the free movement of the European companies after 1992 becomes a reality they will bring their own spray plastering expertise to this country and an increase in this form of plastering may become commonplace.

CHAPTER TWELVE

Geometry of Arches

ARCHES STRUCK FROM ONE CENTRE

Stilted semi-circular (Figure 205)
This is an ordinary semi-circular arch resting on stilts which are a vertical continuation of the arch moulding below the springing line. Parts of the arch are identified. Note that the arch joints and keystone are radial from the centre, which in turn is at the mid point of the springing line.

Segmental arch (Figure 205)
The span is bisected and the position of the rise found by measuring vertically above the springing line on the bisector. One side of the span and the rise are now bisected. The intersection of the two bisectors will determine the centre for the arch.

Horseshoe (Figure 205)
The span and rise are set as described for the segmental arch. Span and rise are bisected to determine the position of the centre on the first bisector.

ARCHES STRUCK FROM TWO CENTRES

Moorish (Figure 206)
The position of the two centres can be found as follows. Firstly an arbitrary line is drawn horizontally about one-third of the distance between the springing line and the rise. Span and rise are bisected from both sides to determine the position of the two centres on the arbitrary line.

Gothic arches (Figure 206)
These are pointed arches, struck from two centres each of which must be on the level of the springing line. When the radius is equal to the span it is termed an Equilateral Gothic. If the radius is less than the span it is Obtuse Gothic and when the radius is greater than the span it is termed a Lancet Gothic.

With the Equilateral Gothic only the span requires setting out, the radius (equal to the span) is struck from each side of the span until the two sides intersect.

ARCHES STRUCK FROM TWO CENTRES

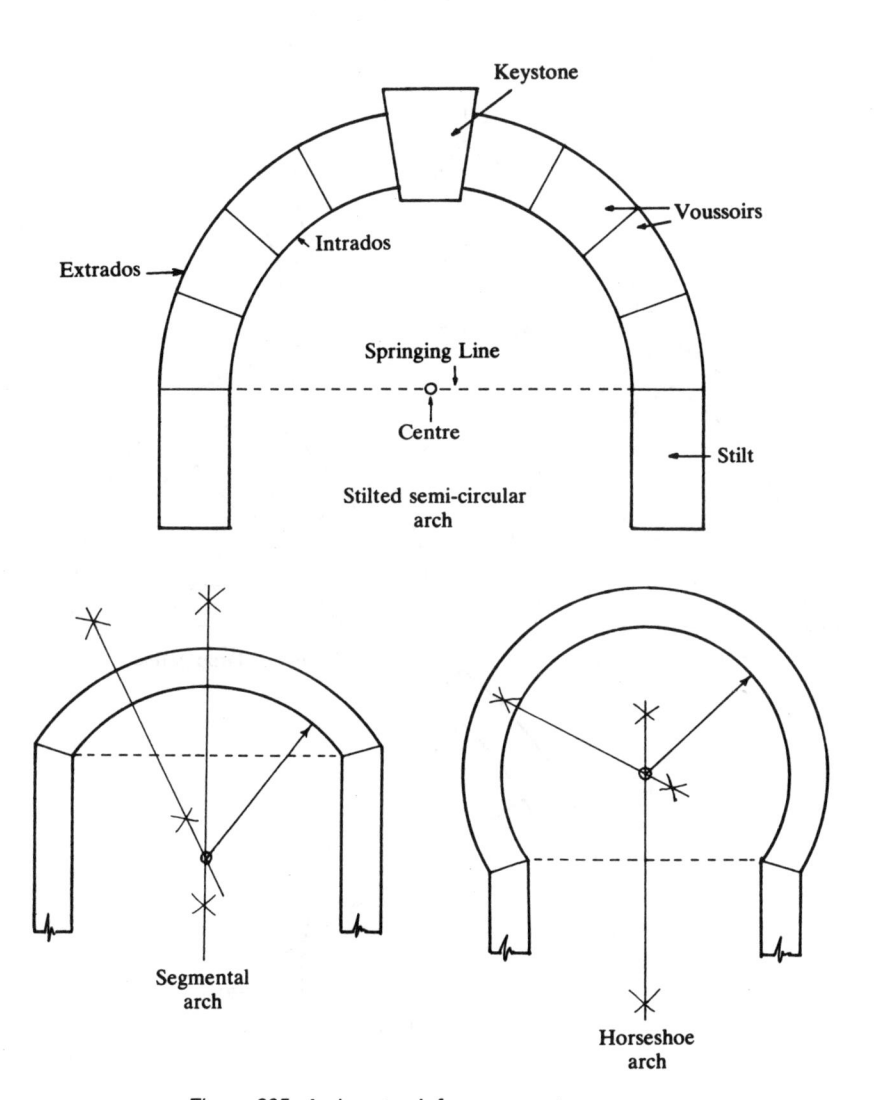

Figure 205: Arches struck from one centre

GEOMETRY OF ARCHES

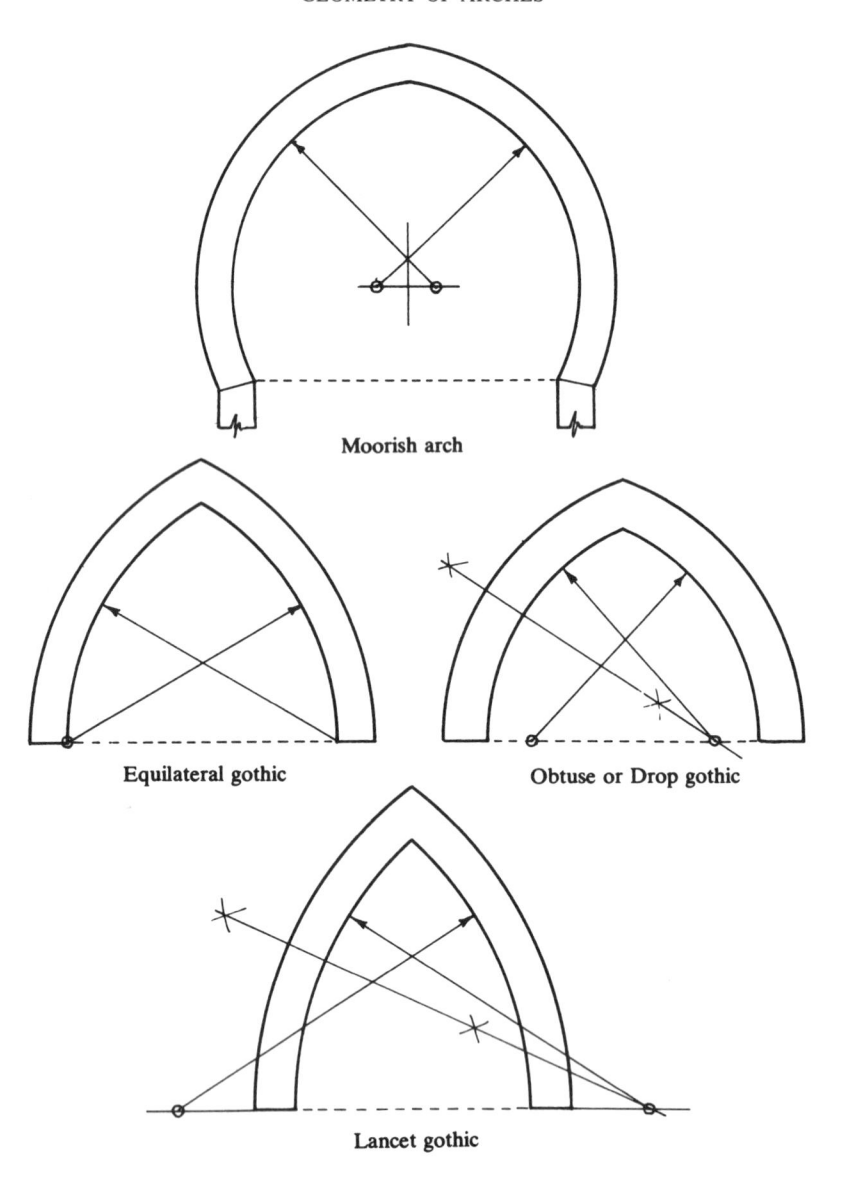

Figure 206: Arches struck from two centres

Drop Gothic and Lancet Gothic arches are formed by setting out span and rise and then bisecting these points. Where the bisectors cross the springing line, or the extension of the springing line, will determine the two centres. (Normally it is sufficient to bisect one side only, the second centre can be found by measuring across an equal distance for the vertical centre line.)

ARCHES STRUCK FROM THREE CENTRES

In the three-centred false *Elliptical Arch*, Figure 207, the span and rise are connected as shown. A semi-circle is drawn from the centre of the span with radius equal to half the span. A circular arc is drawn from the rise with radius equal to the distance between the rise and the crown of the semi-circle, until the arc cuts the connecting line between rise and span at point 'A'. The left hand edge of the span and point 'A' are now bisected. Where this bisector crosses the springing line determines one centre and where it crosses the vertical span bisector gives the position of the second centre. The third centre can be measured away from the centre of the springing line equidistant from the first centre. Common normals are drawn through each centre and the three arcs drawn as shown.

Ogee arch (Figure 207 – rise optional)
The span is bisected to find one centre. The other two centres are determined by drawing construction lines at 60° from the first centre in each direction until they intersect with vertical lines drawn from each end of the span. The four arcs should intersect on the two common normals connecting the centres.

ARCHES STRUCK FROM FOUR CENTRES

Tudor arch (Figure 208 – rise optional)
The span is divided into four equal parts. Construction lines are drawn at 45° through the first and third divisional points as shown. A square is formed centrally beneath the springing line with side equal to half the span. Each corner of the square becomes a centre, the diagonal lines produced are the common normals.

Tudor arch (Figure 208 – fixed rise)
The span and rise A, B and C are positioned. An arbitrary vertical line equal to about two-thirds of the rise is erected from 'A' to establish point 'D'.
'E' is found by measuring the distance 'A' – 'D' along the springing line from 'A'.

GEOMETRY OF ARCHES

3-centred false elliptical

3-centred ogee arch
(rise optional)

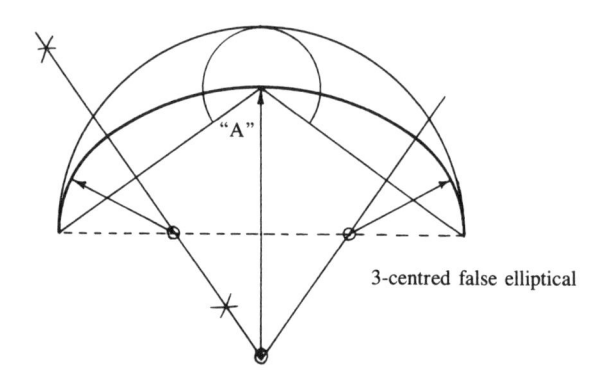

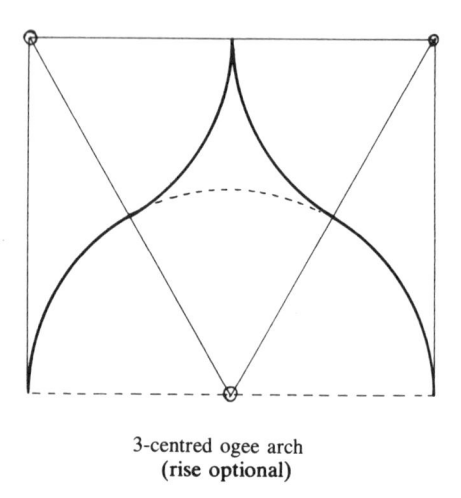

Figure 207: *Arches struck from three centres*

Points 'D' and 'C' are connected and a line is drawn at right angles for 'C', along which the distance 'A' – 'D' is measured from 'C' to position 'F'.

'E' and 'F' are next bisected and this bisector is extended until it meets the line 'C' – 'F' produced to intersect at point 'G'.

'G' and 'E' are two centres and a line through both gives a common normal. The other two centres 'H' and 'J' can be found by direct measurement away from the vertical centre line.

ARCHES STRUCK FROM FOUR CENTRES

Ogee arch (Figure 208 – with stated rise)

Span and rise are set out to find points 'A', 'B' and 'C'. The connecting lines 'A' – 'C' and 'B' – 'C' are drawn, and an arbitrary point 'D' is positioned just over half way along line 'A' – 'C' from 'A'.

Points 'D' and 'A' are bisected to find centre 'E' on the springing line. A line drawn through 'D' from 'E' will establish centre point 'F' horizontally from 'C'. Centres 'H' and 'G' are found by direct measurement.

True elliptical (Figure 209 – auxiliary circles method)

Span and rise are set out. Semi-circles with radius equal to the rise and half the span are drawn. The semi-circles are divided by any number of radial lines, not necessarily evenly spaced. Vertical lines are drawn from the intersection of each radial line with the outer semi-circle. Horizontal lines are drawn from the intersection of each radial line with the inner semi-circle. The paired intersections of vertical and horizontal lines will position a series of points through which the arch outline can be drawn free hand or with the aid of French curves or a flexicurve.

True elliptical (Figure 209 – trammel method)

The major and minor axes are drawn. A template is made by marking the lengths 'A' – 'B' equal to half the major axis and 'B' – 'C' equal to half the minor axis on a suitable staff.

Points around the ellipse can be formed by keeping the point 'C' of the template on the major axis and point 'A' on the minor axis. The other end of the template ('B') will be on the circumference of the ellipse. A series of points are found around the arch by placing the trammel rod at a number of different positions as described. The positions marked from point 'B' on the trammel can now be connected together to form the arch outline.

A continuous trammel can be made for greater accuracy.

Elliptical arch (Figure 210 – pin and string method)

Span and rise are set out. Focal points are marked at two points on the springing line by measuring the distance of half the span from the rise, NB focal point to rise = ½ span.

Pins are driven into the focal points and rise. A strong thin line is passed over the pin at the rise and tied tightly to each pin at the focal points. The pin at the rise is removed and replaced by a pencil which can be traversed around the arch outline, taking care to ensure that the line is kept tight for the whole of the sweep.

Elliptical arch with parallel curves (Figure 210)

No two elliptical curves are parallel but a parallel extrados can be formed to a true elliptical intrados.

To draw a normal to a true ellipse select any point on the perimeter and draw two lines connecting with the focal points. A normal can be found by

GEOMETRY OF ARCHES

bisecting the angle formed.

If a number of normals are drawn and measured equidistant from the original arch outline a parallel curve can be drawn through the upper ends of the normals.

A quicker method is to draw a series of arcs of a suitable radius from a large number of points around the original arch outline. The parallel curve can be drawn through the crown of the arc outlines.

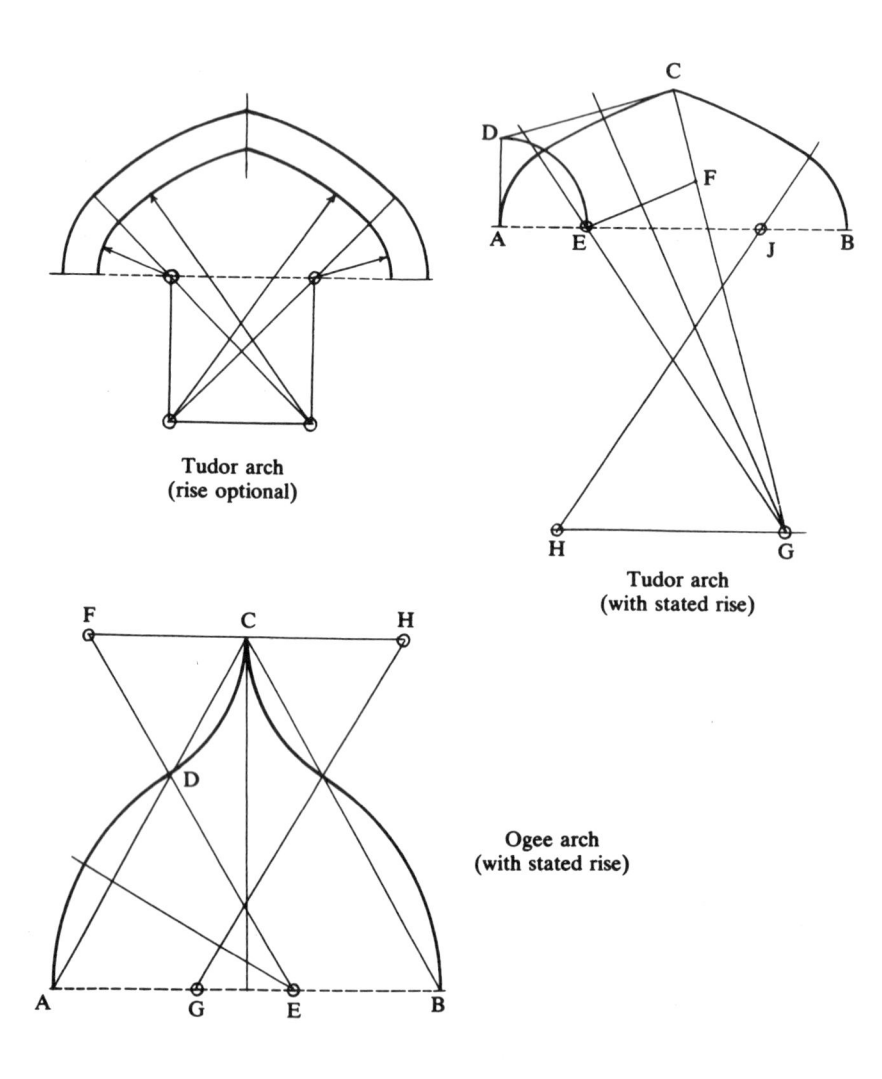

Tudor arch
(rise optional)

Tudor arch
(with stated rise)

Ogee arch
(with stated rise)

Figure 208: *Arches struck from four centres*

ARCHES STRUCK FROM FOUR CENTRES

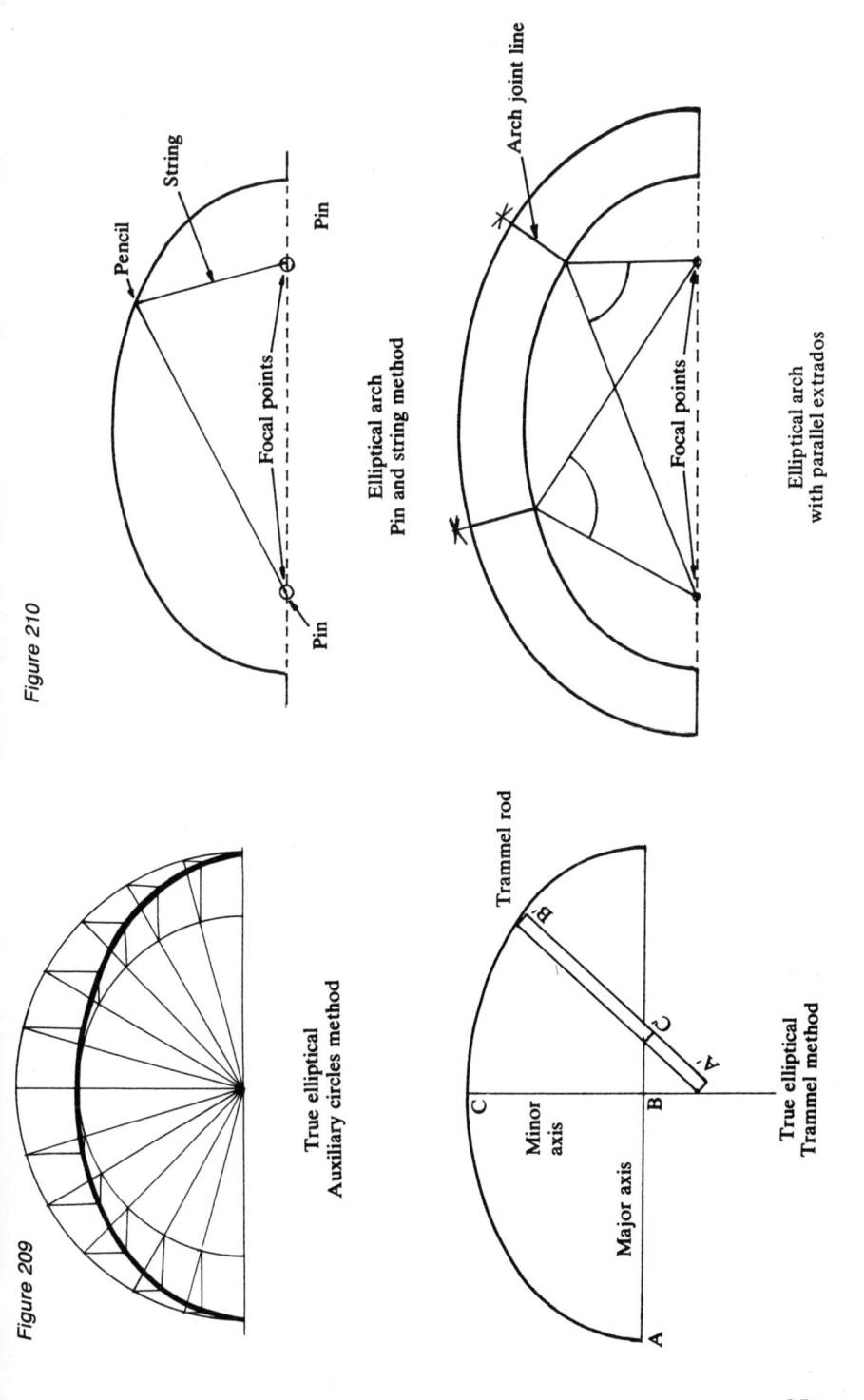

Figure 210

Elliptical arch
Pin and string method

Elliptical arch
with parallel extrados

Figure 209

True elliptical
Auxiliary circles method

True elliptical
Trammel method

251

PLASTERERS' OVAL

This is a simply constructed approximate ellipse an example of which is shown in Figure 211. It is drawn inside a rectangle whose sides (and therefore the major and minor axes of the ellipse) are in the ratio of 3:2. The four centres are the extremities of the minor axis and halfway from the centre to the ends of the major axis. The common normals for the intersection of the curves of different radii are on the diagonals of the two halves of the rectangle as shown.

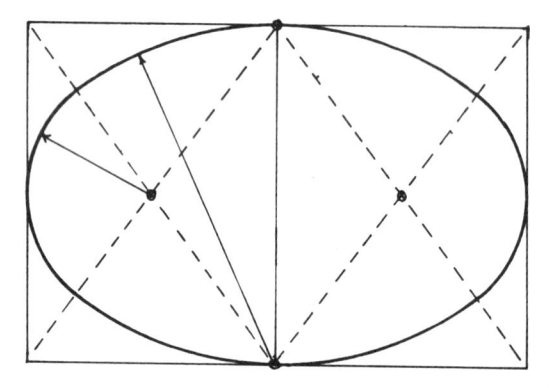

*Figure 211: **Plasterers' oval***

APPENDIX A

Plastering Terms

The following is a list of terms in present day usage connected with the plastering trade. They have been compiled to conform to the Glossary of Terms applicable to Internal Plastering, External Rendering and Floor Screeding; British Standard 4049: 1966.

PLASTERING TERMS

Accelerator. A substance added to a mix to speed up the setting or hardening process

Additive. A material added to the binder in a mix to alter the properties of the mix and/or the set or hardened product

Adhesion. The strength of attachment between two coats or to a background without the use of mechanical key

Admixture. A material added to the aggregate, binder and water in a mix to alter the properties of the mix and/or the ultimate hardened product

Aggregate. That part of the mix which does not set and acts as the filler, such as sand, crushed stone, etc.

Arris. A sharp-edged external angle

Ashlar jointing. The marking of cement rendering to imitate ashlar stonework

Background. The base to which plastering material is applied

Bay. The area of wall, floor or ceiling done in one operation

Beam case. A fibrous plaster cast surrounding a beam

Bed. A recess in a moulding into which an ornament can be planted

Bell Cast. The projection over door or window openings to shed rain clear of the opening

Binder. That part of the mix that sets and binds the aggregate together

Bird's Beak Stop. The feature formed at external angles where a moulded or rounded angle has to be stopped and brought to a square arris

APPENDIX A

Blabbing, Blebbing or Blistering All mean the formation of small swellings on the plastering surface

Blowing. See *Popping*

Bond. The holding strength of a plastering material to the background, or previous coat, due to mechanical key or adhesion or both

Bonding Agent. A substance applied to a smooth surface to increase its adhesive qualities

Brace. A support, often temporary, used in the making of fibrous plaster casts

Bracketing. The provision of a series of wood or metal brackets to which lathing can be applied to form a hollow core for large section cornices or around beams, also termed cradling when used around beams

Bruising. See *Kerfing*

Bulking. The increase in volume of sand due to moisture

Busk. A thin piece of flexible steel used for scraping, cleaning up and finishing fibrous plaster work

Calcination. The heating process used to manufacture plaster from gypsum or quicklime from limestone

Canvas. See *Scrim*

Casing bead. A metal bead used at openings or edges of plaster work

Cast. The reproduction obtained from a mould

Cat's face. See *Gaul*

Coarse stuff. Lime/sand mortar for use in undercoats

Cockling. The buckling of 'firstings' in a fibrous plaster cast

Cold pour. A flexible moulding compound which does not require heating to prepare it.

Collars. Circular screeds used when forming columns *in situ*

Compo. Sand and cement, or cement/lime/sand mortar

Composition. A stiff paste composed of resin, linseed oil and glue, used to reproduce small ornament from reverse moulds

Core. The major sectional portion of a moulding filled out to leave only a suitable finishing thickness to the profile

Cottle. A clay fence used in fibrous plaster work

Cove. A cavetto or concave moulding for internal angles

Cradling. See *Bracketing*

APPENDIX A

Crazing. Hair cracks in a plaster or cement surface

Curing. The process used to ensure hardness of cement work by preventing moisture loss

Dado. The lower part of a wall above the skirting

Darby. A wooden or light alloy rule about 1.1 m long, with two handles; used for straightening floating coats

Devilling. Scratching the floating surface with a nail float to provide key for the subsequent setting coat

Dot. Short pieces of wood lath, or similar, which are bedded then plumbed, levelled or lined in; these positions are used as guide points in the formation of screeds

Drag. A thin steel plate with an undercut toothed edge used for scraping and levelling fibrous plaster surfaces

Draught. A slope given to an otherwise vertical straight member in a plaster reverse mould to assist release of the cast

Drum. A curved framework formed in fibrous plaster to provide the reverse shape for arches, barrel ceilings, domes, etc.

Dry-lining. The technique of surfacing walls with plasterboard instead of traditional plastering

Dubbing out. The filling out of hollow places on a solid background before the rendering or floating coat is applied

Efflorescence. White powdery deposit of salts brought to the surface of building materials on drying

Expansion joint. A joint formed between two different parts of the work made to allow small movement to occur without disrupting the finished work

Fat. The fine residue formed on a plasterer's trowel when trowelling up

Fattening Up. The increase in plasticity and workability when putty lime after the slaking process, or hydrated lime after soaking in water, is allowed to mature

Feather-edged rule. A rule used for ruling in or 'wiping out' internal angles

Fence. A strip of wood, clay, sheet metal or plaster used to form a boundary in some moulding processes

Fire cracking. The crazing of a skimming coat

'Firstings'. The first coat of plaster applied on the face of a model or mould

Flaking. The scaling away of patches of the plaster surface due to insufficient adhesion to the previous coat

Flanking in. The process of filling in the areas between screeds with a plastering mix

Flash set. The sudden set of plaster or cement when mixed with water

Fresco. A painting done on wet lime plaster

Furring. Wood or metal battens fixed to backgrounds and to which plasterboard or lathwork can be fixed

Gaul. A blemish in a plaster surface due to failure to fill in a small depression when tightening in and trowelling up

Gauge box. A bottomless box of given size used to measure materials by volume

Gesso. A mixture used for modelling or casting and made from plaster, glue and linseed oil or boiled oil, glue and whiting

Gig stick. A radius rod attached to a running mould for forming circular mouldings

Glass-reinforced gypsum (GRG). A gypsum plaster mixture reinforced with glass fibres.

Green suction. The early suction of a cement-based backing

Grinning. The appearance on the plaster surface of the pattern of background joints due to differences in suction of these materials

Grounds. Wood battens fixed to backgrounds and to which joinery finishings may be secured; such grounds, where suitable, may be used as screeds

Grout. A slurry, or sloppy mixture, of cement, sand and water or of cement and water

Hacking. The roughening of solid backgrounds to provide key

Hawk. Handboard for holding plastering material ready for application with a trowel

Joggle. A method of positioning adjoining pieces in a plaster mould by means of an interlocking notch and projection

Kerfing. The bruising of reinforcement laths to assume curved shapes in fibrous plaster casts

Killing. Destroying the set of hemi-hydrate plasters by continued mixing after the normal setting time has elapsed

Knocking back. See *Retempering*

Laitance. The scum formed on the surface of freshly laid concrete, due to excessive trowelling, vibration or the use of mixes with a high water content

APPENDIX A

Larry. A steel hoe with wooden handle, used for mixing coarse stuff

Mechanical application. The application of plaster or cement mixes by machine

Mechanical key. Openings or grooves in the surface of an undercoat or background, into or through which the plastering material may pass and when set will be bonded mechanically

Mitre stop. A section of moulding cast from a reverse mould, cut to the desired angle to form internal or external mitres, in the following casts

Model. The original ornament or feature which is to be reproduced

Moulding compounds. Flexible compounds such as rubber, gelatine or PVC which are used in moulds for casting plaster mouldings or enrichments with undercut sections

Muffle. A thickness piece placed beyond the profile of a mould when coring out

Noggin. A reinforcing member between a joist or stud.

Pattern staining. The appearance in light and shade of the pattern of the structure on plaster surfaces, caused by uneven deposits of dust due to temperature differences on both sides of the structure; this occurs mostly in ceilings and the pattern of joists and laths can be observed on the ceiling soffit

Plasticizer. An additive to improve the workability or plasticity of the mix

Polyvinyl acetate (PVAC). A bonding agent.

Polyvinyl chloride (PVC). An artificial rubber-like material used as a hot melt moulding compound

Popping, Pitting or Blowing. Small blow holes in lime plastered surfaces due to expansion of previously unslaked lime particles coming into contact with moisture

Pricking up coat. The first coat on wood or metal lathing

Pugging. Plastering or other insulating material applied above the ceiling between the joists to assist sound and thermal insulation

Radius rod. See *Gig stick*

Releasing agent. A material used on the surface of a mould to assist easy release of the cast such as French chalk, grease, etc.

Retarder. A material added to a mix to delay the setting action

Retempering. Re-mixing stiffened mixes

Ropes. Twined strips of scrims soaked in gauged plaster

Screed. A narrow band of material or batten, used as a guide for ruling off

APPENDIX A

Seasoning. The treatment of models and moulds by sealing to prevent absorption

Scrim. Metal, hessian, canvas or similar material having a mesh wide enough to allow plaster to pass through easily; it is used in the construction of casts and to strengthen and reinforce the joints between plasterboards, casts, etc.

'Seconds'. The retarded second coat of plaster applied over the 'firstings' so that the two coats are bonded together

Sgraffito. Decorative work or lettering formed by scratching through one or more coats of coloured plaster or cement to reveal different colour or colours below

Shelling. The lifting off and loosening of plaster or cement coats due to adhesion failure

Shoes. Sheet metal runners attached to each end of the slipper of a running mould

Size water. A dilute solution of glue or gelatine and water used to retard the setting action of plaster

Slacking or Slaking. The process of adding water to quicklime to form slaked or putty lime

Soffit. The underside of a beam or ceiling

Spatterdash. A mixture of coarse sand or grit and cement, mixed into a slurry and thrown on to dense backgrounds to provide key for subsequent plaster or cement applications

Spline. A wood lath used for reinforcement of fibrous plaster casts

Squeezing. A method of obtaining the reverse shape of an enrichment by use of a clay squeezed impression

Studding. A wood or metal framework for plasterboard or lath attachment; mainly used for forming partitions

Styrene-butadiene rubber (SBR). A bonding agent suitable for use externally, (not affected by moisture).

Suction. The water absorption property of a background

Surface retardants. Special preparations which are applied to the inside faces of concrete formwork to retard the setting action of the adjacent concrete; when the formwork is removed the concrete is easily roughened to provide key for plastering

Sweat out. The appearance of water on a finished surface

Template or Templet. A suitable frame having the desired profile, or curved shape to which the work is to be formed

Turning box. A box or frame with a profile and metal spindle used to form plaster balusters or similar features; also termed a hand lathe

Voids. The spaces in a material occupied by air, water, or both

Wads. Pieces of scrim soaked in gauged plaster and used in the making and fixing of fibrous plaster casts

Working time. The period, after mixing, during which the mix may be applied and finished

X-ray plaster. A plastering mix containing barium sulphate as the aggregate; this type of plaster is used to provide insulation against certain types of radiation

APPENDIX B

Information sources and references

Manufacturers' literature has been a valuable source of information concerning their specialist products. British Gypsum, Cement Marketing Corporation, Cement and Concrete Association, Expanded Metal Company Limited, Tinsley Building Products Ltd, and Newton & Co. Limited have been particularly helpful and have freely supplied details of their products. I am also grateful to the Cement Marketing Corporation, the Cement and Concrete Association and the Blue Circle Group for the supply of, and permission to use, the photographs included in this book.

Government publications concerning plastering include digests, bulletins and advisory leaflets issued from the Building Research Establishment (Department of the Environment), together with British Standards and Codes of Practice from the British Standards Institution.

Relevant publications from the Department of the Environment include the following:

Sands for plasters, Mortars and External Renderings, A. D. Cowper, BRE Bulletin No 7

BRE Digests
    49 *Choosing Specifications for Plastering*
    104 *Floor Screeds*

Relevant publications from the British Standards Institution include:

British Standards
BS 12   *Portland cement*
BS 410 *Test sieves*
BS 812 *Testing aggregates*
BS 882 *Aggregates from natural sources for concrete*
BS 890 *Building limes*

APPENDIX B

BS 1191 *Gypsum building plasters*
          Part 1 Excluding premixed lightweight plasters
          Part 2 Premixed lightweight plasters
BS 1192 *Construction drawing practice*
BS 1199–BS 1200 *Building sands from natural sources*
BS 1230 *Gypsum plasterboard*
BS 1369 *Steel lathing for internal plastering and external rendering*
BS 4022 *Prefabricated gypsum wallboard panels*
BS 4049 *Glossary of terms applicable to plastering, external rendering and floor screeding*
BS 4551 *Methods of testing mortars, screeds and plasters*
BS 4721 *Ready mixed building mortars*
BS 4887 *Mortar admixtures*
BS 5075 *Concrete admixtures*
BS 5262 *Code of practice for External Rendered Finishes*
BS 5270 *Specifications for Polyvinyl acetate (PVAC) emulsion bonding agents for internal use with gypsum plasters*
BS 5492 *Code of practice for internal plastering*
BS 6100 *Glossary of building and civil engineering terms*
BS 6452 *Part 1 Beads for internal plastering and dry lining*
BS 6462 *Quick lime, hydrated lime and natural calcium carbonate*
BS 8203 *Installation of sheet and tile flooring*
BS 8204 *In-situ floorings*
BS 8212 *Dry lining and partitioning using gypsum plasterboards*
BS 8000 *Workmanship on building sites*

Codes of Practice
CP 204 *In-situ floor finishes*
CP 290 *Suspended ceilings and linings of dry construction using metal fixing systems*

Department of the Environment publications can be obtained from local government offices or Her Majesty's Stationery Office, situated in many of the larger cities in Britain.

British Standards and Codes of Practice can be obtained from the BSI Sales Department, Lingford Wood, Milton Keynes, MK14 6LE, or at local branches in the larger cities.

Index

Aggregates, 19–21
Arches
  fibrous, 220–22
  geometry, 244–52
  in situ, 166–77

Backgrounds
  preparation, 37, 38
  types, 37
Bell-casting, 122–3
Benchwork
  benches, 190–91
  casting, 210
  running, 191–3
Bonding adhesives, 30, 31

Cement
  manufacture, 17
  types, 18
Cement casting, 236, 237
Chattering, 192
Cold cure rubber moulding compounds, 200, 201
Complaints and remedies, 86–9
Coved skirtings, 75, 76
Curved walls
  concave, 54–7
  convex, 57–8

Damp-proofing, 83, 84
Draught, 194
Dry-lining
  metal furring, 105–7
  thistlebond, 101–3
Dubbing out, 39

External cement work
  modern, 138–43
  traditional, 114–38

Finishing coats
  Keene's cement, 53
  plasters, 51–3
Floating coats
  ceilings, 58–60
  external angles, 45–51
  internal angles, 51
  reveals, 42, 43
  walls, 39–41
Floors
  curing, 69, 70
  dusting, 71
  granolithic, 68

hardening, 71
laying to falls, 72, 73
non-flip finishes, 70, 71
screeds, 72, 73

Glass-reinforced gypsum (GRG), 238, 239
Glass-reinforced plastics (GRP)
  glass fibres, 31, 35, 36
  resins, 35
  techniques, 237–9

Insulation
  radiation, 81, 82
  sound, 81–2
  thermal, 81

Jointing metal angle beads, 48, 49

Lathing
  expanded metal lathing, 24, 93–5
  Newlath, 30, 97, 98
  plasterboarding faults, 99–100
  plasterboards, 24, 90–93
  rib lath and Hy-rib, 25, 94
  Twil-Lath, 29, 95, 96
  wood lath, 96, 97
Limes
  classification, 13, 15
  limestone cycle, 13, 14
  manufacture and setting, 15
Lining in dots, 40, 41, 42

Mechanical plastering
  one coat protection, 243
  types and methods of use, 240–43
Metal trims
  angle beads, 25
  fixing, 48–51
  other types, 27, 28
  stop beads, 26
Mixes
  cement/sand, 23
  lime/sand, 21
  plaster/lightweight aggregate, 22
  uses and coverages, 23
Mixing bankers and troughs, 12
Moulding materials
  cold cure rubber, 200, 201
  gelatine, 33
  grease, 32
  PVC, 33, 34
  shellac, 31, 32
  size, 32

INDEX

wax, 34
Moulding members, 144–6
Moulding methods
  beam casing mould, 214–16
  cornice reverse mould, 211, 212
  gelatine mould, 199, 206
  lighting trough mould, 215, 216
  loose piece mould, 213
  piece mould, 194–6
  plaster reverse, 209
  PVC mould, 199, 200
  waste mould, 196–8
  wax mould, 198–9

Niches
  fibrous, 217–20
  in situ, 176–9

Partitions
  cellular, 103, 105
  laminated, 107, 109, 110
  metal stud, 109, 111
Patching
  ceilings, 60–62
  walls, 53, 54
Pattern staining, 82, 83
Pickling and seasoning
  gelatine moulds, 199
Plain faced casts, 209, 210
Plain interior plastering
  one coat work, 38
  three coat work, 39
  two coat work, 38–41
Plasterboard coves, 111, 113
Plasters
  manufacture, 15
  properties, 16
  types, 15, 16
Plasticizers, 30
Plumb, dot and screed, 40, 41

Rendering coats
  internal, 39
Repair and maintenance of moulded and enriched surfaces, 186–9
Roof screeds, 72–4
Run cases, 208, 209
Run casts, 209
Run cores, 207, 208
Running and turning moulds
  double angle mould, 152
  external cornice mould, 161, 163–5
  flush bead mould, 147, 148
  hand lathe, 216
  hanging mould, 152, 153
  internal cornice mould, 154–61
  panel mould, 147
  peg mould, 172, 173
  raised panel mould, 150
  raking mould, 160–61
  ramp mould, 126
  single angle mould, 151
  thumb mould, 144–6
  trammel, 172–4
  turning moulds, 216, 217
  twin slippered mould, 153

Scaffolding, 11, 114, 115
Setting out flutes, 185
Skirtings (cement)
  coved, 75, 76
  moulded, 149, 150
  plain, 74, 75
Spatterdash, 125
Stairs, 75–9

Taking a squeeze, 193
Tools, 1–10
Trueness of plastering systems, 40

Water level, 86
Waterproofing, 84–6

X-ray plasters, 82